The Clean Water Act Handbook

Fourth Edition

DUKE K. MCCALL, III

Bernan Press

Lanham • Boulder • New York • London

Published by Bernan Press
An imprint of The Rowman & Littlefield Publishing Group, Inc.
4501 Forbes Boulevard, Suite 200, Lanham, Maryland 20706
www.rowman.com
800-865-3457; info@bernan.com

Unit A, Whitacre Mews, 26-34 Stannary Street, London SE11 4AB

Library of Congress Cataloging-in-Publication Data Available

ISBN: 978-1-59888-818-8 (pbk. : alk. paper)
ISBN: 978-1-59888-819-5 (electronic)

∞™ The paper used in this publication meets the minimum requirements of American National Standard for Information Sciences—Permanence of Paper for Printed Library Materials, ANSI/NISO Z39.48-1992.

Printed in the United States of America

The Clean Water Act Handbook

Contents

Preface

This Handbook was designed to provide a comprehensive road map to the requirements, legal theories, and critical issues related to regulation of water pollution under the Clean Water Act. The Clean Water Act, the primary federal statute that addresses water pollution in the United States, is one of the longest-lived of the federal environmental statutes, and while the Clean Water Act is not complicated, it is complex. It contains a wide range of programs and requirements that have developed over the years through statutory changes, regulatory enactments, and court decisions. Only by understanding these programs and requirements at a fundamental level can individuals or companies develop an effective compliance program. Although this Handbook focuses on the federal Clean Water Act, it provides useful information for compliance with state and federal water pollution control programs.

This Fourth Edition includes a comprehensive update on various Clean Water Act programs. In addition, this edition includes significant new information regarding efforts to define "the waters of the United States," the phrase that determines the scope of the Clean Water Act.

This Handbook was written from a legal perspective, but it is intended to be a helpful resource for any environmental professional. I welcome any questions or suggestions as to how to make the Handbook as practical and useful as possible.

Duke K. McCall, III
Morgan, Lewis & Bockius LLP
Washington, D.C.

About the Author

Duke K. McCall, III is a partner in the Environmental Practice Group of Morgan, Lewis & Bockius LLP, resident in the firm's Washington, DC office. He represents clients in environmental regulatory and complex litigation matters, including contribution and cost recovery actions, enforcement proceedings, citizen suits, and toxic tort litigation brought under the Comprehensive Environmental Response, Compensation, and Liability Act (Superfund), the Clean Water Act, the Clean Air Act, the Resource Conservation and Recovery Act, and analogous state laws. He also assists clients with related matters such as the pursuit of insurance coverage for environmental liabilities.

Mr. McCall frequently speaks on environmental law matters, is the author of the Clean Water Act chapter of the *Environmental Law Handbook* published by Government Institutes, serves on the Editorial Board of the *Environmental Liability, Enforcement & Penalties Reporter*, and serves on the Editorial Advisory Board for *Law360 Environmental*. He is a member of the American Bar Association's Section of Environment, Energy, and Resources and Section of Litigation.

Mr. McCall earned his J.D. from Vanderbilt University and a B.A. from Furman University. He clerked for the Honorable William W. Wilkins of the U.S. Court of Appeals for the Fourth Circuit, and the Honorable William B. Traxler, Jr. of the U.S. District Court for the District of South Carolina. He is admitted to the bars of the District of Columbia, South Carolina, the U.S. Supreme Court, and numerous federal courts at the trial and appellate levels.

Email: duke.mccall@morganlewis.com

1 Introduction

1.1 OVERVIEW

The Clean Water Act (hereafter CWA), 33 U.S.C. §§ 1251–1387, is the primary federal statute that addresses water pollution in the United States. It establishes a number of programs designed to restore and protect the quality of our nation's waters by eliminating the discharge of pollutants into surface waters. The programs established under the act include the National Pollutant Discharge Elimination System (NPDES) permit program, the dredge and fill permit program, and municipal wastewater treatment programs. The U.S. Environmental Protection Agency (EPA), together with other federal, state, and local agencies, administers these and the other programs established under the act.

1.2 BRIEF HISTORY OF THE CWA

Today's CWA traces its roots to the Federal Water Pollution Control Act (FWPCA) amendments of 1972.[1] The 1972 amendments to the FWPCA required, for the first time, that EPA set nationwide limits for discharges from industrial sources and publicly owned treatment works into the navigable waters of the United States. The 1972 amendments further required that EPA establish more stringent effluent limitations for specific sources or groups of sources where necessary to protect water quality in a specific portion of the nation's navigable waters. The 1972 amendments also established the NPDES permit program. These provisions—the effluent limitations, water quality requirements, and NPDES permit program—remain the foundation of the CWA.

In implementing the 1972 amendments, EPA initially focused on the control of conventional pollutants, such as biological oxygen demand and suspended solids, rather than toxic pollutants. Dissatisfied with EPA's progress toward reducing toxic water pollution, environmental groups sued the agency for its failure to meet the statutory deadlines for toxics controls. This lawsuit resulted in a consent decree, known as the Flannery Decree, in 1976.[2] The Flannery Decree set out a detailed toxics strategy, requiring EPA to promulgate effluent guidelines, new source performance standards, and pretreatment standards for 65 toxic pollutants (the "priority pollutants") in each of 21 major industrial categories by December 31, 1979.[3] Congress endorsed the Flannery Decree's approach to toxics regulation in its 1977 amendments to the FWPCA, writing several portions of the decree into law. The 1977 amendments to the FWPCA also amended the short title of the statute, noting that it had become "commonly referred to as the Clean Water Act."[4]

The act was amended again in 1987. The 1987 amendments created new programs for toxics control, established a timetable for regulation of storm water, strengthened requirements related to water quality, tightened requirements for certain variances, established a revolving loan fund for construction of sewage treatment plants, and expanded EPA's enforcement tools.[5]

In response to the *Exxon Valdez* oil spill, Congress overhauled the oil spill provisions of the act in the Oil Pollution Act of 1990, sometimes referred to as OPA 90.[6] In addition to its amendments to the existing oil spill provisions in Section 311 of the CWA, the Oil Pollution Act also created a separate statutory program governing oil spill liability and compensation.[7]

1.3 CLEAN WATER ACT GOALS AND POLICIES

Like many federal environmental statutes, the CWA contains a statement of objectives, goals, and policies. The act's stated objective is to "restore and maintain the chemical, physical, and biological integrity of the nation's waters."[8] To achieve this objective, the act declares an interim goal of attaining a level of water quality that "provides for the protection and propagation of fish, shellfish and wildlife and provides for recreation in and on the water."[9] The act further declares an ultimate goal of eliminating the discharge of pollutants into the nation's navigable waters.[10]

The stated policies sought to be advanced by the act include "that the discharge of toxic pollutants in toxic amounts be prohibited" and "that programs for the control of nonpoint sources of pollution be developed and implemented in an expeditious manner so as to enable the goals of [the act] to be met through the control of both point and nonpoint sources of pollution."[11]

While these objectives, goals, and policies are not legal mandates, EPA and the courts rely on them to determine what Congress intended when it enacted the CWA. Thus, for example, the act's stated policy of prohibiting the discharge of "toxic pollutants in toxic amounts" provided the foundation for the Flannery Decree and has been a primary focus of the act's implementation since 1977.

1.4 ELEMENTS OF THE CWA

The CWA contains a broad range of regulatory tools to attain its statutory objectives and goals and to implement its underlying policies. These tools include the following major elements:

- A prohibition on discharges, except as in compliance with the act (Section301);
- A permit program to authorize and regulate discharges in compliance with the act (Section 402);
- A system for determining the limitations to be imposed on authorized and regulated discharges (Sections 301, 306, 307);
- A permit program governing the discharge or placement of dredged or fill material in the nation's waters (Section 404);
- A system for preventing, reporting, and responding to spills into the nation's waters (Section 311);
- A procedure for cooperative federal/state implementation of the act (Sections 401, 402); and
- Strong enforcement mechanisms (Sections 309, 505).

These elements and other aspects of the CWA program will be discussed in some detail in the rest of this Handbook.

1.5 THE DISCHARGE PROHIBITION

Section 301 of the CWA establishes a broad prohibition against "the discharge of any pollutant by any person" except as in compliance with the act's permit requirements, effluent limitations, and other enumerated provisions.[12] The effect of the "except as in compliance" language is to shift the burden of proof in an enforcement action. This language requires the discharger to prove that the discharge was in compliance with the act rather than requiring the government to prove that a discharge was out of compliance.

Because the Section 301 prohibition establishes the act's scope, it is useful to dissect and examine its language. First, the discharge of a pollutant is defined to mean, in relevant part, "any addition of any pollutant to navigable waters from any point source."[13] Thus, the terms "addition," "pollutant," "navigable waters," and "point source" are all of critical importance.

1.5.1 Addition

EPA and the courts have interpreted the term "addition" broadly. Generally, almost any introduction of a pollutant into a body of water is an addition. The scope of the term has been limited only by the requirement that there must be an addition of a new material into an area or an increase in the amount or type of material that is already present.[14]

EPA's regulations make an exception to this principle for pollutants that are present in a discharge only by reason of their presence in the discharger's intake water if the intake water is drawn from the same body of water as the one into which the discharge is made and if the pollutants are not removed by the discharger as part of its normal operations.[15] The Supreme Court also has held that the transfer of polluted water from one part of a water body to another part of the same water body does not constitute an "addition" of pollutants.[16] In addition, discharges of water from dams, even if a dam's operations adversely affect the temperature or dissolved oxygen content of the water, have been determined not to be additions of pollutants.[17]

The EPA Environmental Appeals Board has held that a company can be held liable for exceedances of metals limitations that result from metals in rainfall and from leaching of metal building materials caused by acid rain.[18] Moreover, courts have held that the mere transfer of water from one body of water to another may be an addition of pollutants, even if the water bodies are of the same quality.[19]

1.5.2 Pollutant

Pollutant is defined in the act as "dredged spoil, solid waste, incinerator residue, sewage, garbage, sewage sludge, munitions, chemical wastes, biological materials, radioactive materials, heat, wrecked or discarded equipment, rock, sand, cellar dirt and industrial, municipal, and agricultural waste discharged into water."[20] Despite this seemingly specific definition, courts have broadly interpreted the term to include virtually any material, as well as such material characteristics as toxicity and acidity.[21]

1.5.3 Point Source

Point source is defined in the act as "any discernable, confined and discrete conveyance . . . from which pollutants are or may be discharged."[22] This definition has been interpreted to cover almost any natural or man-made conveyance from which a pollutant may be discharged, including pipes, ditches, erosion channels, and gullies.[23] Vehicles, such as bulldozers or tank trucks, have also been held to be point sources.[24] But human beings have been held not to be point sources, at least for purposes of criminal enforcement of the act.[25]

Point sources include sources from which pollutants "may be" discharged. Thus, the term includes not only sources from which pollutants are routinely discharged but also conveyances from which a pollutant may be discharged only in extreme conditions, such as during an unusual storm event.[26]

1.5.4 Navigable Waters ("Waters of the United States")

Navigable waters are defined by the act as "the waters of the United States, including the territorial seas."[27] This definition has proved difficult to apply to the myriad of different types of water bodies in the United States, resulting in numerous judicial and administrative efforts to determine the scope of the CWA. A joint rule promulgated by EPA and the Army Corps of Engineers in June 2015 defines the term "waters of the United States" to include:

- All waters which are currently used, were used in the past, or may be susceptible to use in interstate or foreign commerce, including all waters which are subject to the ebb and flow of the tide;
- All interstate waters, including interstate wetlands;
- The territorial seas;
- All impoundments of waters otherwise identified as waters of the United States;
- All tributaries of waters identified in paragraphs 1–3;
- All waters adjacent to a water identified in paragraphs 1–5, including wetlands, ponds, lakes, oxbows, impoundments, and similar waters;
- All prairie potholes, Carolina bays and Delmarva bays, pocosins, Western vernal pools, and Texas coastal prairie wetlands, where they are determined, on a case-specific basis, to have a significant nexus to a water identified in paragraphs 1–3;
- All waters located within the 100-year floodplain of a water identified in paragraphs 1–3, and all waters located within 4,000 feet of the high tide line or ordinary high water mark of a water identified in paragraphs 1–5, where they are determined on a case-specific basis to have a significant nexus to a water identified in paragraphs 1–3.[28]

A number of features are also specifically excluded from the definition of "waters of the United States" under the 2015 rule. Excluded features include prior converted cropland, certain ditches, stormwater control features, artificially irrigated areas, and artificial lakes or ponds.[29] Another notable feature that is explicitly excluded from the scope of the Clean Water Act under the 2015 rule is groundwater.[30] Groundwater is nonetheless included in many states' definitions of "waters of the state," and in these states, point-source discharges into groundwater may be covered by the state water permit program, even though they are not considered to be discharges to the "waters of the United States."

The determination of whether relatively isolated, intrastate bodies of water fall within the definition of "waters of the United States" has long been controversial. EPA and the Corps of Engineers have generally supported the broadest possible interpretation of the Act's coverage. But federal courts have not been willing to find that all waters are "waters of the United States." For example, in 2001, the U.S. Supreme Court invalidated a long-standing Corps of Engineers regulation that extended the act's dredge and fill permit program to intrastate waters used by migratory birds.[31]

In the U.S. Supreme Court's most recent case on the jurisdiction of the CWA, *Rapanos v. United States*, the Court reviewed consolidated cases in which the Sixth Circuit held that the CWA afforded federal regulatory jurisdiction over wetlands located some distance from navigable waters.[32] The Supreme Court issued a split decision overruling the Sixth Circuit.[33] The plurality concluded that the phrase "waters of the United States" includes only "relatively permanent, standing or continuously flowing bodies of water 'forming geographical features'" and that federal regulatory jurisdiction extended to "only those wetlands with a continuous surface connection to bodies that are 'waters of the United States' in their own right."[34] Justice Kennedy concurred in the judgment but rejected the plurality's reasoning. He concluded that federal regulatory jurisdiction under the CWA extends to wetlands that "possess a 'significant nexus' to waters that are or were navigable in fact or that could reasonably be so made."[35] He further concluded that wetlands "possess the requisite nexus" if "either alone or in combination with similarly situated lands in the region, [they] significantly affect the chemical, physical, and biological integrity of other covered waters more readily understood as 'navigable.'"[36]

Lower courts have struggled with how to apply the fractured reasoning of the Supreme Court in *Rapanos*. Some courts have concluded that Justice Kennedy's "significant nexus" test is controlling because it resolves the issue on the "narrowest grounds" or is the least restrictive of federal authority to regulate.[37] Other courts have concluded that jurisdiction should be found if either of the tests articulated by the plurality and Justice Kennedy are met.[38] EPA and the Corps of Engineers have applied the "significant nexus" test,[39] and some courts have been deferential to the agencies in determining what constitutes a significant nexus. For example, the Fourth Circuit has upheld a Corps of Engineers determination that a wetland connected by intermittent flow through man-made drainage ditches to a river approximately seven miles away had a significant nexus to the river.[40]

The 2015 rule promulgated by EPA and the Corp of Engineers sought to resolve some of the confusion over the scope of Clean Water Act jurisdiction.[41] Among other changes, the rule added for the first time two "bright-line" categories of waters that are always jurisdictional, "tributaries" and "adjacent waters," thus obviating the need to analyze jurisdiction over those waters on a case-by-case basis. The rule also added new categories of waters that are excluded from the definition of waters of the United States, such as stormwater control features and cooling ponds. The rule then expanded upon the definition of "significant nexus" articulated by Justice Kennedy in *Rapanos* by delineating the set of waters to which the significant nexus test applies and listing factors that may be considered when assessing whether a nexus is significant. While EPA and the Corps of Engineers have described the 2015 rule as a clarification of existing law, some industry members and property rights groups have viewed it as an expansion of the jurisdiction of the CWA that will result in increased permitting costs. Those stakeholders have brought a number of legal challenges to the rule, which have been consolidated for review by the Sixth Circuit.[42] A related issue courts have struggled with is when parties may obtain judicial review of determinations by the Corps of Engineers that specific waters are subject to the CWA.

The Fifth Circuit held in 2014 that such jurisdictional determinations are not final agency actions subject to judicial review.[43] Under the Fifth Circuit's ruling, such jurisdictional determinations may only be challenged through applying for a permit and then appealing the jurisdictional determination or beginning work without a permit and challenging the government's jurisdiction after receiving a compliance order or civil enforcement action. The Eighth Circuit, in contrast, held in 2015 that jurisdictional determinations by the Corps are reviewable final agency actions, in part because of the prohibitive costs of the alternative methods of challenging whether waters are subject to the CWA.[44] The U.S. Supreme Court resolved this circuit split in May 2016, holding in *Army Corps of Engineers v. Hawkes Co.* that determinations by the Corps of Engineers regarding whether a particular property is covered by the CWA are subject to immediate judicial review.[45] The Court found that the alternatives of (1) applying for a permit and appealing the jurisdictional determination or (2) discharging without a permit and contending in any enforcement action that no permit was required did not provide adequate avenues for review due to their high costs and risk. The Court concluded that the lack of alternatives for review, combined with the significant implications of jurisdictional determinations, demonstrated that the determinations were final agency actions for which judicial review should be available. The Supreme Court's decision opens the door for a number of new challenges to jurisdictional determinations by landowners and project proponents, and it could also be used by environmental groups to challenge Corps of Engineers determinations that properties are not jurisdictional.

1.6 OVERVIEW OF THIS HANDBOOK

This Handbook first looks at the NPDES permit program, providing an overview of the program in chapter 2, a discussion of effluent limitations in chapter 3, and a discussion of whole effluent toxicity control in chapter 4. Next, the Handbook describes programs for the control of storm water in chapter 5 and other types of discharges in chapter 6. Chapter 7 discusses the pretreatment program that governs discharges into public water treatment works. Chapter 8 covers nonpoint-source discharges of pollutants. Chapter 9 addresses the U.S. Army Corps of Engineers program under

Section 404 of the Act pertaining to the discharge of dredged or fill material. Chapter 10 discusses the Clean Water Act programs for control of and response to spills of oil and hazardous substances. Finally, chapter 11 discusses the enforcement of the various programs under the Act.

2

The NPDES Permit Program

2.1 WHAT IS AN NPDES PERMIT?

The NPDES permit program implements the CWA's prohibition on unauthorized discharges by requiring a permit for every discharge of pollutants from a point source to waters of the United States. NPDES permits give the permittee the right to discharge specified pollutants from specified outfalls, normally for a period of five years. The permit usually sets numerical limitations on the authorized discharges and imposes other conditions on the permittee.

2.2 WHAT DISCHARGES REQUIRE AN NPDES PERMIT?

Generally, an NPDES permit is required for any discharge of a pollutant from a point source to waters of the United States. Discharges that require an NPDES permit include such waste streams as industrial process water, noncontact cooling water, and collected or channeled storm-water runoff. An NPDES permit is not required for sheet runoff (which is not a point-source discharge), discharges into wastewater treatment systems (which are excluded from the definition of "waters of the United States"), and certain exempted activities.[1]

2.3 STATE AND FEDERAL ROLES

An NPDES permit is issued either by EPA or by a state, if the state has received permitting authority from EPA pursuant to Section 402(b) of the act.[2] Currently, 46 states have received permitting authority.[3] Upon EPA approval of a state program, EPA and the state enter into a memorandum of agreement (MOA) concerning the specific elements of the authorized state program.[4] In states that are not authorized to administer the NPDES program, permits are issued by EPA regional offices.

Permitting procedures for state-issued permits normally follow the EPA procedures described in section 6.4 below. EPA's regulations mandate that state permit

EPA Regional Offices

Region 1 **Boston** Connecticut Maine Massachusetts New Hampshire Rhode Island Vermont	**Region 2** **New York City** New Jersey New York Puerto Rico Virgin Islands	**Region 3** **Philadelphia** Delaware District of Columbia Maryland Pennsylvania Virginia West Virginia	**Region 4** **Atlanta** Alabama Florida Georgia Kentucky Mississippi North Carolina South Carolina Tennessee
Region 5 **Chicago** Illinois Indiana Michigan Minnesota Ohio Wisconsin	**Region 6** **Dallas** Arkansas Louisiana New Mexico Oklahoma Texas	**Region 7** **Kansas City** Iowa Kansas Missouri Nebraska	**Region 8** **Denver** Colorado Montana North Dakota South Dakota Utah Wyoming
	Region 9 **San Francisco** Arizona California Hawaii Nevada American Samoa Guam Trust Territories Northern Mariana Islands	**Region 10** **Seattle** Alaska Idaho Oregon Washington	

programs include certain elements, such as signatory requirements, effluent limitations calculation methods, and issuance procedures.[5] It is not unusual, however, for aspects of state programs that are not mandated by EPA regulations, such as appeal procedures, to differ from federal procedures. Substantively, state programs must be at least as stringent as the federal program. But states are free to implement requirements that are more stringent than the federal program.[6]

Permits issued by authorized states are subject to review by EPA.[7] If EPA objects to a state permit and the state does not change the permit to address EPA's concerns, EPA may issue its own permit for the facility.[8] EPA has the power to withdraw its approval of a state permit program and to take over the program if it finds that the state is not administering the program in accordance with the act's requirements.[9]

Where EPA is the permitting authority, the state in which the discharger is located must certify that the discharge authorized in the permit will comply with state water quality standards and other requirements.[10] If the state does not either certify the permit or deny certification within a reasonable time, it is deemed to have waived the certification requirement.[11] States often use the certification requirement as a means of persuading EPA to adopt more stringent conditions in a permit.

2.4 THE PERMIT PROCESS

2.4.1 The Permit Application

The NPDES permit application, whether for a new discharge or for an existing discharge, requires extensive information about the facility and the nature of its discharges. EPA application forms include Form 1 (general information), Form 2C (detailed information for existing industrial facilities), Form 2D (detailed information for new industrial facilities), Form 2E (for new and existing industrial facilities that discharge only non-process wastewater), Form 2F (for new and existing facilities whose discharge is composed entirely of storm water associated with industrial activity), and other forms for other specific types of discharges.[12] State application forms must, at a minimum, require the information required by EPA's forms.[13]

The permit application must be signed by a responsible corporate officer as defined in EPA's regulations.[14] The person signing the application must certify as follows:

> I certify under penalty of law that this document and all attachments were prepared under my direction or supervision in accordance with a system designed to assure that qualified personnel properly gather and evaluate the information submitted. Based on my inquiry of the person or persons who manage the system, or those persons directly responsible for gathering the information, the information submitted is, to the best of my knowledge and belief, true, accurate and complete. I am aware that there are significant penalties for submitting false information, including the possibility of fine and imprisonment for knowing violations.[15]

Because the person signing the application may have little direct knowledge of the application's contents, some companies have adopted the practice of having the application signed in a "signing ceremony" during which the responsible corporate officer may question those directly responsible for the application in order to ensure the correctness of its contents. This ceremony is then memorialized in a memorandum or other writing so that a record of this inquiry will exist if the accuracy of the application becomes an issue in the future.

If the discharger is located in a state authorized to issue NPDES permits, it must submit its permit application to the state permitting agency. If the discharger's state is not authorized to issue NPDES permits, the application must be submitted to the appropriate EPA regional office. Normally, the application must be submitted at least 180 days prior to the date a proposed discharge is to commence or 180 days prior to the expiration of an existing permit.[16]

If a complete application is filed at least 180 days prior to the expiration of an existing permit, the existing permit will be continued (i.e., will remain in effect) until a new permit is issued by the permitting authority.[17] Because it often takes up to a year for a state or EPA regional office to issue a renewed permit, it is essential that a permittee ensure that its renewal application is both timely and complete.

2.4.2 The Draft Permit and Comment Period

When a permit involves complicated or unusual features, EPA or the state typically will initiate early informal discussions with the permit applicant about the permit terms. Often the permitting agency will issue a pre-draft permit, sometimes accompanied by what is called a 14-day letter, upon which the permit applicant can submit comments, so that the draft permit, when it is issued for public comment, reflects the input of the permit applicant.

The draft permit, when it is issued, must be accompanied by a Fact Sheet or Statement of Basis explaining how the permit terms and conditions were calculated and developed.[18] The permitting agency must publish a notice of the issuance of the draft permit in a local newspaper and must accept comments from the public for at least 30 days.[19] If there is significant public interest in the draft permit, EPA or the state agency will hold a public hearing.[20]

All persons, including the permit applicant, must raise all reasonably ascertainable issues and submit all reasonably available arguments supporting its positions during the comment period on the draft permit.[21] If an issue is not raised during the comment period, the party loses its right to raise that issue during subsequent administrative or judicial challenges to the permit.[22]

2.4.3 Appealing the Final Permit Decision

After accepting comments, the permitting agency will issue a final permit decision, which becomes effective in thirty days, unless a later date is specified or review is requested.[23] Within thirty days after the final permit decision is issued, any person who filed comments on the draft permit or participated in the public hearing may directly petition the EPA Environmental Appeals Board (EAB) for review.[24]

A petition for review must include a statement of the reasons supporting review, including, as appropriate, that the permit condition in question is based on a clearly erroneous finding of fact or conclusion of law or that it is based on an exercise of discretion or important policy consideration which the EAB should review.[25] When the petition for review is denied or a decision on the merits of the appeal is rendered, the EPA regional administrator is required to issue a final permit decision, constituting the final agency action on the permit application.[26]

Where EPA is the permitting authority, the final agency action is subject to judicial review in the federal courts of appeal.[27] Where the state issues the permit, judicial review is in accordance with state procedures.[28] Normally, judicial review is a review of the agency's decision based solely on the administrative record. The court will review the record to determine whether the agency's decision was arbitrary and capricious, an abuse of discretion, contrary to a constitutional right, or otherwise not in accordance with law.[29]

2.5 NPDES PERMIT CONDITIONS

The primary purpose of NPDES permits is to establish enforceable effluent limitations. In addition to effluent limitations, which are discussed in chapter 3, NPDES permits establish a number of other enforceable conditions, such as monitoring and reporting requirements (which are discussed in more detail in section 2.6 below), a duty to properly operate and maintain systems, upset and bypass provisions,[30] recordkeeping, and inspection and entry requirements.[31]

Although these other conditions are often "boilerplate," some may be subject to a degree of negotiation with the permit writer and should not be accepted if they would impose an unreasonable burden upon a facility.[32] State boilerplate may be more stringent than EPA by, for example, omitting the upset defense provision or establishing more stringent standards for the use of the defense.[33]

In addition, NPDES permits may require the permittee to perform best management practices (BMPs).[34] BMPs are procedures designed to prevent or minimize the release of toxic pollutants. BMPs are often simple housekeeping measures such as requirements to store drums in specific locations or to clean up spills promptly. BMPs are especially appropriate for nontraditional NPDES permits, such as storm-water permits or permits for mining operations.

Where it is impossible for a permittee to immediately come into compliance with a permit, the permit may contain a schedule of compliance. Such a schedule may contain interim limitations and dates for the submittal of compliance plans designed to achieve full compliance by a certain date.

2.6 MONITORING REQUIREMENTS

The implementation and enforcement of the NPDES program depend to a large extent on self-monitoring. Permits require dischargers to monitor their compliance with permit limitations on a regular basis and to report the results of this monitoring to the permitting authority on standardized discharge monitoring reports (DMRs).[35] EPA has revised the NPDES program to allow dischargers to obtain a waiver for sampling of pollutants that the discharger can certify are not part of its effluent (i.e., the effluent levels are no higher than background levels), although any permit limitation for the pollutant would remain in place.[36]

Monitoring must take place at the point of discharge into the receiving waters unless monitoring at that location is infeasible. For example, the permit may require the monitoring of internal waste streams where the final discharge point is inaccessible, where wastes at the point of discharge are so diluted as to make compliance monitoring impracticable, or where interference among pollutants at the point of discharge would prevent accurate detection or analysis.[37]

Permits normally contain requirements for the maintenance and proper installation of monitoring equipment and specify the monitoring and analytical methods that must be used.[38] Tampering with monitoring equipment or submitting false monitoring data to the permitting authority are criminal violations of the act.[39] The act gives the permitting authority the right to enter the premises of any permittee to inspect the facility's monitoring and other records, inspect monitoring equipment, and take samples to verify monitoring results.[40]

2.7 REPORTING REQUIREMENTS

EPA's NPDES regulations contain numerous provisions requiring both routine and non-routine reporting. These requirements are listed in the following chart.

Reporting Requirements under the NPDES Program 40 C.F.R. § 122.41(1)

Planned changes—any planned physical alteration or addition to the permitted facility that could result in the creation of a new source or that could significantly change the nature or increase the quantity of the pollutants discharged or that could significantly affect the permittee's sludge use or disposal practices

Anticipated noncompliance—planned changes in the permitted facility or activity that may result in noncompliance

Transfers—transfer of the permit to another "person"[1]

Monitoring results—Discharge Monitoring Reports as required in the permit

Compliance schedules—reports required under compliance schedules

24-hour reporting—any noncompliance that may endanger health or the environment, a violation of a maximum daily discharge limitation, bypasses, upsets[2]

Other noncompliance—any other noncompliance (reported with DMRs)

Discharges exceeding notification levels in § 122.42—discharges in which levels of toxic pollutants exceed levels established in EPA regulations

Other information—when permittee becomes aware that it failed to submit any relevant facts, or submitted incorrect information, in a permit application

1. The permitting authority may require modification or revocation and reissuance of the permit to change the name of the permittee and incorporate any such other requirements as may be necessary.
2. Notice must be given orally within 24 hours of when the permittee becomes aware of the circumstance. Written notice must be provided within five days. The written notice must contain a description of the noncompliance and its cause, the period of noncompliance, and if the noncompliance has not been corrected, the anticipated time it is expected to continue, and the steps taken or planned to reduce, eliminate, and prevent reoccurrence of the noncompliance.

3

Effluent Limitations

3.1 INTRODUCTION

The CWA mandates a two-part approach to establishing effluent limitations. First, all dischargers are required to meet treatment levels based on EPA's assessment of the capabilities of treatment technologies that are technologically and economically achievable in the discharger's particular industry. This technology-based treatment level is considered to be the baseline for dischargers. Second, more stringent treatment requirements must be met where they are found to be necessary to achieve water quality goals for the particular body of water into which a facility discharges. Water quality–based controls may be a combination of chemical-specific limitations, whole effluent toxicity control, and a biological criteria/bioassay and biosurvey approach.[1]

3.2 FORMS OF PERMIT LIMITATIONS

Permit limitations may be expressed in several ways. Most technology-based or water quality–based limitations are expressed either as a mass limitation (e.g., two pounds per day or two pounds per "X" units of production) or a concentration limitation (e.g., 50 parts per million). Other types of permit limitations include visual observations (e.g., no visible sheen, foam, or floating solids), monitor-only requirements, requirements to perform specified tests, limitations on indicator parameters (e.g., biological oxygen demand, total organic carbon), flow limitations, pH range limitations, and temperature limitations.

Most permits impose both maximum limitations (i.e., the discharge may not exceed the limit during any monitoring event during the permit term, known as a daily maximum limitation) and monthly average limitations (i.e., the average of discharge levels as revealed in daily, weekly, or monthly monitoring throughout the month may not exceed the limitation). For some parameters, particularly pH, a permit may require continuous monitoring and virtually continuous compliance.[2]

3.3 TECHNOLOGY-BASED LIMITATIONS

EPA establishes national effluent guidelines (or effluent limitations guidelines) for industrial categories through notice and comment rulemakings.[3] These guidelines establish limitations for all types of dischargers within the industrial category (i.e., direct and indirect dischargers, existing and new sources) and for specific types of discharges (e.g., process water, cooling water, sanitary wastewater). The effluent guidelines are enforceable only through their incorporation into an NPDES permit.

EPA has issued effluent guidelines for more than 50 industrial categories.[4] In a 1992 consent decree,[5] EPA agreed to meet a schedule for issuing 19 additional effluent guidelines over an 11-year period. This schedule has been extended through several amendments to the 1992 agreement.[6]

Industrial Categories for Which Effluent Guidelines Have Been Promulgated

Dairy products	Pulp, paper and paperboard
Grain mills	Meat products
Canned and preserved	Metal finishing
Fruits and vegetables	Coal mining
Canned and preserved	Oil and gas extraction
Seafood processing	Mineral mining and processing
Sugar processing	Centralized Waste Treatment
Textile mills	Pharmaceutical manufacturing
Cement manufacturing	Ore mining and dressing
Feedlots	Transportation Equipment Cleaning
Electroplating	Paving and roofing materials
Organic chemicals, plastics, and synthetic fibers manufacturing	Waste combustors
	Landfills
Inorganic chemicals manufacturing	Paint formulation
Soap and detergent manufacturing	Ink formulation
Fertilizer manufacturing	Gum and wood chemicals manufacturing
Petroleum refining	Pesticide chemicals
Iron and steel manufacturing	Explosives manufacturing
Nonferrous metals manufacturing	Carbon black manufacturing
Phosphate manufacturing	Photographic
Steam electric power generating	Hospitals
Ferroalloy manufacturing	Battery manufacturing
Leather tanning and finishing	Plastics molding and forming
Glass manufacturing	Metal molding and casting
Asbestos manufacturing	Coil coating
Rubber manufacturing	Porcelain enameling
Timber products processing	Aluminum forming
	Copper forming
	Electrical and electronic components
	Nonferrous metals forming

Section 304(m) of the act requires EPA to issue biennial reports, or effluent limitations guideline plans, listing candidate sectors for new or revised effluent limits.[7] In 2006 one district court held that, although EPA historically has opted not to issue effluent limitations guidelines for all sectors identified for possible regulation, Section 304(m) requires EPA to issue effluent limitations guidelines for industry sectors identified for possible regulation in the plan.[8]

For dischargers in industrial categories for which EPA has not yet issued effluent guidelines and for types of discharges not covered by an applicable effluent guideline, permit writers will apply best professional judgment (BPJ) to establish permit limitations.[9] Normally, in applying BPJ, the permit writer will assess potentially applicable technologies applied to similar discharges in other industrial categories and may evaluate effluent treatability and analytical methods to develop limitations roughly equivalent to what an applicable effluent guideline would prescribe.[10]

Permit limitations based on BPJ are subject to EPA's antibacksliding policy. This policy, which was codified in the 1987 amendments to the act, prohibits, with a few exceptions, the relaxation of BPJ limitations in subsequent permits even if subsequently promulgated effluent guidelines would allow less stringent limitations.[11] The 1972 amendments to the act established a two-step program for the development of effluent limitations based on an application of technology-based controls. In the first phase, industrial dischargers were required by July 1, 1977, to meet a level of pollutant control based on the application of the best practicable control technology currently available (BPT).[12] In the second phase, to be achieved by July 1, 1983, industrial dischargers were required to meet differing standards depending on the type of pollutant at issue.[13] For toxic and nonconventional pollutants, industrial dischargers were required to meet a level of pollution control based on the application of best available technology economically achievable (BAT).[14] For conventional pollutants (biological oxygen demand, total suspended solids, fecal coliform, pH, oil, and grease), the act allows the application of a more lenient best conventional pollutant control technology (BCT).[15]

The BPT/BAT/BCT system of standards does not apply to new sources.[16] Direct discharges that are new sources must meet new source performance standards (NSPS), which are based on best available demonstrated control technology.[17]

Industrial dischargers that discharge into publicly owned treatment works (POTWs)—known as indirect dischargers—are not regulated by the standards applicable to direct dischargers. Instead, they are regulated under the Section 307(b) pretreatment program, which is discussed in chapter 7.

3.3.1 BPT

EPA sets BPT standards by surveying the particular industry to determine the types of treatment facilities typical of the industry and, with this information, determining the levels of pollution control achieved by the better-run facilities using the typical technologies. The agency then considers the category-wide or subcategory-wide cost of applying the technology in relation to the effluent reduction benefits.[18] This results in a control level reflecting the average of the best achieved by the industry.[19]

Although the 1977 deadline for meeting BPT limitations is long past, EPA has continued to promulgate or revise BPT standards for conventional pollutants.[20] It is EPA's position that Congress did not intend BCT standards to replace BPT standards for conventional pollutants. Instead, EPA has asserted that BCT, which places cost-effectiveness constraints on incremental technology requirements that exceed BPT, was intended only to supplement BPT.[21]

3.3.2 BAT

BAT controls are intended to represent the maximum feasible pollution reduction for an industry. In making this determination, EPA often looks beyond the technologies usually employed by an industry, basing its standards on technologies used in other industries or on pilot plant data. The act's requirement that BAT be the best technology does not mean that the chosen technology must be the one with the highest level of pollutant removal. The determination of best technology is based on numerous factors—for example, non-water-quality environmental impacts, energy requirements, and the cost of achieving effluent reduction—only one of which is pollution control.[22]

Although no cost-benefit analysis is required, the standards must be economically achievable. EPA generally considers BAT standards to be economically achievable if they would not force the closure of a large portion of the plants in an industrial category or subcategory.[23] In June 2002, EPA proposed to significantly pare back proposed effluent guidelines for the iron and steel manufacturing industry based on a finding that the industry was not financially equipped to comply with the broader proposed guidelines.[24]

3.3.3 BCT

Congress created the BCT category of controls because it was concerned that the costs of moving beyond BPT to BAT for conventional pollutants were likely to be unreasonable considering the nature of conventional pollutants. It therefore required BCT standards to meet a cost-reasonableness test.[25] In most industrial categories, the BCT limitations are no more stringent than those established for BPT.

3.3.4 NSPS

New sources are subject to NSPS. The question of whether a discharge is a new source is often disputed. New source is defined in the act as "any source, the construction of which is commenced after the publication of proposed regulations prescribing a standard of performance . . . which will be applicable to such source, if such standard is thereafter promulgated."[26] EPA regulations modify this definition somewhat, defining a new source as a facility for which construction begins after the promulgation of an applicable final NSPS or after proposal of such an NSPS, if the NSPS is finally promulgated within 120 days of its proposal.[27]

The determination of whether a facility is a new source can be far more complicated than suggested by the regulatory definition. For one thing, the U.S. Courts of Appeals are in conflict regarding the validity of the 120-day provision.[28] Further, it is not always clear when construction has begun. EPA's regulations define commencement of construction to include the placement, assembly, or installation of facilities or equipment, or significant site preparation work, or the entering into of a binding contract for the purchase of facilities or equipment.[29] Controversy may also arise as to whether construction at an existing facility should be considered a new source or simply a facility modification. To answer this question, EPA uses a substantial independence test, which looks at the degree to which the new unit functions independently of the existing facility, the degree of integration, and the extent to which the new unit is engaged in the same general type of activity as the existing source.[30]

NSPS are intended to reflect "the greatest degree of effluent reduction . . . achievable through application of the best available demonstrated control technology, processes, operating methods, or other alternatives, including, where practicable, standards permitting no discharge of pollutants."[31] Although NSPS often are very similar to BAT standards, they can be more stringent than BAT, based on EPA's position that it can require the installation of state-of-the-art treatment technology in new facilities where requiring existing facilities to retrofit their processes and systems to include such technology would be economically unreasonable. In addition, because EPA must consider alternative production processes and operating methods, NSPS can also effectively dictate the choice of production processes used by a new source. For example, the NSPS for the steam electric category prohibit the discharge of fly ash transport water, effectively requiring new source plants in this category to use a dry-ash handling system.[32]

One benefit of being a new source is that any new source facility constructed to meet all applicable NSPS may not be subjected to any more stringent standards for 10 years after the date construction is completed or for the period of depreciation under the Internal Revenue Code, whichever is shorter. This 10-year protection, however, applies only to technology-based limitations; it does not prevent more stringent water quality–based limitations from being imposed on a new source. In addition, at the end of the 10-year period, immediate compliance with the standards in effect at that time will be required.[33]

3.3.5 Variances

The statutory mechanisms for obtaining a variance from a technology-based standard are very limited. One such variance is the fundamentally different factors (FDF) variance.[34] The FDF variance mechanism allows a discharger to obtain a variance from technology-based limitations (including pretreatment limitations) other than NSPS. The variance is not available for water quality–based limitations.

To obtain an FDF variance, a discharger must demonstrate that the factors applicable to its facility are fundamentally different from those considered in the development of the effluent limitations guidelines applicable to the facility. Factors that may be fundamentally different include the following:

1. The nature or quality of the pollutants contained in the discharge;
2. The volume of the discharge;
3. The non-water-quality environmental impacts of control and treatment of the discharge;
4. The energy requirements of the treatment technology; and
5. The age, size, land availability, and configuration as they relate to the discharger's equipment, facilities, and processes.

The variance is not available simply because the cost of compliance with the limitations would force plant closure.[35]

Under the 1987 amendments to the act, a permittee must file a request for an FDF variance within 180 days after the publication of the applicable effluent guideline or standard.[36] In addition, an applicant for an FDF variance must show that it raised the fundamentally different factors during the development of the effluent guideline or show why it did not have a reasonable opportunity to raise such issues. This requirement makes it imperative that affected industries monitor and participate in the development and revision of effluent guidelines.

Several other variances are available under the act, although they are rarely, if ever, granted. Under Section 301(c), EPA may modify BAT requirements or pretreatment requirements affecting nonconventional nontoxic pollutants if it can be shown that the economic capability of the discharger requires less stringent limitations. The modified limitation must be shown to result in further progress toward elimination of the discharge of pollutants. Under Section 301(g), EPA may grant a variance from BAT for several nonconventional nontoxic pollutants (ammonia, chlorine, color, iron, and total phenols) where the applicable BAT limits are unnecessarily stringent.

3.4 WATER QUALITY–BASED LIMITATIONS

3.4.1 Water Quality Standards

Water quality standards are established by the states and consist of two elements: (1) use classifications and (2) criteria that, if not exceeded, will protect those uses.[37] The CWA requires all states to classify the waters within the state according to intended use (e.g., public drinking water supplies, propagation of fish and wildlife, recreational purposes, and industrial, agricultural, and other uses).[38]

Water quality criteria quantitatively describe the physical, chemical, and biological characteristics of waters necessary to support the designated uses. State water quality standards thus normally consist of a numeric level of a pollutant that cannot be exceeded in the ambient water in order to protect the designated use. For example, the standard may state that the level of arsenic in a stream designated for trout propagation may not exceed 0.2 milligrams per liter.[39]

The state water quality standards must attain the act's goal of fishable, swimmable waters wherever possible and, under EPA's anti-degradation policy, must maintain both the uses designated in the standards and the current uses unless the state can demonstrate that the designated use is unattainable or infeasible for reasons specified in EPA's regulations.[40]

State criteria normally are based on federal water quality criteria, which have been published for more than 150 pollutants.[41] The federal criteria are not enforceable standards but are guidance that can be used by states in determining appropriate numerical criteria for water bodies within the state. A state may, however, choose to set site-specific criteria based on the characteristics of the local water body. For example, site-specific criteria may be appropriate in water bodies with species different from those used in the development of the federal criteria or where adaptive processes have enabled a viable, balanced aquatic community to exist with levels of pollutants that exceed the national criteria.[42]

Courts have held that both states and citizen groups may bring actions for violations of state water quality standards, even if those standards are not translated into permit limitations.[43]

3.4.2 Translating Standards into Chemical-Specific Permit Limitations

EPA's regulations require permits to include water quality–based limitations for all pollutants that "are or may be" discharged at levels that cause, have a "reasonable potential to cause, or contribute to an excursion above any State water quality standard."[44] In setting a permit limitation to meet a water quality standard, the permit writer will calculate how much of the pollutant the permittee may discharge without causing the ambient standard to be exceeded.

To calculate an appropriate permit limitation based on a numerical state standard, the permit writer will take into account the dilution provided by the receiving water. Where mixing with the receiving water occurs rapidly, the permit writer typically will consider the flow of the receiving stream (usually the 7Q10 low flow[45]) as compared to the flow of the outfall and will develop a dilution factor.[46] For example, an outfall that discharges 50,000 (0.05 million) gallons per day into a receiving stream with a low flow of 1.0 million gallons per day would be assigned a dilution factor of 20. The ambient water standards would then be multiplied by this dilution factor to determine the permit limitations.[47]

Where mixing occurs less rapidly, a permit writer typically will set permit limitations based on a mixing zone. A mixing zone is that portion of the receiving water that immediately receives an effluent discharge and in which the initial dilution of the discharge takes place.[48] Normally, permit writers will allow water quality standards to be exceeded within the mixing zone so long as acutely toxic conditions do not exist.[49] Some states will take mixing into account through the use of dynamic modeling which models the fate of a discharge in the receiving waters.

3.4.2.1 Total Maximum Daily Loads

Section 303(d) of the act requires that states identify and establish a priority ranking for water bodies for which technology-based effluent limitations required by Section 301 are not stringent enough to attain and maintain applicable water quality standards.[50] For these priority waters, the states must establish total maximum daily loads (TMDLs) for the pollutants causing the impairment in those water bodies and submit the list of impaired water bodies and TMDLs to EPA for approval.[51] If a state fails to do so, EPA will establish a list of impaired water bodies and TMDLs for the state.[52]

A TMDL is a quantitative assessment of pollutants that cause water quality impairments in a particular water body. A TMDL specifies the total amount of a particular pollutant that a segment of water may receive (from point, nonpoint, and natural background sources)[53] without exceeding applicable water quality criteria, allocates allowable pollutant loads among the sources contributing the pollutant to the water body, and provides the basis for attaining or maintaining water quality standards. Where a TMDL has been established for a water body, water quality–based effluent limits for point-source discharges to the water body are based on allocations of pollutants assigned to the point source through the TMDL.

Despite the long-standing statutory mandate, states have generally been slow to submit lists of impaired water bodies to EPA and even slower in developing TMDLs. To jump-start the TMDL program, EPA issued a final rule revising the federal TMDL regulations on July 13, 2000.[54] The rule expanded the program's reach to include the regulation of nonpoint sources of pollution as well as point sources and set rigid schedules for submitting impaired waters lists and TMDLs. The July 2000 rule was extremely controversial. The agency received more than 34,000 comments on the proposed TMDL rule. To counter congressional threats to eliminate funding for implementing the rule, EPA established the effective date of the final rule as October 1, 2001.

More than a dozen groups, including farm groups, environmental groups, municipal wastewater treatment officials, and state water pollution agencies, filed challenges to the revised TMDL rule.[55] On July 16, 2001, EPA filed a proposal to stay the litigation for 18 months to allow the agency to revise the TMDL rule. EPA issued a *Federal Register* notice in October 2001 that extended the effective date of the final rule to April 30, 2003.[56] Ultimately, however, EPA withdrew the July 2000 rule. As a result, the 1992 TMDL amendments establish the current program rules.

On December 5, 2013, EPA released a new "collaborative framework" for TMDL implementation.[57] This framework provides long-term goals for EPA and the states to work together to achieve. The chief goal is to, by 2020, have states identify and characterize the extent of healthy and impaired waters in priority watersheds or waters through site-specific assessments. The framework does not alter existing responsibilities or authorities under Section 303(d) or its regulations.

Apart from the challenges to the federal TMDL regulation, there have also been lawsuits filed in at least 38 states over the delay in issuance of TMDLs under the existing TMDL program. In most cases, courts have established mandated timetables for the establishment of TMDLs (generally five years).

A 2006 decision by the District of Columbia Court of Appeals has proved significant in this regard.[58] At issue in the case before the court was EPA's approval of TMDLs for the Anacostia River, an impaired waterway, that limited the annual discharge of oxygen-depleting substances and limited the seasonal discharge of pollutants contributing to turbidity. An environmental group petitioned for review of EPA's approval of the TMDLs, arguing that the CWA requires the establishment of "total maximum daily loads," not seasonal or annual loads.[59] The Court of Appeals agreed that the plain language of the act requires the establishment of daily limits and remanded the case to the district court with instructions to vacate EPA's approval of the TMDLs.[60] EPA published guidance in light of the court's decision, noting that load allocations should be expressed in daily time increments.[61]

EPA has considered an alternative approach known as the Watershed Rule, which contains many of the elements of the TMDL rule, but also varies in some important ways. For example, the Watershed Rule would include a pollutant trading program (see section below) and require states to submit a Continuing Planning Process program that would provide blueprints for implementing the states' clean water programs rather than TMDL implementation plans. A draft of the Watershed Rule was issued in 2003, but has never been finalized.

3.4.2.2 Pollutant Trading

In June 1996, EPA introduced the concept of watershed-based trading of pollutants in effluents.[62] Under the proposed effluent trading approach, facilities that can reduce pollution to meet water quality standards at lower costs could accumulate credits by going beyond their permit requirements. Once these reductions are achieved, a trading program would allow companies to sell or barter credits to other facilities within the same watershed. Small dischargers are expected to be able to purchase these credits at a lower price than they would pay to install the technology needed to reduce pollution to meet water quality standards. A trade will not be approved if it would result in any violation of water quality standards. In addition, trades would require a net reduction in pollution. Pollutant trading may not be used to meet federal technology-based permit limitations (e.g., BPT, BAT, NSPS).[63]

State and local permitting authorities have included pollutant trading in their permit programs. Point/nonpoint trading programs were originally established for the Tar-Pamlico estuary in North Carolina, the Dillon Creek Reservoir in Colorado, and Cherry Creek in Colorado. More recently, planned or pilot trading programs have been developed for the state of Michigan, the Chesapeake Bay watershed, the Long Island Sound, the Lower Boise River in Idaho, and the Fox-Wolf, Rock River, and Red Cedar basins in Wisconsin. Other states are considering trading programs. EPA has published materials for permit writers and stakeholders to assist in determining whether trading programs are appropriate for specific watersheds and how such programs could be designed and implemented.[64]

On October 13, 2006, EPA and the Agriculture Department's Natural Resources Conservation Service (NRCS) entered into a partnership agreement intended to eliminate barriers to and to develop standards for water quality credit trading. In addition to coordinating programs and activities, the agreement provides for the development of a water quality credit trading pilot project (or related market-based project) in a watershed within the Chesapeake Bay basin. This partnership between EPA and the NRCS is intended to increase the use of market-based approaches to improve water quality.

3.4.2.3 Watershed Permits

In 2003, EPA adopted a new approach to NPDES permits that allows states to issue NPDES permits for an entire watershed instead of specific facilities. Such permits are intended, among other objectives, to facilitate pollutant trading and ease the significant backlog of permits to be issued. EPA issued guidance in 2003 and again in 2007 on the watershed-based approach, setting forth a number of case studies implementing the NPDES watershed framework, including some involving multisource watershed-based permits.[65]

3.5 TOXICITY-BASED LIMITATIONS

Because of the enormous difficulties associated with setting individual water quality–based permit limitations, EPA and the states rely extensively on limitations based on whole effluent toxicity (WET).[66] WET limitations are imposed in the form of a permit requirement to perform toxicity testing on the permittee's effluent. Effluent toxicity tests involve exposing selected species of aquatic life to one or more concentrations of an effluent in a laboratory setting to determine the short-term and/or long-term effects of exposure to the effluent.[67]

Most permits now require dischargers to conduct toxicity tests on a regular basis (normally monthly or quarterly).[68] For example, a permit may require the permittee to conduct a monthly 24-hour acute toxicity test on a species of water flea known as Daphnia magna using 100 percent effluent. In some cases, the permit may require longer-term chronic toxicity testing on a species of fish, such as the fathead minnow. This toxicity testing can be very expensive, in some cases more expensive than all of the other monitoring and testing required by a permit.

In addition, a permit may establish a limitation on the results of the testing, such as providing that the mortality rate in an acute toxicity test may be no more than 10 percent. With this type of permit, the failure of a toxicity test will be considered a violation of the permit and could subject the permittee to penalties. In other permits, the failure of a toxicity test is not a permit violation, but it triggers a requirement to perform a toxicity identification evaluation or toxicity reduction evaluation to identify and eliminate the source of the effluent toxicity.[69]

3.6 BIOLOGICAL CRITERIA

In the early 1990s, EPA began to focus on a third approach for controlling water pollution. In addition to chemical-specific and WET requirements, EPA turned its attention to biological assessments—that is, evaluating the integrity of receiving waters directly by comparing various in-stream characteristics, such as species diversity and number, to the characteristics representative of "unaffected" waters.[70] It is now EPA's policy that states must fully integrate chemical-specific limitations, toxicity testing, and biological criteria, or biocriteria, into their water quality programs.[71]

3.7 NUTRIENT CRITERIA

Eutrophication of water bodies has been a significant long-standing problem in the United States, causing fish kills, low dissolved oxygen, murky water, and depletion of desirable flora and fauna. In February 2002, EPA published nine nutrient water quality criteria documents for water bodies within certain geographical "eco-regions."[72] These criteria documents supplemented the 17 eco-regional nutrient criteria documents announced by EPA in January 2001.[73] The criteria documents present recommended criteria for causal parameters (total phosphorus and total nitrogen) and response variables (chlorophyll and turbidity). In 2008 EPA released a report describing progress made by states in the preceding decade in adopting numeric nutrient criteria.[74] Because states have been slow to adopt numeric nutrient criteria for all their waters, EPA has continued to offer guidance and technical assistance.[75]

4

Whole Effluent Toxicity Control

4.1 BACKGROUND

Since 1984, EPA and the states have employed an integrated strategy consisting of both biological and chemical means to control toxic effects on water quality beyond the Act's technology-based requirements to achieve and maintain state water quality standards.[1] One method for measuring the biological effects of toxic pollutants upon aquatic life is whole effluent toxicity (WET) testing. EPA and the states have used the data derived from WET testing to assess compliance with state water quality standards and to establish NPDES permit limitations necessary to attain and maintain those standards.

The WET approach to toxics control for the protection of aquatic life involves the use of acute and chronic WET testing to measure the toxicity of wastewaters. Whole effluent toxicity tests typically use standardized, surrogate freshwater or marine plants, vertebrates, or invertebrates to measure the aggregate toxic effect of an effluent. An acute WET test is typically a test of 96 hours or less in duration in which lethality is the measured endpoint. A chronic WET test is typically a longer-term test in which sublethal effects, such as effects on growth and reproduction, can be measured in addition to lethality.[2] Most chronic tests required today are seven-day tests, called short-term chronic tests, that do not last for the entire life cycle of the test organism, but instead focus on the most sensitive life cycle stages.[3]

In the past ten years, the Agency has published regulations, policy statements, and guidance documents that address a variety of issues associated with WET controls in NPDES permits. In 1984, EPA issued the "Policy for the Development of Water Quality-Based Permit Limitations for Toxic Pollutants."[4] This policy addresses the use of biological and chemical methods to assure that toxic whole effluent discharges are regulated consistent with federal and state requirements.

In 1989, EPA revised existing 40 C.F.R. § 122.44(d)(1), which previously required NPDES permits to contain any more stringent requirements necessary to achieve state water quality standards.[5] The revised regulation describes in greater detail requirements for NPDES permitting authorities to follow in developing NPDES effluent limitations to assure compliance with state water quality standards.[6]

4.2 WHEN ARE WET LIMITATIONS REQUIRED?

Under EPA's regulations, when the permit writer determines that a discharge causes, or has the reasonable potential to cause, or contributes to an in-stream excursion of a state water quality criterion for WET, the permit must contain effluent limits for WET.[7] Limits on WET are not necessary where the permit writer demonstrates that chemical-specific limits for the effluent are sufficient to attain and maintain applicable numeric and narrative water quality standards.[8]

Several states have adopted numeric criteria for WET. Most control of WET, however, is based on state narrative standards, such as "all state waters must, at all times and all flows, be free from substances that are toxic to humans or aquatic life." EPA regards such narrative criteria to cover both short-term and long-term WET effects.[9] Most effluent limitations to control WET result from the permit writer's case-by-case determination of what concentration of WET meets the narrative criterion.

State WET criteria or the procedures implementing state criteria typically describe the magnitude, duration, and return frequency for WET. Magnitude is the maximum allowable concentration of WET, which is typically expressed as a concentration of toxicity in-stream. Duration is the period of time over which the in-stream concentration is averaged for comparison with criteria concentrations, in order to limit the durations of concentrations above the criterion. Return frequency is a designation of how often the criterion may be exceeded without affecting the organisms in the water body. The permit writer uses the magnitude, duration, and return frequency provisions of WET criteria to develop wasteload allocations and effluent limitations to control WET in a discharge.

WET criteria as adopted by the state or as implemented by the permit writer on a case-by-case basis normally consist of two expressions of magnitude of toxicity to test species, one to protect against acute toxicity effects and one to protect against chronic toxicity effects, along with a duration and return frequency for each. EPA has not developed a recommended Section 304(a) criterion for WET. The Agency's *Technical Support Document for Water Quality-Based Toxics Control* (hereinafter referred to as the *TSD*), however, contains recommended magnitudes, durations, and return frequencies for assessing acute WET (.3 acute toxic unit [TU_a][10] to the most sensitive of at least three species, with an averaging period of one hour and a once in three years return frequency) and chronic WET (1.0 chronic toxic unit [TU_c] to the most sensitive of at least three species, with an averaging period of four days and a once in three years return frequency).

4.3 WET-BASED PERMIT LIMITATIONS

4.3.1 Types of Limitations

A limit on WET in an NPDES permit may be expressed in a number of ways, such as toxic units, no observed effect concentration (NOEC), lethal concentration, or percent mortality. For acute toxicity tests, the endpoint is normally a lethal concentration which is the concentration of effluent at which a certain percentage of the test organisms die. For example, the LC_{50} is the effluent concentration that results in the mortality of 50 percent of the test organisms.[11] Endpoints for chronic testing are usually the "no observed effect concentration" (NOEC), the "lowest observed effects concentration" (LOEC), or the "effect concentration." The NOEC is the highest concentration of effluent at which no adverse effects of the effluent (e.g., decreases in reproduction and growth, or lethality) are observed in the test species. The LOEC is the lowest concentration to which test organisms are exposed that causes an observed effect. The EC is the concentration that would cause an adverse effect upon a certain percentage of the test organisms (e.g., the EC_{10} or EC_{50}).

Because there is an inverse relationship between toxicity and the EC (i.e., the lower the EC, the higher the toxicity of the effluent), EPA recommends that toxicity be measured in terms of toxic units (TUs). An acute toxic unit (TU_a) is 100 divided by the LC_{50}; a chronic toxic unit (TU_c) is 100 divided by the NOEC.[12] Acute and chronic TUs make it easy to quantify the toxicity of an effluent and to specify water quality criteria based on toxicity. For example, an effluent that contains 20 TU_cs is twice as toxic as an effluent that contains 10 TU_cs.

4.3.2 Elements of WET Testing

Permit requirements relating to WET may differ in a number of factors, including test organism, dilution water, sampling frequency, exposure method (flow-through, static, renewal), sampling method (grab, composite), and testing location (off-site, on-site). Each of these elements is described below.

4.3.2.1 Test Organism

The typical test organisms used in WET testing are listed on the following chart.

To provide sufficient information for making permitting decisions, EPA recommends that a minimum of three species representing three different phyla (e.g., a fish, an invertebrate, and a plant) be used to test an effluent for toxicity. In some cases, such as where the effluent is highly consistent, the use of the one or two most sensitive species may be allowed on a case-by-case basis.[13]

Sometimes, state agencies require testing on representative resident species under the assumption that such tests are needed to assess impact to local aquatic life. EPA considers this unnecessary since standard test species have been shown to represent

Standard Test Species

Acute Tests	*Chronic Tests*
Freshwater Fathead minnow Bluegill Channel catfish Trout/Salmon Daphnia magna Daphnia pulex Ceriodaphnia	**Freshwater** Fathead minnow Daphnia Ceriodaphnia Selenastrum capricornutum (algae)
Marine Mummichog Sheepshead minnow Silverside Flounder Mysid shrimp Grass shrimp Oysters	**Marine** Silverside Sheepshead minnow Mysid shrimp Sea urchin Red macroalgae

the sensitive range of all ecosystems. EPA strongly discourages toxicity testing using resident species unless it is required by a state statute, or the state agency determines that using a unique resident species would be far more protective of the receiving water than testing using the EPA surrogate species.[14]

4.3.2.2 Dilution Water

The permit may require any of several different sources of dilution water, including the receiving water near the discharge point, clean receiving water, or laboratory tap water. Use of receiving water for dilution most closely simulates the real interactions between the effluent and the receiving waters. This is particularly important where the receiving water has unique salinity or hardness characteristics.

4.3.2.3 Testing Frequency

Testing frequency requirements differ greatly among states and EPA regions. The testing frequency often depends on the variability of the plant processes. For example, if the plant processes involve short-term variability, a permit may require four to six samples monthly for at least one year. Where the plant processes show long-term variability, EPA recommends that tests be scheduled to coincide with expected process changes.

4.3.2.4 Flow-Through v. Static and Renewal Tests

Two basic types of testing systems are available to measure effluent toxicity: flow-through systems and static systems. A flow-through test is conducted using a diluter system and a continuous feed of effluent and dilution water. A static test is conducted in test chambers into which effluent and diluent are added only at the beginning of the test. A variation of the static procedure is the renewal toxicity test. This test uses the same delivery system as a static test, but the test solutions are changed, or renewed, on a predetermined schedule (e.g., every 24 hours).

If the effluent is highly variable, either a flow-through or renewal test would be appropriate. If the effluent is not considered variable, such as a discharge from a 30-day retention basin, then a static or renewal test procedure. Because flow-through tests are significantly more expensive and complicated to use, cost may also be a factor in choosing an appropriate test method.[15]

4.3.2.5 Effluent Concentrations

Permits may require the permittee to conduct either screening toxicity tests or more definitive tests. Normally, a screening test involves only a single concentration of effluent (usually 100%). A definitive test will normally include five concentrations (e.g., 100%, 50%, 25%, 12.5%, 6.25%). Both tests require the use of a control concentration as well as the test concentrations.

4.3.2.6 On-Site v. Off-Site Testing

For the most part, comparisons of toxicity data from tests conducted on-site and tests conducted off-site have shown little variation. However, if the presence of non-persistent compounds (e.g., chlorine or other volatiles) in the effluent is suspected or known, the regulatory authority may require on-site testing. In addition, when conducting flow-through toxicity tests which require a continuously pumped sample, on-site testing is strongly recommended.[16]

4.3.2.7 Grab v. Composite Sampling

The use of a grab sample or a composite sample depends on the objectives of the test and the long-term operations of the discharger. If the toxicity of the effluent is variable, grab samples collected during the peaks of toxicity provide a measure of maximum toxic effect. Collection of grab samples may also be necessary if there is little dispersion or mixing of the effluent in the receiving water. Because grab samples must be collected at the time of the toxicity spike in order to measure peak toxicity, a 24-hour composite sample may be used to catch such spikes. The compositing, however, may tend to dilute the toxicity of the spikes, leading to an inaccurate measure of the maximum toxicity of the effluent. Composite samples are thus more appropriate for chronic tests where peak toxicity of short duration is of a lesser concern.[17]

4.3.2.8 Acute-to-Chronic Ratio

The acute-to-chronic ratio (ACR) expresses the relationship between the concentration of WET or a toxicant causing acute toxicity to a species (e.g., the LC_{50}) and the concentration causing chronic toxicity to the same species (e.g., the NOEC). In other words, the ACR can be defined as LC_{50}/NOEC.

The ACR is commonly used to extrapolate to a chronic toxicity concentration using available acute toxicity data coupled with exposure considerations when chronic toxicity data is not available. The ACR varies between species for a given chemical, and, for any one species, between different toxicants. The ACR may be used in developing chronic toxicity limits where chronic toxicity testing is not measured directly, in order to minimize testing costs. Where acute and chronic toxicity data are available, the ACR should be calculated directly for that specific effluent.

Collected data show that ACRs for WET range from less than one to more than 50, with the majority falling below 20. In the absence of data to develop an ACR, EPA recommends that an ACR of 10 be used.[18]

4.3.3 WET Permit Requirements

The following is an example of a typical permit provision requiring WET testing:

WET Permit Requirements

The Permittee shall determine by use of the "Effluent Toxicity Screening Test" set out in Attachment A if the effluent from outfall 002 is lethal to more than 10 percent of a culture of test organisms, as specified in Attachment A, when the effluent is tested as follows:

Effluent sample—100% by volume
Dilution water—0% by volume

The Permittee shall conduct the first test specified above within sixty (60) days of the issuance of this permit. The tests will be conducted monthly thereafter for twelve (12) consecutive months. If, at the end of twelve months, no tests have shown the effluent to be lethal to more than 10 percent of the test organisms, the Permittee shall conduct the test every six months during the remaining term of the permit.

Within ninety (90) days after any test that caused lethality of more than 10 percent, the Permittee must submit to the Department an explanation and evaluation of the cause of the toxicity. If the effluent continues to show toxicity, the Permittee will submit to the Department a plan for reducing toxicity to a safe level.

4.3.4 Toxicity Identification and Reduction Evaluations

Where monitoring indicates unacceptable effluent toxicity, the principal mechanism for bringing a discharger into compliance with a water quality–based whole effluent toxicity requirement is a Toxicity Identification and Reduction Evaluation (TI/RE). The purpose of a TI/RE is to investigate the causes and to identify corrective actions for difficult effluent toxicity problems. A permit may require a discharger to conduct a TI/RE whenever a toxicity test fails permit criteria, or may require a TI/RE only where the permittee is unable to explain adequately and immediately correct exceedances of a whole effluent toxicity permit limit or requirement.

Sample Permit Language Requiring a TI/RE

The discharger shall demonstrate that effluent toxicity-based permit limits described in Part I.A of the permit are being attained and maintained through the application of all reasonable treatment and/or source control measures. Upon identifying noncompliance with those limits, the discharger shall initiate corrective action according to the following schedule:

Task	*Deadline*
1. Take all reasonable measures necessary to reduce toxicity immediately.	Within 24 hours
2. Submit a plan and schedule to attain continued compliance with the effluent toxicity-based permit limits in Part I.A., where the source of toxicity is known, if immediate compliance is not attained.	Within 30 days
3. Submit a TI/RE study plan detailing the toxicity reduction procedures to be employed where the source is unknown and toxicity cannot be immediately controlled through operational changes.	Within 45 days
4. Initiate TI/RE plan.	Within 45 days
5. Comply with approved TI/RE schedule.	Immediately upon approval
6. Submit results of TI/RE.	Per schedule
7. Implement TI/RE controls as described in final report.	On due date of report
8. Complete TI/RE implementation to meet permit limits and conditions.	Per schedule

5

Storm-Water Discharges

5.1 THE STORM-WATER PROGRAM

The problem of how to regulate storm-water discharges has plagued EPA for decades. Although the CWA has prohibited unpermitted point-source discharges of any pollutant to navigable waters since 1972, EPA and the courts struggled long and hard to develop a feasible program to reconcile the statutory requirement with the practical problems of regulating possibly millions of diverse point-source discharges of storm water. At the same time, EPA continued to identify storm-water runoff as a significant source of water quality impairment. Ultimately, the storm-water program required congressional action.

In the 1987 amendments to the Act, Congress established a schedule for the regulation of, and issuance of permits for, storm-water discharges. The amendments required EPA to issue regulations setting forth permit requirements for storm-water discharges "associated with industrial activity" and for discharges from municipal separate storm-water sewer systems serving populations of 250,000 or more by February 4, 1989; to issue regulations setting forth permit requirements for municipal separate storm-water sewer systems serving populations of 100,000 or more but less than 250,000 by February 4, 1991; and to issue regulations designating other storm-water discharges to be regulated to protect water quality by October 1, 1992.[1]

In the regulations EPA issued, storm water is defined as "storm water runoff, snow melt runoff, and surface runoff and drainage."[2] This definition does not include infiltration (water that enters a sewer system from below the surface of the ground through defective pipes, pipe joints, connections, or manholes) or street wash waters.

The definition of storm-water discharge associated with industrial activity in the regulations is long and complex and requires careful reading.[3] In general, the term is defined to include discharges from any point source used for collecting and conveying storm water that is directly related to manufacturing, processing, or materials storage areas at an industrial plant.[4] The definition excludes storm water discharged from areas that are separate from industrial activities, such as office buildings and parking lots, unless the drainage is combined with storm water drained from areas used for industrial activities.[5] The definition also excludes discharges from facilities engaged in wholesale, retail, service, or commercial activities.[6] EPA and permitting states retain the authority to require a permit for discharges falling outside this definition that contribute to water quality violations or are significant contributors of pollutants to water bodies.[7]

Industrial facilities with no exposure of industrial materials and activities to storm water may be able to obtain a conditional "no exposure" exemption from the storm-water permit requirement.[8] To obtain the exemption, the facility operator must submit a written certification that a condition of no exposure exists at the facility. The operator must provide a storm-resistant shelter to protect industrial materials and activities from exposure to rain, snow, snow melt, and runoff.[9]

The regulations define a municipal separate storm sewer as a conveyance or system of gutters, ditches, man-made channels, or storm drains that is owned by a state, county, municipality, or other public entity; is designed or used for conveying storm water; and is not a combined sewer or part of a publicly owned treatment works.[10] Discharges into municipal separate storm sewers are considered discharges to waters of the United States and require a permit to the same extent as direct discharges.

EPA may issue permits for municipal storm sewer systems either for a particular system or on a jurisdiction-wide basis. Permits for municipal storm sewer systems must control pollutants "to the maximum extent practicable."[11] Such permits must include a requirement to effectively prohibit non-storm-water discharges into the system's storm sewers. A federal appeals court has upheld EPA's policy that allows states to issue storm-water permits to municipal storm-water systems that rely on BMPs rather than numeric discharge limits.[12]

In 1999, EPA issued regulations requiring small municipalities (populations under 100,000) and construction sites of between one and five acres to obtain a discharge permit by March 10, 2003, and to implement BMPs to meet water quality standards.[13] The regulation was challenged in three separate actions and consolidated in the Ninth Circuit, which invalidated parts of the regulations that allowed small municipalities to obtain permits without typical regulatory review and public participation.[14] EPA has since issued guidance to update the permitting process for municipal separate storm sewer systems.[15]

5.2 THE STORM-WATER PERMIT PROCESS

Dischargers initially had three options for obtaining coverage under a storm-water permit:

- Coverage under a general permit
- Application for a permit through a group application
- Obtaining an individual storm-water permit.[16]

Only the general permit and individual permit options remain.[17]

5.2.1 General Permits

A storm-water discharger may apply for an NPDES permit by filing a notice of intent to be covered by a general permit issued by EPA or an authorized state. EPA has issued several general permits for storm-water discharges associated with industrial activity and discharges associated with construction activity. Although EPA's general permits apply only in the states and territories where EPA is the permitting authority, most authorized states have adopted EPA's general permits or have developed general permits of their own.

In September 1995, EPA issued a multi-sector general storm-water permit (MSGP) that would apply to 11,000 facilities in 29 industrial segments in the unauthorized states and territories.[18]

Industrial Categories Covered By the Multi-Sector General Permit—2000
Timber Products
Paper and Allied Products Manufacturing
Chemical and Allied Products Manufacturing
Asphalt Paving and Roofing Materials Manufacturing and Lubricant Manufacturing
Glass, Clay, Cement, Concrete, and Gypsum Product Manufacturing
Primary Metals
Metal Mining (Ore Mining and Dressing)
Coal Mines and Coal Mining-Related Facilities
Oil and Gas Extraction and Refining
Mineral Mining and Dressing
Hazardous Waste Treatment, Storage or Disposal
Landfills and Land Application Sites
Automobile Salvage Yards
Scrap Recycling Facilities
Steam Electric Power Generating Facilities
Land Transportation Facilities
Water Transportation
Ship and Boat Building or Repairing Yards
Air Transportation Facilities
Treatment Works
Food and Kindred Products
Textile Mills, Apparel, and other Fabric Product Manufacturing
Furniture and Fixture Manufacturing
Printing and Publishing
Rubber, Miscellaneous Plastic Products, and Miscellaneous Manufacturing Industries
Leather Tanning and Finishing
Fabricated Metals Products
Transportation Equipment, Industrial, or Commercial Machinery
Electronic and Electrical, Photographic and Optical Goods Manufacturing
Facilities not covered under other sectors and designated by the permitting authority

Under the multi-sector program, any facility that falls into one of the 29 industrial sectors, including facilities already covered by one of the baseline general permits, is allowed to apply for coverage under the five-year MSGP permit. Under the permit, storm-water dischargers are required to develop site-specific pollution prevention plans based on industry-specific BMPs specified in the permit. For most facilities, the requirements under the more tailored multi-sector permit are less stringent than those under the baseline general permit. Most authorized states have used the EPA multi-sector permit as a model for state-issued general permits.

5.2.2 Individual Storm-Water Permits

If a facility cannot qualify for coverage under a general permit, it must submit an individual storm-water permit application. The preparation of the individual application is very burdensome, requiring not only detailed information about the facility, but also quantitative data based on sampling of storm-water discharges collected during storm events.[19]

Information required in an application for an individual storm-water permit (Form 2F):

- Topographic map and/or site drainage map, showing: all drainage and discharge structures; the drainage area of each storm water outfall; paved areas and buildings within the drainage area of each outfall; past and present areas used for outdoor storage or disposal of significant materials; structures used to reduce pollutants in storm water runoff; materials loading and access areas; areas where pesticides, herbicides, soil conditioners and fertilizers are applied; hazardous waste treatment, storage, or disposal facilities; all wells used for underground injection of fluids; and all receiving waters from the facility;
- An estimate of the area of impervious surfaces and the total area drained by each outfall;
- A narrative description of materials that in the past three years have been exposed to storm water, all materials management practices and control measures employed to reduce contact with storm waters and other information;
- A certification that all storm water outfalls have been tested for non-storm water discharges and a description of testing methods employed;
- Information about significant leaks or spills of toxic or hazardous pollutants within the last three years;
- Quantitative data, based on samples collected during representative storm events, for parameters including pollutants limited in an effluent guideline or in the facility's NPDES permit, conventional pollutants, flow rates, and timing of sampling.

5.2.3 Storm-Water Management Plans and Pollution Prevention Plans

EPA contends that pollution prevention is the most environmentally sound and cost-effective way to control the discharge of pollutants in storm-water runoff from industrial facilities.[20] Thus, a primary component of all state and federal general permits is the requirement to develop and implement a storm-water pollution prevention plan, also referred to as a storm-water management plan, to control the pollutants carried by storm-water discharges into surface waters. These plans generally contain various BMPs that may be general, industry specific, or site specific.

The major objectives of such plans are (1) to identify sources of pollution potentially affecting the quality of storm-water discharges associated with industrial activity from the facility and (2) to describe and ensure implementation of practices to minimize and control pollutants in storm-water discharges from the facility and to ensure compliance with the terms and conditions of the general permit.[21]

6

Other Types of Discharge

6.1 COMBINED SEWER OVERFLOWS AND SANITARY SEWER OVERFLOWS

Two other wet weather problems that are largely the responsibility of municipalities are combined sewer overflows (CSOs) and sanitary sewer overflows (SSOs). CSOs usually occur when combined sanitary and storm sewer systems are overwhelmed during heavy rains. EPA estimates that approximately 770 communities, primarily in the Northeast and Great Lakes regions, still have combined sewer systems. EPA issued a CSO control strategy in April 1994 that requires municipalities to take several short-term and long-term measures to control CSOs.[1] SSOs occur when a separate sanitary sewer system is overwhelmed, generally as a result of increased flows or structural, operational, or maintenance problems.[2] EPA has identified CSOs and SSOs, particularly those occurring in dry weather, as top priorities for enforcement and has issued an enforcement policy to address such overflows.[3]

6.2 THERMAL DISCHARGES

Section 316 of the act establishes special criteria for the discharge of heat, which is defined as a pollutant under the act and is therefore subject to technology-based limitations. Under Section 316(a), if a discharger can demonstrate that the applicable technology-based limitation is more stringent than necessary to ensure protection and propagation of a balanced, indigenous population of shellfish, fish, and wildlife in and on the receiving water, EPA or the state may adjust the limitation to a less stringent level. The Supreme Court has also held that EPA may utilize a cost-benefit analysis in setting Section 316(b) performance standards.[4] Section 316 is of particular importance to electric-generating facilities because heat is a significant part of their discharge.[5]

The Section 316 process is quite complex and requires the discharger to develop substantial amounts of scientific data. Today, most Section 316 applications are supported by extensive computer modeling of the effects of a thermal discharge on the receiving waters.

6.3 OCEAN DISCHARGES

No NPDES permit may be issued for a discharge into the territorial seas, the contiguous zone, or the oceans unless the permittee complies with special criteria.[6] The permitting authority may issue a permit for such a discharge only if it determines that the discharge is in the public interest and will not result in the unreasonable degradation of the marine environment. Unreasonable degradation of the marine environment is defined as any one of the following conditions:

- a significant adverse change in ecosystem diversity, productivity, and stability of the biological community within the discharge area and surrounding biological communities;
- a threat to human health through direct exposure to pollutants or through consumption of exposed aquatic organisms; or
- a loss of aesthetic, recreational, scientific, or economic values that is unreasonable in relation to the benefit to be derived from the discharge.[7]

Permit conditions may be imposed to ensure that such degradation does not occur.

The dumping of materials into the oceans (i.e., discharges that are not through outfalls) is regulated under the Marine Protection Research and Sanctuaries Act (MPRSA).[8] Under the MPRSA, transporting material from the United States for ocean dumping is prohibited except as authorized by a permit issued under the act.[9] The only ocean dumping activity of any significance that is currently permitted is the disposal of dredged spoil.[10] Under the MPRSA, the U.S. Army Corps of Engineers is authorized to issue permits for the transportation of dredged materials for ocean disposal.[11] EPA may designate ocean sites for dumping and may designate other sites as protected from ocean dumping. Approximately 100 dredged material disposal sites have been designated.[12] The Ocean Dumping Ban Act of 1988, which amended the MPRSA, imposed strict prohibitions on the types of materials that can be dumped in the oceans. Under that act, it is unlawful to dump sewage sludge or industrial waste into ocean waters.[13]

Ocean dumping is also regulated by the Coast Guard under Annex V of the International Convention for the Prevention of Pollution from Ships, known as MARPOL V. This treaty restricts the discharge of garbage and plastics into the ocean from ships of the signatory countries. Provisions of MARPOL V were codified as part of the Marine Plastic Pollution Research and Control Act of 1987.[14] In addition, the discharge of sewage from vessels is controlled under the Marine Sanitation Device Program.[15]

7

The Pretreatment Program

7.1 INTRODUCTION

Industrial discharges that do not discharge directly into waters of the United States but instead discharge into a public sanitary sewer system are regulated under the CWA pretreatment program.[1] POTWs receive about 80 percent of the nation's wastewater flow (approximately 30 billion gallons per day). Under the pretreatment program, limitations are imposed on industrial users (IUs) of a POTW through a permit, order, or contract issued by the POTW or municipality rather than by the state or EPA.

7.2 PRETREATMENT STANDARDS

The pretreatment program involves a three-part system for controlling the pollution caused by IUs of POTWs. This system includes (1) national general and specific discharge prohibitions, (2) national categorical standards, and (3) local limits developed by POTWs.

7.2.1 National General Prohibitions

Discharges of pollutants that may interfere with a POTW's operations are regulated through the pretreatment regulations found at 40 C.F.R. Part 403.[2] The general prohibitions in the regulations prohibit an IU from introducing into a POTW any pollutant that causes pass through or interference.[3] Pass through is defined as a discharge that exits the POTW into waters of the United States in quantities or concentrations that, alone or in conjunction with a discharge or discharges from other sources, cause a violation of any requirement of the POTW's NPDES permit (including an increase in the magnitude or duration of a violation).[4] Interference is defined as a discharge that, alone or in conjunction with a discharge or discharges from other sources, (1) inhibits or disrupts the POTW, its treatment processes or operations, or its sludge processes, use, or disposal and (2) causes a violation of any requirement of the POTW's NPDES permit (including an increase in the magnitude or duration of a violation) or prevents sewage sludge use or disposal in compliance with applicable statutory provisions and regulations.[5]

7.2.2 National Specific Prohibitions

The pretreatment regulations also establish eight specific prohibitions that are generally intended to prevent interference with the POTW's operations. These prohibitions include prohibitions on the discharge of pollutants that may be fire or explosion hazards, pollutants that will cause corrosive structural damage to the POTW, solid or viscous pollutants that would cause an obstruction in flow, heat that would inhibit the biological activity at the POTW, and pollutants that could cause acute worker health and safety problems.[6]

7.3 NATIONAL CATEGORICAL STANDARDS

The prohibition on discharges that pass through POTWs is implemented through categorical effluent guidelines. Effluent guidelines for an industrial category will normally include pretreatment standards for existing sources and pretreatment standards for new sources (PSNS). The standards establish specific numerical limitations on pollutants considered incompatible pollutants—that is, pollutants other than biological oxygen demand, suspended solids, pH, and fecal coliform bacteria.

The categorical standards are intended to result in the same level of treatment prior to discharge from the POTW as would have been required if the industrial facility had discharged those pollutants directly into the receiving waters. Thus, the user must meet the equivalent of BAT control unless the stringency of the standard is reduced through the mechanism of removal credits, which gives an industrial discharger credit for the actual level of removal of a pollutant consistently achieved by the POTW.[7] In other words, the industrial discharger may meet a less stringent limitation if treatment of the pollutant occurs at the POTW. Removal does not include dilution or volatilization that occurs at or on the way to the POTW.[8]

In 1986, the U.S. Court of Appeals for the Third Circuit interpreted the act to require EPA to promulgate comprehensive sewage sludge regulations before any removal credits could be authorized.[9] Congress codified this interpretation in the 1987 amendments to the act.[10] Removal credits were thus not available for several years because EPA had failed to promulgate final regulations governing the disposal and use of POTW sludge.

Regulations governing sludge that is co-disposed in municipal solid waste landfills were promulgated in October 1991[11] and were upheld by the D.C. Circuit Court of Appeals in December 1993.[12] EPA promulgated Phase I of its comprehensive sludge use and disposal regulations in February 1993.[13] Phase I established numerical pollutant limits for certain metals when sewage sludge is applied to the land, disposed of at surface disposal sites, or incinerated.

Removal credits are thus now available for (1) pollutants sent to POTWs that co-dispose of their sludge in municipal landfills and (2) the pollutants listed in 40 C.F.R. § 403 Appendix G that are discharged to other POTWs that comply with the sludge regulations.[14] EPA has considered proposing a rule that would expand the universe of pollutants for which removal credits may be authorized.[15]

7.4 LOCAL LIMITS

POTWs may establish local limits that are more stringent than the federal categorical standards.[16] For example, the local limits may impose more stringent limitations on pollutants regulated under the federal standards or may establish limits on pollutants that are not limited in the federal standards. Two factors that often require local limits are the prevention of fume toxicity to workers and the reduction of POTW air emissions. Where the local limit is more stringent than the federal standard, the local limit supersedes the federal standard. A local limit that is less stringent than a federal standard, however, does not relieve the IU of its obligation to meet the federal standard.

7.5 PRETREATMENT PROGRAM ENFORCEMENT

Although the pretreatment program is implemented primarily by the municipality operating the POTW, EPA and the states retain enforcement authority.[17] If EPA becomes aware of a pass through or interference at a POTW and sends notice of the incident to the POTW and the POTW does not take appropriate enforcement action within 30 days, EPA may bring an enforcement action for the violation.[18] Similarly, an authorized state has the power to bring an action for penalties even though a POTW has sought penalties for the same noncompliance, if the state determines that the penalty sought by the POTW is insufficient.[19]

POTWs with approved pretreatment programs may take enforcement action against violations of pretreatment permits, including violations of the national pretreatment standards.[20] POTWs, to be approved under the federal pretreatment program, must have the authority to seek injunctive relief for noncompliance and to seek or assess civil or criminal penalties of at least $1,000 per day for each violation of a pretreatment requirement.

8

Nonpoint-Source Discharges

8.1 INTRODUCTION

Nonpoint sources of pollution (i.e., discharges other than those defined as point-source discharges) are a major source of pollution of our nation's waters. It is estimated that agricultural runoff alone is responsible for 70 percent of the nation's impaired rivers and streams. Siltation, salinity, pesticides, and nutrient discharges are the primary impacts on surface waters caused by nonpoint sources.

Several sections of the act establish various programs for controlling nonpoint sources of pollution. In addition, states and EPA have targeted nonpoint sources through the TMDL program, discussed above.

8.2 THE SECTION 319 PROGRAM

The 1987 amendments to the act established a program for the control of nonpoint sources of water pollution, codified in Section 319, that depends primarily on state implementation. Section 319 requires states to submit to EPA for approval an assessment of waters within the state that, without additional action to control nonpoint sources of pollution, cannot reasonably be expected to attain or maintain applicable water quality standards.[1] Section 319 further requires states to obtain EPA approval of state management programs that identify the measures to be undertaken by the state to reduce the pollutant loadings resulting from nonpoint-source discharges, the state programs to achieve these measures, and a schedule for implementing the measures.[2] State programs must be developed in cooperation with local, substate regional, and interstate entities involved in the nonpoint-source issue. Congress authorized up to $400 million over four years to fund the state management programs.[3] This statutory provision has been greatly weakened, however, by Congress's failure to appropriate the money. Additionally, the statute provides no sanction against states for failing to submit an adequate plan.[4]

8.3 COASTAL ZONE MANAGEMENT PROGRAM

The Coastal Zone Management (CZM) program provides for the protection of coastal areas from nonpoint sources of pollution.[5] The CZM program is a voluntary partnership between the federal government and the 28 coastal states and five U.S. territories. Under the program, which is jointly administered by EPA and the National Oceanic and Atmospheric Administration, the participating states and territories must develop Coastal Nonpoint Pollution Control (CNPC) programs. The CNPC programs describe how the state will implement nonpoint-source pollution controls in accordance with technical guidance issued by EPA.[6] States were scheduled to implement the first phase of their approved programs by 2004 and, if necessary, the second phase by 2009.

8.4 NATIONAL ESTUARY PROGRAM

The 1987 amendments to the CWA also created the National Estuary Program (NEP) to promote long-term planning and management for nationally significant estuaries threatened by pollution, development, or overuse.[7] This program is a major vehicle for the implementation of nonpoint-source controls. The NEP requires the preparation

28 Estuaries in the National Estuary Program
Puget Sound, WA
Buzzards Bay, MA
Narragansett Bay, RI
Casco Bay, ME
Massachusetts Bays, MA
Long Island Sound, NY and CT
NY-NJ Harbor, NY and NJ
Delaware Estuary, DE, PA and NJ
Delaware Inland Bays, DE
Albemarle-Pamlico Sounds, NC
Indian River Lagoon, FL
Tampa Bay, FL
Sarasota Bay, FL
Barataria-Terrebonne Estuarine Complex, LA
Galveston Bay, TX
Santa Monica Bay, CA
San Francisco Estuary, CA
Peconic Bay, NY
San Juan Bay, PR
Coastal Bend Bays and Estuaries, TX
Tillamook Bay, OR
Mobile Bay, AL
Morro Bay, CA
Charlotte Harbor, FL
Maryland Coastal Bays, MD
New Hampshire Estuaries, NH
Barnegat Bay, NJ
Lower Columbia River Estuary, OR/WA

of Comprehensive Conservation and Management Plans (CCMP), which recommend approaches for correcting and preventing problems for estuaries nominated by state governors or the EPA administrator. A CCMP is prepared through a Management Conference, to which federal, state, and local governments and representatives of industry and the general public may be invited. There are currently 28 estuaries included in the NEP.

9

Dredge and Fill Permits

9.1 INTRODUCTION

The CWA's broad prohibition on the discharge of pollutants, except in compliance with the act, applies to dredged or fill material. Section 404 of the CWA provides for the issuance of permits for the discharge of dredge or fill material into the navigable waters of the United States.[1] The Section 404 dredge and fill permit program is administered by the U.S. Army Corps of Engineers.

9.2 WATERS WITHIN THE SCOPE OF THE PROGRAM

Questions regarding the scope of the Section 404 program often arise with respect to the dredging and filling of wetlands. Wetlands may be jurisdictional waters if they are adjacent to or have a significant nexus with other waters of the U.S., or if they cross state lines. But it can be very difficult to determine whether certain areas are wetlands covered by the CWA, or if they are even wetlands at all. In 1987, the EPA, Corps of Engineers, Fish and Wildlife Service, and Soil Conservation Service therefore issued a guidance manual for identifying wetlands that constitute waters of the U.S.[2] This document describes technical criteria, field indicators, and other relevant information for identifying and delineating jurisdictional wetlands.

The 2015 CWA jurisdiction rule does not alter the definition of "wetlands" or the procedures for wetlands delineation set out in the 1987 manual.[3] The proposed rule does, however, modify the definition of adjacency, so that waters adjacent to other waters of the U.S. do not necessarily need to be wetlands to be jurisdictional. Prior to the proposed rule, only "adjacent wetlands" had been specifically recognized as jurisdictional;[4] under the 2015 rule, all waters adjacent to another water of the U.S. are jurisdictional.[5] Nonetheless, the 1987 manual will likely continue to be important in determining whether certain areas are wetlands or non-jurisdictional dry lands.

9.3 COVERED ACTIVITIES

The Section 404 permit program covers only the actual discharge or placement of dredged and fill material into waters of the United States. EPA's and the Corps' Section 404 jurisdiction does not extend to incidental fallback of dredged material during wetlands dredging operations.[6] However, regulations create a rebuttable presumption that all mechanized ditching, channelization, and excavation results in discharges greater than incidental fallback and is therefore within Section 404 jurisdiction.[7]

Redeposit of dredged materials into a wetland or water body is subject to the Section 404 program. For example, in one case, the court held that the rearrangement of indigenous materials in a riverbed, undertaken to cut off high water channels to protect riverbanks from erosion, required a Section 404 permit.[8] In another case, the court held that Section 404 applied to owners of a tugboat where the boat's propellers cut into a river bottom, uprooting and destroying sea grass and depositing bottom sediment on adjacent sea grass beds.[9] Some fill activities are exempted from Section 404 if specified effects on navigable waters are avoided. These exempted activities include maintenance of dams, dikes, and similar structures; the construction of temporary sedimentation basins and temporary farm, forest, and mining roads; and some agricultural activities.[10]

In *Southeast Alaska Conservation Council v. U.S. Army Corps of Engineers*, the Ninth Circuit ruled that Section 404 also does not apply to a discharge of material that meets the regulatory definition of "fill material" where EPA previously has adopted, under the NPDES permit program, a performance standard applicable to such discharges.[11] Conservation groups challenged the Corps' issuance of a Section 404 permit to Coeur Alaska, Inc. for the disposal of mine tailings.[12] Coeur Alaska planned to use a froth flotation system to remove gold fines from tailings, and then discharge the process wastewater, containing the residual tailings, into a nearby lake.[13] Because the discharge would raise the bottom of the lake by approximately fifty feet, the discharge met the regulatory definition of "fill material."[14] EPA, however, previously had promulgated a performance standard under Sections 301 and 306 of the CWA that prohibits discharges from froth-flotation mills into waters of the United States.[15] The plaintiffs argued that the Corps' issuance of the Section 404 permit and the proposed discharge of process wastewater thus violated Sections 301 and 306 of the CWA.[16]

The Ninth Circuit held that "[e]ven though the discharge . . . facially qualifies for the permitting scheme under § 404 of the Clean Water Act . . . the discharge is nevertheless prohibited by the clearly applicable and specific performance standard."[17] The court reasoned that the plain language of Sections 301 and 306 requires that all discharges comply with applicable effluent and performance standards, with no exceptions for discharges that would otherwise qualify for regulation under Section 404.[18] The court further reasoned that the regulatory history demonstrated that EPA and the Corps "intended for effluent limitations and performance standards to apply even to discharges that facially meet the definition of the term 'fill material.'"[19]

Coeur Alaska sought certiorari from the Ninth Circuit's decision, arguing that "[i]t changes the Section 404 permit program from a distinct permitting scheme . . . , into a secondary permitting scheme that has only residual application."[20] The Supreme Court reversed the Ninth Circuit, holding that the Corps acted in accordance with law in issuing the permit.[21]

9.4 INDIVIDUAL PERMITS

The Corps has issued national regulations governing the issuance of Section 404 permits.[22] Corps district engineers also develop and implement local policies and procedures. The local procedures must allow potential applicants to contact the local office for pre-application consultation for major projects.[23]

In reviewing an application for an individual Section 404 permit, the Corps will conduct a "Public Interest Review."[24] This review involves "evaluation of the probable impacts . . . of the proposed activity and its intended use on the public interest" and consideration and balancing of "conservation, economics, aesthetics, general environmental concerns, wetlands, historic properties, fish and wildlife values," flood damage prevention, water supply, water quality, energy, and other factors.[25] The Corps cannot deny a permit solely on the basis of economics unrelated to environmental impacts.[26]

In reviewing a permit application, the Corps must apply EPA's Section 404 guidelines.[27] Under EPA's guidelines, "no discharge of dredged or fill material shall be permitted if there is a practicable alternative to the proposed discharge which would have less adverse impact on the aquatic ecosystem."[28] An alternative is "practicable" if it is "capable of being done," taking into account "cost, existing technology, and logistics in light of overall project purposes."[29] There is a presumption that a permit will not be granted for work in a wetland unless there are no practicable, less environmentally damaging alternatives.[30] A heavy burden is placed on the applicant to overcome this presumption.[31] EPA's guidelines also contain a policy statement that a discharge may not cause or contribute to significant degradation of the aquatic ecosystem.[32]

No Section 404 permit may be granted by the Corps unless the affected state (or states) certifies that the permitted activities will not violate applicable laws and regulations.[33] The Corps must consult with EPA[34] and the federal Fish and Wildlife Service[35] before issuing a permit. Section 404(c) allows EPA to veto such activities. The D.C. Circuit has held that EPA can veto a permit years after issuance, even when the permittee is in full compliance with the permit.[36]

In addition, the requirements of the National Environmental Policy Act[37] apply to Section 404 permits.[38] Thus, if a project authorized by a Section 404 permit would result in significant effects on the human environment, the Corps of Engineers must prepare an environmental impact statement before issuing the permit.[39] This requirement can significantly delay development and construction projects.

The Corps' regulations require permit applications to be concurrently processed at federal, state, and local levels. Some states and Corps districts are working to streamline the permit application process. For example, the Pennsylvania Department of Conservation and Natural Resources (DCNR) and the Corps have created a general wetlands permit, called the Pennsylvania State Programmatic General Permit, for activities that have a limited environmental impact on wetlands less than one acre in size or on less than 250 linear feet of streams. This general permit will allow developers to submit only one application for authorization and will allow the DCNR to approve an activity without it undergoing extensive federal review.

9.5 THE MITIGATION POLICY

The guidelines issued under Section 404(b)(1) require applicants to take all practicable steps to minimize the adverse effects of proposed filling activities. Once the amount of wetland damage has been reduced to its barest minimum, the remaining damage must be mitigated.

Under a February 1990 MOA between the EPA and the Corps of Engineers, the Corps committed to minimize the loss of wetlands resulting from its permit decisions.[40] This no-net-loss policy requires the Corps to determine first whether potential impacts have been avoided to the maximum extent practicable, then whether any remaining unavoidable impacts have been mitigated "to the extent appropriate and practicable," and then to compensate financially for aquatic values. Unavoidable wetlands impacts often may be offset by wetlands restoration or creation.[41]

Compensatory mitigation—for example, restoration of existing degraded wetlands or creation of man-made wetlands—can be accomplished by three methods: (i) permittee-responsible mitigation, where the permittee retains responsibility for the mitigation; (ii) mitigation banks, sites where a third party conducts mitigation activities and sells credits to permittees to satisfy their mitigation obligations; (iii) or in-lieu fee programs, where a governmental or non-profit resource management entity conducts mitigation and provides credits to permittees.[42] On March 31, 2008, EPA and the Corps of Engineers issued a final rule setting forth performance standards for compensatory mitigation projects and establishing a preference for mitigation banks over in-lieu fee programs and permittee-responsible mitigation.[43] The rule also provides that all mitigation should be located in the same watershed as the impacted resource and must be sufficient to replace the lost aquatic resource functions.[44]

9.6 NATIONWIDE PERMITS

The Corps is authorized to issue nationwide and general (state, regional, or nationwide) Section 404 permits for specific categories of activities involving the discharge of dredged or fill materials determined to have minimal adverse environmental effects.[45]

An activity covered by a nationwide or general permit may be performed without an individual Section 404 permit so long as the general conditions[46] of the nationwide or general permit are observed.[47] The general and nationwide permits are intended to allow certain minor-impact activities to take place without delay or paperwork.[48]

General Conditions Applicable to Nationwide Permit Activities

1. That the activity may not cause more than a minimal adverse effect on navigation;
2. That the structure or fill will be properly maintained;
3. That appropriate erosion and siltation controls will be used and maintained during construction;
4. That any activity shall not substantially disrupt the movement of indigenous species unless the primary purpose of the fill is to impound water;
5. That heavy equipment must be placed on mats or other measures must be taken to minimize soil disturbance;
6. That the activity complies with regional and case-specific conditions;
7. That the activity will not occur in a component of the National Wild and Scenic River System, nor in a river officially designated by Congress as a "study river" for possible inclusion in the System while the river is in an official study status;
8. That the activity will not impair reserved tribal rights, including reserved water rights and treaty fishing and hunting rights;
9. That in certain states an individual state water quality certification must be obtained or waived;
10. That an individual state coastal zone management consistency concurrence must be obtained or waived in certain states;
11. That the activity will not jeopardize a threatened or endangered species or destroy or adversely modify the critical habitat of such species;
12. If the activity may adversely affect historic properties listed on or eligible for the National Register of Historic Places, that the permittee shall notify the Corps and wait for its approval; and
13. Where notification is required, that the permittee has complied with the specific notification provisions.

In addition to these general conditions, activities performed under a nationwide permit must also comply with the following "Section 404 Only Conditions."

Section 404 Only Conditions

1. No discharge may occur in proximity to a public water supply intake except for repair purposes;
2. No discharge shall occur in areas of concentrated shellfish production except when harvesting;
3. No discharge may consist of unsuitable material and all discharges must be free from toxic pollutants in toxic amounts;
4. Discharges must be minimized or avoided to the maximum extent practicable unless there is a Corps-approved mitigation plan;
5. Discharges into spawning areas during spawning season must be avoided to the maximum extent practicable;
6. Discharges must not permanently restrict or impede the passage of normal or expected high flows or cause the relocation of water unless the primary purpose of the fill is the impoundment of water;
7. Impoundments of water shall not cause adverse impacts caused by the accelerated passage of water and/or the restriction of its flow;
8. Discharges into migratory waterfowl breeding areas must be avoided; and
9. Temporary fills must be removed and the affected areas returned to their preexisting condition.

Nationwide permits have been issued for approximately 50 activities, including some survey activities, backfilling of utility lines, minor road crossings, construction of outfall structures, bank stabilization, oil and gas structures, hydropower projects, and maintenance activities.[49] A number of the nationwide permits require that special notice be given to the Corps prior to proceeding with an activity. The remaining nationwide permits are self-executing,[50] although it is prudent to contact the Corps when planning an activity within a possible wetland area.[51]

9.7 POTENTIAL LIABILITIES UNDER THE SECTION 404 PROGRAM

Like violations of other parts of the act, violations of the Section 404 permit requirement may result in penalties of up to $37,500 per day.[52] In addition, the Corps may bring an action to compel the restoration of areas that have been filled without obtaining the required permit or dredged in violation of permit conditions. Extensive restoration, including replication of natural topography and hydrology, has been ordered in some cases.[53]

Discharging dredged or fill material without a Section 404 permit can also lead to criminal penalties. Criminal penalties will normally be imposed only for cases of extreme conduct, such as refusing to obey a cease and desist order or causing severe damage to a wetland.[54]

10

Preventing, Reporting, and Responding to Spills

10.1 SPILL PREVENTION

Section 311 of the CWA establishes a national policy that "there should be no discharges of oil or hazardous substances into or upon the navigable waters of the United States, adjoining shorelines, or into or upon the waters of the contiguous zone" and creates a comprehensive scheme of prohibitions, reporting requirements, penalties, and cleanup obligations to implement this policy.[1]

The act requires many facilities to develop and maintain plans for preventing and responding to spills of oil and hazardous substances, called Spill Prevention Control and Countermeasure (SPCC) Plans. As amended by the Oil Pollution Act of 1990 (OPA), the act now also requires owners and operators of some facilities to prepare and submit a more extensive plan, called a Facility Response Plan, for responding to a worst-case spill of oil.

10.1.1 SPCC Plans

A facility must prepare an SPCC plan if it drills, produces, gathers, stores, processes, refines, transfers, distributes, or consumes oil and if, because of its location, it could reasonably be expected to discharge[2] oil in harmful quantities, as defined in Part 110 of the regulations, into or upon surface waters, adjoining shorelines, or the contiguous zone.[3] Exempted from this requirement are facilities where (1) the total underground storage capacity is 42,000 gallons or less of oil and (2) the aboveground storage capacity is 1,320 gallons or less of oil.[4] Also exempted are completely buried underground tanks already regulated under the Underground Storage Tank Program.[5]

The SPCC plan describes steps the facility will take to prevent spills and to minimize the risk of harm to surface waters in the event of a release of oil. The SPCC plan need not be submitted to the regulatory agency but must be maintained at the facility at all times and may be reviewed during state or federal inspections.[6] Owners or operators of new facilities must prepare and implement an SPCC plan within six months after the date the facility begins operations.[7]

The SPCC plan must be amended whenever there is a change in facility design, operation, or maintenance that materially affects the facility's potential to discharge oil. Notwithstanding this requirement, the plan must be reviewed and evaluated for adequacy at least once every five years.[8] Plans must be reviewed and certified by a registered professional engineer.[9]

If a facility discharges more than 1,000 gallons of oil into the waters of the United States or upon adjoining shorelines or discharges more than 42 gallons of oil to waters of the United States or adjoining shorelines twice within a 12-month period, the owner or operator of the facility must submit additional information to the EPA regional administrator within 60 days.[10] After reviewing this information, the regional administrator may require the facility to amend the SPCC plan if he or she finds that the existing plan does not meet the requirements of the regulations or that amendment of the plan is necessary to prevent or contain discharges from the facility.[11]

EPA's regulations set forth detailed guidelines for the preparation of SPCC plans.[12] The plan must contain a description of recent spill events and, in some circumstances, a prediction of the direction, rate of flow, and total quantity of oil that could be discharged from the facility as a result of a failure in containment.[13] The regulations specify the types of containment structures and other spill control mechanisms that may—or in some cases must—be included in the plan.[14]

The plan must also designate a person who will be accountable for oil spill prevention and who will report to facility management.[15] Owners and operators are responsible for properly instructing their personnel in the operation and maintenance of equipment to prevent oil discharges, and in applicable pollution control laws and regulations.[16]

10.1.2 Facility Response Plans

Under the OPA, a non-transportation-related onshore facility is required to prepare a Facility Response Plan if it handles, transports, or stores oil and if it, "because of its location, could reasonably be expected to cause substantial harm to the environment by discharging into or on the navigable waters, adjoining shorelines, or the exclusive economic zone."[17] Facilities that do not meet EPA's substantial harm criteria must complete and maintain a certification verifying that the criteria do not apply.[18]

EPA has established criteria for determining which facilities may cause substantial harm in the event of a discharge of oil.[19] Facility Response Plans must be prepared for two classes of facilities:

1. Facilities that transfer oil over water to or from vessels that have a total oil storage capacity greater than or equal to 42,000 gallons, and

2. Facilities with total oil storage capacity of at least one million gallons, where one or more of the following is true:
   - The facility does not have secondary containment for each aboveground storage area sufficiently large to contain the capacity of the largest tank plus sufficient freeboard for precipitation,
   - The facility is located in the vicinity of fish and wildlife or sensitive environments such that a discharge could cause injury to them,
   - The facility is located in an area where a discharge would shut down operations at a public drinking water intake, or
   - The facility has had a reportable spill greater than or equal to 10,000 gallons within the past five years.[20]

The regional administrator may require facilities other than substantial harm facilities to prepare and submit a Facility Response Plan under certain circumstances.[21]

The Facility Response Plan is intended to be a plan for responding to a worst-case release of oil. EPA's regulations include a worksheet that is to be used to calculate what this worst-case event would be.[22] The regulations also set forth a model plan that is to be followed in developing facility plans.[23] The plan must, among other things, identify the person with authority to implement the plan, require immediate communication to appropriate federal officials, ensure that adequate private personnel and equipment will be available to respond to the discharge, describe the immediate measures that will be taken in the event of a spill to secure the source of the discharge and provide containment, contain plans for evacuation, and require training, drills, and equipment testing.[24] The plan must be consistent with the National Oil and Hazardous Substance Pollution Contingency Plan.[25]

Facilities that have not submitted such plans are not permitted to operate.[26] Facilities may operate for up to two years pending federal approval of submitted plans if the owner of the facility certifies that he or she has ensured by contract or other means the availability of private personnel and equipment necessary to respond to a worst-case discharge. If that period expires or if approval is denied, the facility must discontinue its storage, transportation, and handling of oil.[27]

10.2 SPILL NOTIFICATION

Section 311(b)(5) of the act requires persons in charge of facilities to immediately notify the National Response Center of discharges of harmful quantities of oil or a hazardous substance to navigable waters or adjoining shorelines.[28] It is a criminal offense to fail to make such a report, punishable by up to five years in prison.[29]

For spills of oil,[30] EPA has determined that a harmful quantity is any quantity causing a film or sheen on the receiving waters, any quantity causing a sludge or emulsion to be deposited beneath the surface of the water or upon adjoining shorelines, or any quantity that violates an applicable water quality standard.[31]

EPA has designated over 300 substances as hazardous substances subject to the Section 311 reporting requirement[32] and has identified the reportable quantity for each of these substances.[33] Any release of more than the reportable quantity of a hazardous substance within a 24-hour period must be reported to the National Response Center.[34] This reporting requirement is in addition to reporting requirements under the Comprehensive Environmental Response, Compensation, and Liability Act, the Emergency Planning and Community Right-to-Know Act, and other federal or state requirements.

Under certain conditions, an NPDES permit may insulate a permittee from the statutory notification requirement. For purposes of Section 311, discharge is defined to exclude the following:

- discharges in compliance with an NPDES permit;
- discharges resulting from circumstances identified, reviewed, and made a part of the public record with respect to an NPDES permit, and subject to a condition in such permit;
- continuous or anticipated intermittent discharges from a point source, identified in an NPDES permit or permit application, that are caused by events occurring within the scope of the relevant operating or treatment systems; and
- discharges incidental to authorized mechanical removal.[35]

10.3 SPILL RESPONSE AND LIABILITY

The discharge of harmful quantities of oil or a hazardous substance into waters of the United States or onto adjoining shorelines or the contiguous zone is prohibited.[36] Owners and operators of facilities from which oil or a hazardous substance is discharged to navigable waters, to shorelines, or into the contiguous zone in quantities greater than the reportable quantity are strictly liable for (1) penalties, (2) the costs of cleaning up the spill, and (3) natural resource damages caused by the spill.[37] Liability is strict, although several limited defenses are available against a claim for liability for the costs of the cleanup of a spill.[38]

The OPA increased the penalties to which a discharger may be subject. Any owner, operator, or person in charge may be fined up to $37,500 per day for a discharge, or up to $2,100 per barrel of oil discharged.[39] Where a discharge results from gross negligence or willful misconduct, the minimum penalty is $150,000, and the maximum penalty is $5,300 per barrel of oil or unit of reportable quantity of hazardous substance discharged.[40]

11

Enforcement

11.1 INTRODUCTION

11.1.1 Federal and State Roles

The CWA's enforcement mechanisms are structured to allow states (and local governments in the case of the pretreatment program) to assume an active role in the act's enforcement. To obtain approval of its NPDES permit program, a state must demonstrate that its environmental control agency has adequate powers of enforcement, roughly equivalent to those exercised by EPA under the act.[1] At a minimum, authorized states must have civil and criminal enforcement authority, and virtually all have some form of administrative enforcement authority. State enforcement programs are not required to be identical to the federal program. For example, states may impose maximum civil penalties as low as $5,000 per day per violation (compared to the daily maximum of $32,500 under the federal program) and may impose maximum criminal penalties as low as $10,000 per day per violation (compared to $50,000 per day under the federal program).[2] In addition, EPA currently does not require authorized states to provide a mechanism for citizen suits under the state program. States are required only to allow intervention of citizens in enforcement actions, to investigate and respond in writing to citizen complaints, and to provide notice and opportunity for comment on proposed settlements of state enforcement actions.[3]

In states with authorized permit programs, EPA retains the right to initiate an enforcement action even if the state determines that no action is warranted. EPA must first notify the alleged violator and the state of its intent to bring an enforcement action and must allow the state 30 days to commence a state enforcement action.[4]

Moreover, EPA may bring a federal enforcement action even though an authorized state has brought a parallel action.[5] Where a final judgment has been issued in a state judicial action, however, EPA is generally foreclosed from relitigating the factual issues decided in the state proceeding.[6]

In addition, if EPA finds that permit violations within a state are widespread and appear to result from the state's failure to enforce its permits, EPA must so notify the state. If the state's failure to enforce extends more than 30 days beyond the notice, EPA must assume responsibility for enforcing permits within the state. This period of federally assumed enforcement ends only after EPA determines that the state will adequately enforce its permit program.[7]

In states where EPA retains permitting authority, EPA (usually through the regional offices) is the primary enforcement authority. Nonauthorized states may enforce in state courts any state statutes and regulations that are not inconsistent with or duplicative of the federal scheme.

11.2 ENFORCEMENT THEORIES

The CWA is viewed by many as the easiest of the federal environmental statutes to enforce. This is because persons regulated under the act normally must report their own compliance and noncompliance to the regulating agency. For example, holders of NPDES permits must file periodic discharge monitoring reports (or DMRs), which must contain the results of all monitoring of discharges, and must indicate where those discharges exceeded permit limitations.[8] In addition, permittees must report any noncompliance that may endanger health or the environment within 24 hours of the time the permittee becomes aware of the circumstances.[9] Permittees are also required to report any anticipated noncompliance,[10] any noncompliance not required to be reported under any other specific provision,[11] and any noncompliance that the permittee failed to report as required elsewhere by the regulations or the permit.[12]

The most common substantive basis for a federal enforcement action is Section 301(a) of the act, the discharge prohibition discussed earlier in this chapter.[13] Because this provision effectively prohibits the discharge of any pollutant except as in compliance with the act, it imparts broad enforcement authority. For example, any discharge in excess of a permit limitation is not only a potential violation of Section 402 (which sets out the permit program) but also a violation of Section 301(a). Likewise, any discharge without a permit would potentially violate Section 301(a).

For civil enforcement purposes, and arguably for some criminal violations, the CWA is a strict liability statute. This means that the intent of the violator is irrelevant; once the violation is established, liability attaches. A person need not have acted negligently or with any intent to violate the statute to be found liable.[14] Thus, enforcement actions may be brought based on little, if anything, more than the DMRs and other reports submitted by the permittee itself. In addition, EPA or the state may generate additional evidence of violations through inspections of the permitted facility.

EPA often enforces in cycles, focusing its enforcement efforts on a particular issue or industrial group in a coordinated manner. For example, in October 1997, the EPA water enforcement program announced that it planned to focus its enforcement efforts on combined sewer overflows, storm water, NPDES permitting, deep well injection, and wetlands. That is not to say that when EPA is focusing on a particular type of activity, it will ignore all other types of violations. It is probably safe to say, however, that when the agency is targeting a certain type of activity, all entities participating in that activity should be aware that they are likely to be scrutinized. In 2009 EPA outlined its strategic goals for developing future enforcement priorities.[15]

11.3 DEFENSES

Because liability under the act is strict, there are few defenses available to a permittee accused of violating its permit. For example, good faith, the pendency of a permit modification request, and data errors have all been rejected as defenses to liability under the act.[16] Courts have differed on the question of whether impossibility may be a defense under the act.[17]

Two defenses that exist under EPA's regulations are theories of upset and bypass. A third defense, known as permit-as-a-shield, may also be available in some situations.

11.3.1 Upset

Several courts have ruled that, since the equipment underlying the technology-based permit limitations is inherently subject to failure for reasons beyond the control of the operator, EPA must allow for upsets in applying those standards.[18] EPA's regulations define an upset as "an exceptional incident in which there is unintentional and temporary noncompliance with technology-based permit effluent limitations because of factors beyond the reasonable control of the permittee."[19] The term does not include noncompliance "caused by operational error, improperly designed treatment facilities, inadequate treatment facilities, lack of preventive maintenance, or careless or improper operation."[20] An upset constitutes an affirmative defense in an enforcement action for violations of technology-based permit limitations.[21]

To claim upset as a defense, the permittee must submit notice of the upset within 24 hours of the event.[22] The permittee must be able to show, through properly signed contemporaneous operating logs or other relevant evidence, the cause of the upset, that the facility was being operated properly at the time of the upset, and that appropriate remedial measures were taken.[23] The permittee seeking to establish the upset defense has the burden of proving that the defense applies.[24]

The upset defense is available to a permittee only if it is incorporated into the permit expressly or by reference to the relevant regulatory provisions.[25] It is also important to note that, since state permit programs can be more stringent than the federal program, a state can choose not to allow the upset defense.[26]

11.3.2 Bypass

A more limited defense may be available through EPA's regulations governing bypasses. A bypass is defined as "the intentional diversion of waste streams from any portion of a treatment facility."[27] Bypasses are allowed only in very limited circumstances. Even a bypass that does not cause effluent limitations to be exceeded is allowed only for essential maintenance to ensure efficient operation.[28] Essential maintenance is not routine maintenance that can be performed during periods of non-process operations but includes only repairs and maintenance that cannot wait until the production process is not in operation to be performed; for example, if a seal on a valve malfunctions or a pipe bursts during production hours, the facility operator may bypass that particular process unit to perform corrective maintenance.[29]

Bypasses that cause effluent limitations to be exceeded are prohibited except in circumstances where they are necessary to avoid severe property damage, personal injury, or loss of life.[30] Under these circumstances, bypass is permitted only if there are no feasible alternatives, such as the use of auxiliary treatment facilities, retention of untreated wastewaters, or maintenance during normal periods of downtime.[31] In addition, if the permittee knows in advance of the need for a bypass, it must submit prior notice, if possible, at least 10 days before the date of the anticipated bypass.[32] The regulatory agency must be notified within 24 hours of any unanticipated bypass.[33] Where all of these conditions are met, bypass may be used as an affirmative defense in an enforcement action.[34]

11.3.3 Permit-as-a-Shield

Under Section 402(k) of the act, compliance with an NPDES permit acts as a shield against enforcement.[35] In other words, so long as a permittee is in compliance with the limitations and conditions in its NPDES permit, neither the state nor EPA can bring an enforcement action against it for violation of the act, such as for discharging pollutants not limited in the permit.[36] The Supreme Court has noted that the purpose of Section 402(k) "seems to be . . . to relieve [permit holders] of having to litigate in an enforcement action the question whether their permits are sufficiently strict."[37] The shield does not apply to violations of the act outside the scope of the NPDES program (e.g., spill reporting, Section 404 violations).

In *Atlantic States Legal Foundation, Inc. v. Eastman Kodak Co.*, 12 F.3d 353 (2d Cir.), *cert. denied*, 115 S. Ct. 62 (1994), a citizens group sued an NPDES discharger for discharging, over a four-year period, more than a million pounds of pollutants that were not limited in the discharger's permit. The court held that Section 402(k) prohibited an enforcement action against the discharger since the discharger was in compliance with its permit at all relevant times. The court stated,

> Viewing the regulatory scheme as a whole, . . . it is clear that the permit is intended to identify and limit the most harmful pollutants while leaving the control of the vast number of other pollutants to disclosure requirements. Once within the NPDES or SPDES scheme, therefore, polluters may discharge pollutants not specifically listed in their permits so long as they comply with the appropriate reporting requirements and abide by any new limitations when imposed on such pollutants.[38]

However, the permit shield defense is not absolute. A permittee must strictly comply with disclosure obligations; if a likely pollutant was not disclosed to and considered by the regulator, it can fall outside the scope of the permit.[39] And the Ninth Circuit has held that the permit shield does not apply when the express terms of the permit prohibit discharges not explicitly authorized.[40]

11.4 ENFORCEMENT OPTIONS

Federal enforcement may take the form of (1) an administrative order requiring compliance and/or assessing an administrative penalty, (2) an action for civil penalties and/or an injunction, or (3) an action for criminal penalties. State enforcement schemes generally follow the federal system.

EPA typically will choose the least resource-consuming enforcement option that is appropriate for the violation, which in most cases will be an administrative order.[41] A civil judicial action may be appropriate when there is a need for a court order directing immediate or long-term compliance measures (i.e., a temporary restraining order or injunction), such as where the noncompliance is serious and continuing and the violator is uncooperative. In addition, civil judicial action is needed to assess a penalty of more than $187,500. Criminal enforcement actions will be brought for serious violations that are knowing or negligent.[42]

11.5 ADMINISTRATIVE ORDER

The 1987 amendments to the act authorized EPA to issue administrative orders assessing penalties for CWA violations, including permit violations. Class I penalties may not exceed $16,000 per violation, up to a maximum of $37,500.[43] Before assessing a Class I penalty, EPA must give the alleged violator written notice of the proposed assessment and an opportunity to request an informal hearing.[44] A Class II penalty may not exceed $16,000 per day for each violation, up to a maximum of $187,500.[45] Class II penalties may be imposed only after notice and an opportunity for a full adjudicatory hearing.[46]

EPA may also issue an administrative order requiring compliance with the act.[47] Administrative orders for permit violations often include a compliance schedule and may also include interim limitations that must be met while the scheduled activities are being performed. Administrative compliance orders are administrative commands and do not impose any sanctions for the underlying violations or for a violation of the compliance order itself. Nevertheless, in *Sackett v. EPA* a unanimous Supreme Court held that a compliance order was a final agency action with respect to EPA's jurisdictional decision that certain property contained "navigable waters" and, therefore, the court granted pre-enforcement judicial review.[48] Failure to comply with an administrative order could also form the basis of a criminal prosecution for a knowing violation or of a civil action where a claim of bad faith is made.

11.6 CIVIL JUDICIAL ENFORCEMENT

EPA, through its U.S. Department of Justice attorneys, may bring an action in federal district court seeking civil penalties for violations of the act.[49] Civil penalties may be imposed without a showing of negligence or fault on the part of the defendant and may be assessed at up to $37,500 per day for each violation.[50]

An alleged violator is entitled to a jury trial to determine liability for civil penalties, and a citizen suit plaintiff can also insist on a jury determination of liability.[51] Once liability is established, however, the court retains the power to determine the amount of the penalty.[52]

EPA's CWA Penalty Policy[53] establishes a method for calculating an appropriate penalty as part of a settlement of a CWA enforcement action.[54] The penalty is to consist of an economic benefit component plus a gravity component, plus or minus adjustments. It is EPA's policy that penalties should recover the full economic benefit of noncompliance, calculated from the beginning of the noncompliance until the point when the facility was or will be in compliance. Under the Penalty Policy, EPA normally calculates this benefit using the BEN computer program.[55] The gravity component is based on four considerations:

1. The significance of the violation,
2. The actual or potential harm to human health or the environment,
3. The number of violations, and
4. The duration of noncompliance.

Adjustments may be made for the following circumstances:

- a history of recalcitrance,
- ability to pay,
- litigation considerations (e.g., the potential for protracted litigation and the maximum penalty likely to be awarded by a court), and
- other equitable considerations.

EPA may also initiate a civil action to obtain an injunction. The district court has the power to enter preliminary and permanent injunctions to restrain and abate violations of the act, regulations, and permits, including state permits. If an injunction is violated, the violator is subject to the criminal and civil penalty provisions of the act as well as to the criminal and civil contempt powers of the court.

11.7 CRIMINAL ENFORCEMENT

EPA may refer a matter to the Department of Justice for the institution of a criminal action against any discharger who knowingly or negligently violates the act.[56] The penalties for an initial conviction for a negligent violation include a fine of $2,500 to $25,000 per day, imprisonment for not more than one year, or both. The penalties for a knowing violation are a fine of $5,000 to $50,000 per day, imprisonment for not more than three years, or both.[57] In both cases, the maximum penalties for subsequent convictions are doubled.[58]

In an important case, the Ninth Circuit Court of Appeals held that the manager and assistant manager of a sewage treatment plant were criminally liable under the act for knowing violations even though the two men did not know that their discharges of sludge violated the facility's permit. The court held that criminal sanctions may be imposed on an individual who knowingly engages in conduct that results in a permit violation, regardless of whether he is cognizant of the requirements or even the existence of the permit.[59]

A person who knowingly violates a permit or other requirement of the act and knows at the time that he thereby places another person in imminent danger of death or serious bodily injury is subject to a fine of up to $250,000 ($1 million for an organization) and imprisonment for up to 15 years. The punishment, with respect to both fines and imprisonment, is doubled for second offenses.[60]

An action for such knowing endangerment may, for example, be brought if a person knowingly contaminates a water supply or deliberately dumps hazardous materials into sewers or waterways.[61]

Criminal penalties may also be imposed against any person who makes a false statement, representation, or certification to the government or any person who falsifies, tampers with, or knowingly renders inaccurate any monitoring device required under the act.[62] These violations may subject a violator to a fine of not more than $10,000, imprisonment for not more than two years, or both, with penalties doubled for a subsequent violation.

Criminal penalties may be imposed against corporations, persons directly involved in a violation, and responsible corporate officers.[63] Corporate officers who deliberately shield themselves from knowledge of violations are likely to be considered responsible under the act.

11.8 CITIZEN SUITS

Section 505 of the act allows any person "having an interest which is or may be adversely affected"[64] to commence a civil action against any person for violation of any effluent standard, limitation, or order or against EPA for failure to perform a nondiscretionary duty.[65] Citizen groups have frequently used this citizen suit provision, particularly in actions against dischargers for violations of NPDES permits. These suits are often rather straightforward since NPDES permittees normally must report all exceedances of permit limitations to the permitting agency on a monthly or quarterly basis. A citizen group may therefore bring an action for a permit violation using the permittee's own reports as evidence. In a 2004 decision, the U.S. Court of Appeals for the Sixth Circuit allowed a citizen group to assume an even greater role in enforcement, holding that the group had standing to sue a water commission for failing to submit monitoring reports.[66] And numerous courts have allowed citizen suits even when a discharger was told by the regulator that it did not need a permit.[67]

A citizen suit may be brought only if neither EPA nor a state is "diligently prosecuting" the violation.[68] An EPA or state enforcement action that results in a compliance order and not a penalty assessment, however, has been held not to constitute a diligent prosecution that would foreclose a citizen suit.[69]

The plaintiff must give the alleged violator, EPA, and the state 60 days' notice prior to initiation of the lawsuit, unless the action involves a violation of Sections 306 or 307(a) of the act.[70] The notice must include a parameter-by-parameter description of the alleged violations[71] and must specify a time period in which the alleged violations occurred.[72]

Most citizen suits settle under a consent agreement providing for some combination of the following penalties and/or agreements:

- payment of a civil penalty,[73]
- payment of attorney's fees and costs to the plaintiffs,[74]
- a compliance schedule to bring the permittee into compliance,
- stipulated penalties for failure to meet the compliance schedule, and
- payment of money to support an environmental activity selected by the plaintiff.

EPA has the right to review and object to any consent agreement entered in a citizen suit.[75] A decision in the *Exxon Valdez* case held that, although the CWA does not specifically provide for punitive damages in citizen suits, Section 505 does not prohibit the award of such damages under common law.[76]

Citizen suits may be brought only for continuing or intermittent violations.[77] To meet this requirement, however, a plaintiff need only make a good-faith allegation of a continuous or intermittent violation at the time the statutory 60-day notice of intent to sue is given. A violation will be considered to be continuing for jurisdictional purposes unless it is "absolutely clear that the allegedly wrongful behavior could not reasonably be expected to recur."[78] A court has jurisdiction only over violations included in the plaintiff's 60-day notice letter and post-complaint continuing violations of the same type.[79]

Appendix

FEDERAL WATER POLLUTION CONTROL ACT

[As Amended Through P.L. 107–303, November 27, 2002]

FEDERAL WATER POLLUTION CONTROL ACT

(33 U.S.C. 1251 et seq.)

AN ACT To provide for water pollution control activities in the Public Health Service of the Federal Security Agency and in the Federal Works Agency, and for other purposes.

Be it enacted by the Senate and House of Representatives of the United States of America in Congress assembled,

TITLE I—RESEARCH AND RELATED PROGRAMS

DECLARATION OF GOALS AND POLICY

SEC. 101. (a) The objective of this Act is to restore and maintain the chemical, physical, and biological integrity of the Nation's waters. In order to achieve this objective it is hereby declared that, consistent with the provisions of this Act—

(1) it is the national goal that the discharge of pollutants into the navigable waters be eliminated by 1985;

(2) it is the national goal that wherever attainable, an interim goal of water quality which provides for the protection and propagation of fish, shellfish, and wildlife and provides for recreation in and on the water be achieved by July 1, 1983;

(3) it is the national policy that the discharge of toxic pollutants in toxic amounts be prohibited;

(4) it is the national policy that Federal financial assistance be provided to construct publicly owned waste treatment works;

(5) it is the national policy that areawide treatment management planning processes be developed and implemented to assure adequate control of sources of pollutants in each State;

(6) it is the national policy that a major research and demonstration effort be made to develop technology necessary to eliminate the discharge of pollutants into the navigable waters, waters of the contiguous zone and the oceans; and

(7) it is the national policy that programs for the control of nonpoint sources of pollution be developed and implemented in an expeditious manner so as to enable the goals of this Act to be met through the control of both point and nonpoint sources of pollution.

(b) It is the policy of the Congress to recognize, preserve, and protect the primary responsibilities and rights of States to prevent, reduce, and eliminate pollution, to plan the development and use (including restoration, preservation, and enhancement) of land and water resources, and to consult with the Administrator in the exercise of his authority under this Act. It is the policy of Congress that the States manage the construction grant program under this Act and implement the permit programs under sections 402 and 404 of this Act. It is further the policy of the Congress to support and aid research relating to the prevention, reduction, and elimination of pollution, and to provide Federal technical services and financial aid to State and interstate agencies and municipalities in connection with the prevention, reduction, and elimination of pollution.

(c) It is further the policy of Congress that the President, acting through the Secretary of State and such national and international organizations as he determines appropriate, shall take such action as may be necessary to insure that to the fullest extent possible all foreign countries shall take meaningful action for the prevention, reduction, and elimination of pollution in their waters and in international waters and for the achievement of goals regarding the elimination of discharge of pollutants and the improvement of water quality to at least the same extent as the United States does under its laws.

(d) Except as otherwise expressly provided in this Act, the Administrator of the Environmental Protection Agency (hereinafter in this Act called "Administrator") shall administer this Act.

(e) Public participation in the development, revision, and enforcement of any regulation, standard, effluent limitation, plan, or program established by the Administrator or any State under this Act shall be provided for, encouraged, and assisted by the Administrator and the States. The Administrator, in cooperation with the States, shall develop and publish regulations specifying minimum guidelines for public participation in such processes.

(f) It is the national policy that to the maximum extent possible the procedures utilized for implementing this Act shall encourage the drastic minimization of paperwork and interagency decision procedures, and the best use of available manpower and funds, so as to prevent needless duplication and unnecessary delays at all levels of government.

(g) It is the policy of Congress that the authority of each State to allocate quantities of water within its jurisdiction shall not be superseded, abrogated or otherwise impaired by this Act. It is the further policy of Congress that nothing in this Act shall be construed to supersede or abrogate rights to quantities of water which have been established by any State. Federal agencies shall co-operate with State and local agencies to develop comprehensive solutions to prevent, reduce and eliminate pollution in concert with programs for managing water resources.

(33 U.S.C. 1251)

COMPREHENSIVE PROGRAMS FOR WATER POLLUTION CONTROL

SEC. 102. (a) The Administrator shall, after careful investigation, and in cooperation with other Federal agencies, State water pollution control agencies, interstate agencies, and the municipalities and industries involved, prepare or develop comprehensive programs for preventing, reducing, or eliminating the pollution of the navigable waters and ground waters and improving the sanitary condition of surface and underground waters. In the development of such comprehensive programs due regard shall be given to the improvements which are necessary to conserve such waters for the protection and propagation of fish and aquatic life and wildlife, recreational purposes, and the withdrawal of such waters for public water supply, agricultural, industrial, and other purposes. For the purpose of this section, the Administrator is authorized to make joint investigations with any such agencies of the condition of any waters in any State or States, and of the discharges of any sewage, industrial wastes, or substance which may adversely affect such waters.

(b)(1) In the survey or planning of any reservoir by the Corps of Engineers, Bureau of Reclamation, or other Federal agency, consideration shall be given to inclusion of storage for regulation of streamflow, except that any such storage and water releases shall not be provided as a substitute for adequate treatment or other methods of controlling waste at the source.

(2) The need for and the value of storage for regulation of streamflow (other than for water quality) including but not limited to navigation, salt water intrusion, recreation, esthetics, and fish and wildlife, shall be determined by the Corps of Engineers, Bureau of Reclamation, or other Federal agencies.

(3) The need for, the value of, and the impact of, storage for water quality control shall be determined by the Administrator, and his views on these matters shall be set forth in any report or presentation to Congress proposing authorization or construction of any reservoir including such storage.

(4) The value of such storage shall be taken into account in determining the economic value of the entire project of which it is a part, and costs shall be allocated to the purpose of regulation of streamflow in a manner which will insure that all project purposes, share equitable in the benefits of multiple-purpose construction.

(5) Costs of regulation of streamflow features incorporated in any Federal reservoir or other impoundment under the provisions of this Act shall be determined and the beneficiaries identified and if the benefits are widespread or national in scope, the costs of such features shall be nonreimbursable.

(6) No license granted by the Federal Power Commission for a hydroelectric power project shall include storage for regulation of streamflow for the purpose of water quality control unless the Administrator shall recommend its inclusion and such reservoir storage capacity shall not exceed such proportion of the total storage required for the water quality control plan as the drainage area of such reservoir bears to the drainage area of the river basin or basins involved in such water quality control plan.

(c)(1) The Administrator shall, at the request of the Governor of a State, or a majority of the Governors when more than one State is involved make a grant to pay not to exceed 50 per centum of the administrative expenses of a planning agency for a period not to exceed three years, which period shall begin after the date of enactment of the Federal Water Pollution Control Act Amendments of 1972, if such agency provides for adequate representation of appropriate State, interstate, local, or (when appropriate) international interests in the basin or portion thereof involved and is capable of developing an effective, comprehensive water quality control plan for a basin or portion thereof.

(2) Each planning agency receiving a grant under this subsection shall develop a comprehensive pollution control plan for the basin or portion thereof which—

(A) is consistent with any applicable water quality standards, effluent and other limitations, and thermal discharge regulations established pursuant to current law within the basin;

(B) recommends such treatment works as will provide the most effective and economical means of collection, storage, treatment, and elimination of pollutants and recommends means to encourage both municipal and industrial use of such works;

(C) recommends maintenance and improvement of water quality within the basin or portion thereof and recommends methods of adequately financing those facilities as may be necessary to implement the plan; and

(D) as appropriate, is developed in cooperation with, and is consistent with any comprehensive plan prepared by the Water Resources Council, any areawide waste management plans developed pursuant to section 208 of this Act, and any State plan developed pursuant to section 303(e) of this Act.

(3) For the purposes of this subsection the term "basin" includes, but is not limited to, rivers and their tributaries, streams, coastal waters, sounds, estuaries, bays, lakes, and portions thereof, as well as the lands drained thereby.

(d) [Repealed by section 2021(a) of Public Law 104–66 (109 Stat. 726).]

(33 U.S.C. 1252)

INTERSTATE COOPERATION AND UNIFORM LAWS

SEC. 103. (a) The Administrator shall encourage cooperative activities by the States for the prevention, reduction, and elimination of pollution, encourage the enactment of improved and, so far as practicable, uniform State laws relating to the prevention, reduction, and elimination of pollution; and encourage compacts between States for the prevention and control of pollution.

(b) The consent of the Congress is hereby given to two or more States to negotiate and enter into agreements or compacts, not in conflict with any law or treaty of the United States, for (1) cooperative effort and mutual assistance for the prevention and control of pollution and the enforcement of their respective laws relating thereto, and (2) the establishment of such agencies, joint or otherwise, as they may deem desirable for making effective such agreements and compacts. No such agreement or compact shall be binding or obligatory upon any State a party thereto unless and until it has been approved by the Congress.

(33 U.S.C. 1253)

RESEARCH, INVESTIGATIONS, TRAINING, AND INFORMATION

SEC. 104. (a) The Administrator shall establish national programs for the prevention, reduction, and elimination of pollution and as part of such programs shall—

(1) in cooperation with other Federal, State, and local agencies, conduct and promote the coordination and accelera-

tion of, research, investigations, experiments, training, demonstrations, surveys, and studies relating to the causes, effects, extent, prevention, reduction, and elimination of pollution;

(2) encourage, cooperate with, and render technical services to pollution control agencies and other appropriate public or private agencies, institutions, and organizations, and individuals, including the general public, in the conduct of activities referred to in paragraph (1) of this subsection;

(3) conduct, in cooperation with State water pollution control agencies and other interested agencies, organizations and persons, public investigations concerning the pollution of any navigable waters, and report on the results of such investigations;

(4) establish advisory committees composed of recognized experts in various aspects of pollution and representatives of the public to assist in the examination and evaluation of research progress and proposals and to avoid duplication of research;

(5) in cooperation with the States, and their political subdivisions, and other Federal agencies establish, equip, and maintain a water quality surveillance system for the purpose of monitoring the quality of the navigable waters and ground waters and the contiguous zone and the oceans and the Administrator shall, to the extent practicable, conduct such surveillance by utilizing the resources of the National Aeronautics and Space Administration, the National Oceanic and Atmospheric Administration, the Geological Survey, and the Coast Guard, and shall report on such quality not later than 90 days after the date of convening of each session of Congress; and

(6) initiate and promote the coordination and acceleration of research designed to develop the most effective practicable tools and techniques for measuring the social and economic costs and benefits of activities which are subject to regulations under this Act; and shall transmit a report on the results of such research to the Congress not later than January 1, 1974.

(b) In carrying out the provisions of subsection (a) of this section the Administrator is authorized to—

(1) collect and make available, through publications and other appropriate means, the results of and other information, including appropriate recommendations by him in connection therewith, pertaining to such research and other activities referred to in paragraph (1) of subsection (a);

(2) cooperate with other Federal departments and agencies, State water pollution control agencies, interstate agencies, other public and private agencies, institutions, organizations, industries involved, and individuals, in the preparation and conduct of such research and other activities referred to in paragraph (1) of subsection (a);

(3) make grants to State water pollution control agencies, interstate agencies, other public or nonprofit private agencies, institutions, organizations, and individuals, for purposes stated in paragraph (1) of subsection (a) of this section;

(4) contract with public or private agencies, institutions, organizations, and individuals, without regard to sections 3648

and 3709 of the Revised Statutes (31 U.S.C. 529; 41 U.S.C. 5), referred to in paragraph (1) of subsection (a);

(5) establish and maintain research fellowships at public or nonprofit private educational institutions or research organizations;

(6) collect and disseminate, in cooperation with other Federal departments and agencies, and with other public or private agencies, institutions, and organizations having related responsibilities, basic data on chemical, physical, and biological effects of varying water quality and other information pertaining to pollution and the prevention, reduction, and elimination thereof; and

(7) develop effective and practical processes, methods, and prototype devices for the prevention, reduction, and elimination of pollution.

(c) In carrying out the provisions of subsection (a) of this section the Administrator shall conduct research on, and survey the results of other scientific studies on, the harmful effects on the health or welfare of persons caused by pollutants. In order to avoid duplication of effort, the Administrator shall, to the extent practicable, conduct such research in cooperation with and through the facilities of the Secretary of Health, Education, and Welfare.

(d) In carrying out the provisions of this section the Administrator shall develop and demonstrate under varied conditions (including conducting such basic and applied research, studies, and experiments as may be necessary):

(1) Practicable means of treating municipal sewage, and other waterborne wastes to implement the requirements of section 201 of this Act;

(2) Improved methods and procedures to identify and measure the effects of pollutants, including those pollutants created by new technological developments; and

(3) Methods and procedures for evaluating the effects on water quality of augmented streamflows to control pollution not susceptible to other means of prevention, reduction, or elimination.

(e) The Administrator shall establish, equip, and maintain field laboratory and research facilities, including, but not limited to, one to be located in the northeastern area of the United States, one in the Middle Atlantic area, one in the southeastern area, one in the midwestern area, one in the southwestern area, one in the Pacific Northwest, and one in the State of Alaska, for the conduct of research, investigations, experiments, field demonstrations and studies, and training relating to the prevention, reduction and elimination of pollution. Insofar as practicable, each such facility shall be located near institutions of higher learning in which graduate training in such research might be carried out. In conjunction with the development of criteria under section 403 of this Act, the Administrator shall construct the facilities authorized for the National Marine Water Quality Laboratory established under this subsection.

(f) The Administrator shall conduct research and technical development work, and make studies, with respect to the quality of the waters of the Great Lakes, including an analysis of the present and projected future water quality of the Great Lakes under varying conditions of waste treatment and disposal, an evaluation of the water quality needs of those to be served by such waters, an evaluation of municipal, industrial, and vessel waste treatment and disposal practices with respect to such waters, and a study of alternate means of solving pollution problems (including additional waste treatment measures) with respect to such waters.

(g)(1) For the purpose of providing an adequate supply of trained personnel to operate and maintain existing and future treatment works and related activities, and for the purpose of enhancing substantially the proficiency of those engaged in such activities, the Administrator shall finance pilot programs, in cooperation with State and interstate agencies, municipalities, educational institutions, and other organizations and individuals, of manpower development and training and retraining of persons in, on entering into, the field of operation and maintenance of treatment works and related activities. Such program and any funds expended for such a program shall supplement, not supplant, other manpower and training programs and funds available for the purposes of this paragraph. The Administrator is authorized, under such terms and conditions as he deems appropriate, to enter into agreements with one or more States, acting jointly or severally, or with other public or private agencies or institutions for the development and implementation of such a program.

(2) The Administrator is authorized to enter into agreements with public and private agencies and institutions, and individuals to develop and maintain an effective system for forecasting the supply of, and demand for, various professional and other occupational categories needed for the prevention, reduction, and elimination of pollution in each region, State, or area of the United States and, from time to time, to publish the results of such forecasts.

(3) In furtherance of the purposes of this Act, the Administrator is authorized to—

(A) make grants to public or private agencies and institutions and to individuals for training projects, and provide for the conduct of training by contract with public or private agencies and institutions and with individuals without regard to sections 3648 and 3709 of the Revised Statutes;

(B) establish and maintain research fellowships in the Environmental Protection Agency with such stipends and allowances, including traveling and subsistence expenses, as he may deem necessary to procure the assistance of the most promising research fellows; and

(C) provide, in addition to the program established under paragraph (1) of this subsection, training in technical matters relating to the causes, prevention, reduction, and elimination of pollution for personnel of public agencies and other persons with suitable qualifications.

(4) The Administrator shall submit, through the President, a report to the Congress not later than December 31, 1973, summarizing the actions taken under this subsection and the effectiveness of such actions, and setting forth the number of persons trained, the occupational categories for which training was provided, the effectiveness of other Federal, State, and local training programs in

this field, together with estimates of future needs, recommendations on improving training programs, and such other information and recommendations, including legislative recommendations, as he deems appropriate.

(h) The Administrator is authorized to enter into contracts, with, or make grants to, public or private agencies and organizations and individuals for (A) the purpose of developing and demonstrating new or improved methods for the prevention, removal, reduction, and elimination of pollution in lakes, including the undesirable effects of nutrients and vegetation, and (B) the construction of publicly owned research facilities for such purpose.

(i) The Administrator, in cooperation with the Secretary of the department in which the Coast Guard is operating, shall—

(1) engage in such research, studies, experiments, and demonstrations as he deems appropriate, relative to the removal of oil from any waters and to the prevention, control, and elimination of oil and hazardous substances pollution;

(2) publish from time to time the results of such activities; and

(3) from time to time, develop and publish in the Federal Register specifications and other technical information on the various chemical compounds used in the control of oil and hazardous substances spills.

In carrying out this subsection, the Administrator may enter into contracts with, or make grants to, public or private agencies and organizations and individuals.

(j) The Secretary of the department in which the Coast Guard is operating shall engage in such research, studies, experiments, and demonstrations as he deems appropriate relative to equipment which is to be installed on board a vessel and is designed to receive, retain, treat, or discharge human body wastes and the wastes from toilets and other receptacles intended to receive or retain body wastes with particular emphasis on equipment to be installed on small recreational vessels. The Secretary of the department in which the Coast Guard is operating shall report to Congress the results of such research, studies, experiments, and demonstrations prior to the effective date of any regulations established under section 312 of this Act. In carrying out this subsection the Secretary of the department in which the Coast Guard is operating may enter into contracts with, or make grants to, public or private organizations and individuals.

(k) In carrying out the provisions of this section relating to the conduct by the Administrator of demonstration projects and the development of field laboratories and reseach facilities, the Administrator may acquire land and interests therein by purchase, with appropriated or donated funds, by donation, or by exchange for acquired or public lands under his jurisdiction which he classifies as suitable for disposition. The values of the properties so exchanged either shall be approximately equal, or if they are not approximately equal, the values shall be equalized by the payment of cash to the grantor or to the Administrator as the circumstances require.

(l)(1) The Administrator shall, after consultation with appropriate local, State, and Federal agencies, public and private organizations, and interested individuals, as soon as practicable but not later than January 1, 1973, develop and issue to the States for the purpose of carrying out this Act the latest scientific knowledge available in indicating the kind and extent of effects on health and welfare which may be expected from the presence of pesticides in the water in varying quantities. He shall revise and add to such information whenever necessary to reflect developing scientific knowledge.

(2) The President shall, in consultation with appropriate local, State, and Federal agencies, public and private organizations, and interested individuals, conduct studies and investigations of methods to control the release of pesticides into the environment which study shall include examination of the persistency of pesticides in the water environment and alternative thereto. The President shall submit reports, from time to time, on such investigations to Congress together with his recommendations for any necessary legislation.

(m)(1) The Administrator shall, in an effort to prevent degradation of the environment from the disposal of waste oil, conduct a study of (A) the generation of used engine, machine, cooling, and similar waste oil, including quantities generated, the nature and quality of such oil, present collecting methods and disposal practices, and alternate uses of such oil; (B) the long-term, chronic biological effects of the disposal of such waste oil; and (C) the potential market for such oils, including the economic and legal factors relating to the sale of products made from such oils, the level of subsidy, if any, needed to encourage the purchase by public and private nonprofit agencies of products from such oil, and the practicability of Federal procurement, on a priority basis, of products made from such oil. In conducting such study, the Administrator shall consult with affected industries and other persons.

(2) The Administrator shall report the preliminary results of such study to Congress within six months after the date of enactment of the Federal Water Pollution Control Act Amendments of 1972, and shall submit a final report to Congress within 18 months after such date of enactment.

(n)(1) The Administrator shall, in cooperation with the Secretary of the Army, the Secretary of Agriculture, the Water Resources Council, and with other appropriate Federal, State, interstate, or local public bodies and private organizations, institutions, and individuals, conduct and promote, encourage contributions to, continuing comprehensive studies of the effects of pollution, including sedimentation, in the estuaries and estuarine zones of the United States on fish and wildlife, on sport and commercial fishing, on recreation, on water supply and water power, and on other beneficial purposes. Such studies shall also consider the effect of demographic trends, the exploitation of mineral resources and fossil fuels, land and industrial development, navigation, flood and erosion control, and other uses of estuaries and estuarine zones upon the pollution of the waters therein.

(2) In conducting such studies, the Administrator shall assemble, coordinate, and organize all existing pertinent information on the Nation's estuaries and estuarine zones; carry out a program of investigations and surveys to supplement existing information in

representative estuaries and estuarine zones; and identify the problems and areas where further research and study are required.

(3) For the purpose of this subsection, the term "estuarine zones" means an environmental system consisting of an estuary and those transitional areas which are consistently influenced or affected by water from an estuary such as, but not limited to, salt marshes, coastal and intertidal areas, bays, harbors, lagoons, inshore waters, and channels, and the term "estuary" means all or part of the mouth of a river or stream or other body of water having unimpaired natural connection with open sea and within which the sea water is measurably diluted with fresh water derived from land drainage.

(o)(1) The Administrator shall conduct research and investigations on devices, systems, incentives, pricing policy, and other methods of reducing the total flow of sewage, including, but not limited to, unnecessary water consumption in order to reduce the requirements for, and the costs of, sewage and waste treatment services. Such research and investigations shall be directed to develop devices, systems, policies, and methods capable of achieving the maximum reduction of unnecessary water consumption.

(2) The Administrator shall report the preliminary results of such studies and investigations to the Congress within one year after the date of enactment of the Federal Water Pollution Control Act Amendments of 1972, and annually thereafter not later than 90 days after the date of convening of each session of Congress. Such report shall include recommendations for any legislation that may be required to provide for the adoption and use of devices, systems, policies, or other methods of reducing water consumption and reducing the total flow of sewage. Such report shall include an estimate of the benefits to be derived from adoption and use of such devices, systems, policies, or other methods and also shall reflect estimates of any increase in private, public, or other cost that would be occasioned thereby.

(p) In carrying out the provisions of subsection (a) of this section the Administrator shall, in cooperation with the Secretary of Agriculture, other Federal agencies, and the States, carry out a comprehensive study and research program to determine new and improved methods and the better application of existing methods of preventing, reducing, and eliminating pollution from agriculture, including the legal, economic, and other implications of the use of such methods.

(q)(1) The Administrator shall conduct a comprehensive program of research and investigation and pilot project implementation into new and improved methods of preventing, reducing, storing, collecting, treating, or otherwise eliminating pollution from sewage in rural and other areas where collection of sewage in conventional, community-wide sewage collection systems is impractical, uneconomical, or otherwise infeasible, or where soil conditions or other factors preclude the use of septic tank and drainage field systems.

(2) The Administrator shall conduct a comprehensive program of research and investigation and pilot project implementation into new and improved methods for the collection and treatment of sew-

age and other liquid wastes combined with the treatment and disposal of solid wastes.

(3) The Administrator shall establish, either within the Environmental Protection Agency, or through contract with an appropriate public or private non-profit organization, a national clearinghouse which shall (A) receive reports and information resulting from research, demonstrations, and other projects funded under this Act related to paragraph (1) of this subsection and to subsection (e)(2) of section 105; (B) coordinate and disseminate such reports and information for use by Federal and State agencies, municipalities, institutions, and persons in developing new and improved methods pursuant to this subsection; and (C) provide for the collection and dissemination of reports and information relevant to this subsection from other Federal and State agencies, institutions, universities, and persons.

(4) SMALL FLOWS CLEARINGHOUSE.—Notwithstanding section 205(d) of this Act, from amounts that are set aside for a fiscal year under section 205(i) of this Act and are not obligated by the end of the 24-month period of availability for such amounts under section 205(d), the Administrator shall make available $1,000,000 or such unobligated amount, whichever is less, to support a national clearinghouse within the Environmental Protection Agency to collect and disseminate information on small flows of sewage and innovative or alternative wastewater treatment processes and techniques, consistent with paragraph (3). This paragraph shall apply with respect to amounts set aside under section 205(i) for which the 24-month period of availability referred to in the preceding sentence ends on or after September 30, 1986.

(r) The Administrator is authorized to make grants to colleges and universities to conduct basic research into the structure and function of fresh water aquatic ecosystems, and to improve understanding of the ecological characteristics necessary to the maintenance of the chemical, physical, and biological integrity of freshwater aquatic ecosystems.

(s) The Administrator is authorized to make grants to one or more institutions of higher education (regionally located and to be designated as "River Study Centers") for the purpose of conducting and reporting on interdisciplinary studies on the nature of river systems, including hydrology, biology, ecology, economics, the relationship between river uses and land uses, and the effects of development within river basins on river systems and on the value of water resources and water related activities. No such grant in any fiscal year shall exceed $1,000,000.

(t) The Administrator shall, in cooperation with State and Federal agencies and public and private organizations, conduct continuing comprehensive studies of the effects and methods of control of thermal discharges. In evaluating alternative methods of control the studies shall consider (1) such data as are available on the latest available technology, economic feasibility including cost-effectiveness analysis, and (2) the total impact on the environment, considering not only water quality but also air quality, land use, and effective utilization and conservation of fresh water and other natural resources. Such studies shall consider methods of minimizing adverse effects and maximizing beneficial effects of thermal discharges. The results of these studies shall be reported by the Administrator as soon as practicable, but not later than 270 days after enactment of this subsection, and shall be made available to the public and the States, and considered as they become available by the Administrator in carrying out section 316 of this Act and by the State in proposing thermal water quality standards.

(u) There is authorized to be appropriated (1) not to exceed $100,000,000 per fiscal year for the fiscal year ending June 30, 1973, the fiscal year ending June 30, 1974, and the fiscal year ending June 30, 1975, not to exceed $14,039,000 for the fiscal year ending September 30, 1980, not to exceed $20,697,000 for the fiscal year ending September 30, 1981, not to exceed $22,770,000 for the fiscal year ending September 30, 1982, such sums as may be necessary for fiscal years 1983 through 1985, and not to exceed $22,770,000 per fiscal year for each of the fiscal years 1986 through 1990, for carrying out the provisions of this section, other than subsections (g)(1) and (2), (p), (r), and (t), except that such authorizations are not for any research, development, or demonstration activity pursuant to such provisions; (2) not to exceed $7,500,000 for fiscal years 1973, 1974, and 1975, $2,000,000 for fiscal year 1977, $3,000,000 for fiscal year 1978, $3,000,000 for fiscal year 1979, $3,000,000 for fiscal year 1980, $3,000,000 for fiscal year 1981, $3,000,000 for fiscal year 1982, such sums as may be necessary for fiscal years 1983 through 1985, and $3,000,000 per fiscal year for each of the fiscal years 1986 through 1990, for carrying out the provisions of subsection (g)(1); (3) not to exceed $2,500,000 for fiscal years 1973, 1974, and 1975, $1,000,000 for fiscal year 1977, $1,500,000 for fiscal year 1978, $1,500,000 for fiscal year 1979, $1,500,000 for fiscal year 1980, $1,500,000 for fiscal year 1981, $1,500,000 for fiscal year 1982, such sums as may be necessary for fiscal years 1983 through 1985, and $1,500,000 per fiscal year for each of the fiscal years 1986 through 1990, for carrying out the provisions of subsection (g)(2); (4) not to exceed $10,000,000 for each of the fiscal years ending June 30, 1973, June 30, 1974, and June 30, 1975, for carrying out the provisions of subsection (p); (5) not to exceed $15,000,000 per fiscal year for the fiscal years ending June 30, 1973, June 30, 1974, and June 30, 1975, for carrying out the provisions of subsection (r); and (6) not to exceed $10,000,000 per fiscal year for the fiscal years ending June 30, 1973, June 30, 1974, and June 30, 1975, for carrying out the provisions of subsection (t).

(v) STUDIES CONCERNING PATHOGEN INDICATORS IN COASTAL RECREATION WATERS.—Not later than 18 months after the date of the enactment of this subsection, after consultation and in cooperation with appropriate Federal, State, tribal, and local officials (including local health officials), the Administrator shall initiate, and, not later than 3 years after the date of the enactment of this subsection, shall complete, in cooperation with the heads of other Federal agencies, studies to provide additional information for use in developing—

(1) an assessment of potential human health risks resulting from exposure to pathogens in coastal recreation waters, including nongastrointestinal effects;

(2) appropriate and effective indicators for improving detection in a timely manner in coastal recreation waters of the presence of pathogens that are harmful to human health;

(3) appropriate, accurate, expeditious, and cost-effective methods (including predictive models) for detecting in a timely manner in coastal recreation waters the presence of pathogens that are harmful to human health; and

(4) guidance for State application of the criteria for pathogens and pathogen indicators to be published under section 304(a)(9) to account for the diversity of geographic and aquatic conditions.

(33 U.S.C. 1254)

GRANTS FOR RESEARCH AND DEVELOPMENT

SEC. 105. (a) The Administrator is authorized to conduct in the Environmental Protection Agency, and to make grants to any State, municipality, or intermunicipal or interstate agency for the purpose of assisting in the development of—

(1) any project which will demonstrate a new or improved method of preventing, reducing, and eliminating the discharge into any waters of pollutants from sewers which carry storm water or both storm water and pollutants; or

(2) any project which will demonstrate advanced waste treatment and water purification methods (including the temporary use of new or improved chemical additives which provide substantial immediate improvement to existing treatment processes), or new or improved methods of joint treatment systems for municipal and industrial wastes;

and to include in such grants such amounts as are necessary for the purpose of reports, plans, and specifications in connection therewith.

(b) The Administrator is authorized to make grants to any State or States or interstate agency to demonstrate, in river basins or portions thereof, advanced treatment and environmental enhancement techniques to control pollution from all sources, within such basins or portions thereof, including nonpoint sources, together with in stream water quality improvement techniques.

(c) In order to carry out the purposes of section 301 of this Act, the Administrator is authorized to (1) conduct in the Environmental Protection Agency, (2) make grants to persons, and (3) enter into contracts with persons, for research and demonstration projects for prevention of pollution of any waters by industry including, but not limited to, the prevention, reduction, and elimination of the discharge of pollutants. No grant shall be made for any project under this subsection unless the Administrator determines that such project will develop or demonstrate a new or improved method of treating industrial wastes or otherwise prevent pollution by industry, which method shall have industrywide application.

(d) In carrying out the provisions of this section, the Administrator shall conduct, on a priority basis, an accelerated effort to develop, refine, and achieve practical application of:

(1) waste management methods applicable to point and nonpoint sources of pollutants to eliminate the discharge of pollutants, including, but not limited to, elimination of runoff of pollutants and the effects of pollutants from inplace or accumulated sources;

(2) advanced waste treatment methods applicable to point and nonpoint sources, including inplace or accumulated sources of pollutants, and methods for reclaiming and recycling water and confining pollutants so they will not migrate to cause water or other environmental pollution; and

(3) improved methods and procedures to identify and measure the effects of pollutants on the chemical, physical, and biological integrity of water, including those pollutants created by new technological developments.

(e)(1) The Administrator is authorized to (A) make, in consultation with the Secretary of Agriculture, grants to persons for research and demonstration projects with respect to new and improved methods of preventing, reducing, and eliminating pollution from agriculture, and (B) disseminate, in cooperation with the Secretary of Agriculture, such information obtained under this subsection, section 104(p), and section 304 as will encourage and enable the adoption of such methods in the agricultural industry.

(2) The Administrator is authorized, (A) in consultation with other interested Federal agencies, to make grants for demonstration projects with respect to new and improved methods of preventing, reducing, storing, collecting, treating, or otherwise eliminating pollution from sewage in rural and other areas where collection of sewage in conventional, community-wide sewage collection systems is impractical, uneconomical, or otherwise infeasible, or where soil conditions or other factors preclude the use of septic tank and drainage field systems, and (B) in cooperation with other interested Federal and State agencies, to disseminate such information obtained under this subsection as will encourage and enable the adoption of new and improved methods developed pursuant to this subsection.

(f) Federal grants under subsection (a) of this section shall be subject to the following limitations:

(1) No grant shall be made for any project unless such project shall have been approved by the appropriate State water pollution control agency or agencies and by the Administrator;

(2) No grant shall be made for any project in an amount exceeding 75 per centum of cost thereof as determined by the Administrator; and

(3) No grant shall be made for any project unless the Administrator determines that such project will serve as a useful demonstration for the purpose set forth in clause (1) or (2) of subsection (a).

(g) Federal grants under subsections (c) and (d) of this section shall not exceed 75 per centum of the cost of the project.

(h) For the purpose of this section there is authorized to be appropriated $75,000,000 per fiscal year for the fiscal year ending June 30, 1973, the fiscal year ending June 30, 1974, and the fiscal year ending June 30, 1975, and from such appropriations at least

10 per centum of the funds actually appropriated in each fiscal year shall be available only for the purposes of subsection (e).

(i) The Administrator is authorized to make grants to a municipality to assist in the costs of operating and maintaining a project which received a grant under this section, section 104, or section 113 of this Act prior to the date of enactment of this subsection so as to reduce the operation and maintenance costs borne by the recipients of services from such project to costs comparable to those for projects assisted under title II of this Act.

(j) The Administrator is authorized to make a grant to any grantee who received an increased grant pursuant to section 202(a)(2) of this Act. Such grant may pay up to 100 per centum of the costs of technical evaluation of the operation of the treatment works, costs of training of persons (other than employees of the grantee), and costs of disseminating technical information on the operation of the treatment works.

(33 U.S.C. 1255)

GRANTS FOR POLLUTION CONTROL PROGRAMS

SEC. 106. (a) There are hereby authorized to be appropriated the following sums, to remain available until expended, to carry out the purposes of this section—

(1) $60,000,000 for the fiscal year ending June 30, 1973; and

(2) $75,000,000 for the fiscal year ending June 30, 1974, and the fiscal year ending June 30, 1975, $100,000,000 per fiscal year for the fiscal years 1977, 1978, 1979, and 1980, $75,000,000 per fiscal year for the fiscal years 1981 and 1982, such sums as may be necessary for fiscal years 1983 through 1985, and $75,000,000 per fiscal year for each of the fiscal years 1986 through 1990;

for grants to States and to interstate agencies to assist them in administering programs for the prevention, reduction, and elimination of pollution, including enforcement directly or through appropriate State law enforcement officers or agencies.

(b) From the sums appropriated in any fiscal year, the Administrator shall make allotments to the several States and interstate agencies in accordance with regulations promulgated by him on the basis of the extent of the pollution problem in the respective States.

(c) The Administrator is authorized to pay to each State and interstate agency each fiscal year either—

(1) the allotment of such State or agency for such fiscal year under subsection (b), or

(2) the reasonable costs as determined by the Administrator of developing and carrying out a pollution program by such State or agency during such fiscal year,

whichever amount is the lesser.

(d) No grant shall be made under this section to any State or interstate agency for any fiscal year when the expenditure of non-Federal funds by such State or interstate agency during such fiscal year for the recurrent expenses of carrying out its pollution control program are less than the expenditure by such State or interstate agency of non-Federal funds for such recurrent program expenses during the fiscal year ending June 30, 1971.

(e) Beginning in fiscal year 1974 the Administrator shall not make any grant under this section to any State which has not provided or is not carrying out as a part of its program—

(1) the establishment and operation of appropriate devices, methods, systems, and procedures necessary to monitor, and to compile and analyze data on (including classification according to eutrophic condition), the quality of navigable waters and to the extent practicable, ground waters including biological monitoring; and provision for annually updating such data and including it in the report required under section 305 of this Act;

(2) authority comparable to that in section 504 of this Act and adequate contingency plans to implement such authority.

(f) Grants shall be made under this section on condition that—

(1) Such State (or interstate agency) filed with the Administrator within one hundred and twenty days after the date of enactment of this section:

(A) a summary report of the current status of the State pollution control program, including the criteria used by the State in determining priority of treatment works; and

(B) such additional information, data, and reports as the Administrator may require.

(2) No federally assumed enforcement as defined in section 309(a)(2) is in effect with respect to such State or interstate agency.

(3) Such State (or interstate agency) submits within one hundred and twenty days after the date of enactment of this section and before July 1 of each year thereafter for the Administrator's approval of its program for the prevention, reduction, and elimination of pollution in accordance with purposes and provisions of this Act in such form and content as the Administrator may prescribe.

(g) Any sums allotted under subsection (b) in any fiscal year which are not paid shall be reallotted by the Administrator in accordance with regulations promulgated by him.

(33 U.S.C. 1256)

MINE WATER POLLUTION CONTROL DEMONSTRATIONS

SEC. 107. (a) The Administrator in cooperation with the Appalachian Regional Commission and other Federal agencies is authorized to conduct, to make grants for, or to contract for, projects to demonstrate comprehensive approaches to the elimination or control of acid or other mine water pollution resulting from active or abandoned mining operations and other environmental pollution affecting water quality within all or part of a watershed or river basin, including siltation from surface mining. Such projects shall demonstrate the engineering and economic feasibility and practicality of various abatement techniques which will contribute substantially to effective and practical methods of acid or other mine water pollution elimination or control, and other pollution affecting water quality, including techniques that demonstrate the engineering and economic feasibility and practicality of using sewage sludge materials and other municipal wastes to diminish or prevent pollution affecting water quality from acid, sedimentation, or other pollutants and in such projects to restore affected lands to usefulness for forestry, agriculture, recreation, or other beneficial purposes.

(b) Prior to undertaking any demonstration project under this section in the Appalachian region (as defined in section 403 of the Appalachian Regional Development Act of 1965, as amended), the Appalachian Regional Commission shall determine that such demonstration project is consistent with the objectives of the Appalachian Regional Development Act of 1965, as amended.

(c) The Administrator, in selecting watersheds for the purposes of this section, shall be satisfied that the project area will not be affected adversely by the influx of acid or other mine water pollution from nearby sources.

(d) Federal participation in such projects shall be subject to the conditions—

(1) that the State shall acquire any land or interests therein necessary for such project; and

(2) that the State shall provide legal and practical protection to the project area to insure against any activities which will cause future acid or other mine water pollution.

(e) There is authorized to be appropriated $30,000,000 to carry out the provisions of this section, which sum shall be available until expended.

(33 U.S.C. 1257)

POLLUTION CONTROL IN GREAT LAKES

SEC. 108. (a) The Administrator, in cooperation with other Federal departments, agencies, and instrumentalities is authorized to enter into agreements with any State, political subdivision, interstate agency, or other public agency, or combination thereof, to carry out one or more projects to demonstrate new methods and techniques and to develop preliminary plans for the elimination or control of pollution, within all or any part of the watersheds of the Great Lakes. Such projects shall demonstrate the engineering and economic feasibility and practicality of removal of pollutants and prevention of any polluting matter from entering into the Great Lakes in the future and other reduction and remedial techniques which will contribute substantially to effective and practical methods of pollution prevention, reduction, or elimination.

(b) Federal participation in such projects shall be subject to the condition that the State, political subdivision, interstate agency, or other public agency, or combination thereof, shall pay not less than 25 per centum of the actual project costs, which payment may be in any form, including, but not limited to, land or interests therein that is needed for the project, and personal property or services the value of which shall be determined by the Administrator.

(c) There is authorized to be appropriated $20,000,000 to carry out the provisions of subsections (a) and (b) of this section, which sum shall be available until expended.

(d)(1) In recognition of the serious conditions which exist in Lake Erie, the Secretary of the Army, acting through the Chief of

Engineers, is directed to design and develop a demonstration waste water management program for the rehabilitation and environmental repair of Lake Erie. Prior to the initiation of detailed engineering and design, the program, along with the specific recommendations of the Chief of Engineers and recommendations for its financing, shall be submitted to the Congress for statutory approval. This authority is in addition to, and not in lieu of, other waste water studies aimed at eliminating pollution emanating from select sources around Lake Erie.

(2) This program is to be developed in cooperation with the Environmental Protection Agency, other interested departments, agencies, and instrumentalities of the Federal Government, and the States and their political subdivisions. This program shall set forth alternative systems for managing waste water on a regional basis and shall provide local and State governments with a range of choice as to the type of system to be used for the treatment of waste water. These alternative systems shall include both advanced waste treatment technology and land disposal systems including aerated treatment-spray irrigation technology and will also include provisions for the disposal of solid wastes, including sludge. Such program should include measures to control point sources of pollution, area sources of pollution, including acid-mine drainage, urban runoff and rural runoff, and in place sources of pollution, including bottom loads, sludge banks, and polluted harbor dredgings.

(e) There is authorized to be appropriated $5,000,000 to carry out the provisions of subsection (d) of this section, which sum shall be available until expended.

(33 U.S.C. 1258)

TRAINING GRANTS AND CONTRACTS

SEC. 109. (a) The Administrator is authorized to make grants to or contracts with institutions of higher education, or combinations of such institutions, to assist them in planning, developing, strengthening, improving, or carrying out programs or projects for the preparation of undergraduate students to enter an occupation which involves the design, operation, and maintenance of treatment works, and other facilities whose purpose is water quality control. Such grants or contracts may include payment of all or part of the cost of programs or projects such as—

(A) planning for the development or expansion of programs or projects for training persons in the operation and maintenance of treatment works;

(B) training and retraining of faculty members;

(C) conduct of short-term or regular session institutes for study by persons engaged in, or preparing to engage in, the preparation of students preparing to enter an occupation involving the operation and maintenance of treatment works;

(D) carrying out innovative and experimental programs of cooperative education involving alternate periods of full-time or part-time academic study at the institution and periods of full-time or part-time employment involving the operation and maintenance of treatment works; and

(E) research into, and development of, methods of training students or faculty, including the preparation of teaching materials and the planning of curriculum.

(b)(1) The Administrator may pay 100 per centum of any additional cost of construction of treatment works required for a facility to train and upgrade waste treatment works operation and maintenance personnel and for the costs of other State treatment works operator training programs, including mobile training units, classroom rental, specialized instructors, and instructional material.

(2) The Administrator shall make no more than one grant for such additional construction in any State (to serve a group of States, where, in his judgment, efficient training programs require multi-State programs), and shall make such grant after consultation with and approval by the State or States on the basis of (A) the suitability of such facility for training operation and maintenance personnel for treatment works throughout such State or States; and (B) a commitment by the State agency or agencies to carry out at such facility a program of training approved by the Administrator. In any case where a grant is made to serve two or more States, the Administrator is authorized to make an additional grant for a supplemental facility in each such State.

(3) The Administrator may make such grant out of the sums allocated to a State under section 205 of this Act, except that in no event shall the Federal cost of any such training facilities exceed $500,000.

(4) The Administrator may exempt a grant under this section from any requirement under section 204(a)(3) of this Act. Any grantee who received a grant under this section prior to enactment of the Clean Water Act of 1977 shall be eligible to have its grant increased by funds made available under such Act.

(33 U.S.C. 1259)

APPLICATION FOR TRAINING GRANT OR CONTRACT; ALLOCATION OF GRANTS OR CONTRACTS

SEC. 110. (1) A grant or contract authorized by section 109 may be made only upon application to the Administrator at such time or times and containing such information as he may prescribe, except that no such application shall be approved unless it—

(A) sets forth programs, activities, research, or development for which a grant is authorized under section 109 and describes the relation to any program set forth by the applicant in an application, if any, submitted pursuant to section 111;

(B) provides such fiscal control and fund accounting procedures as may be necessary to assure proper disbursement of and accounting for Federal funds paid to the applicant under this section; and

(C) provides for making such reports, in such form and containing such information, as the Administrator may require to carry out his functions under this section, and for keeping such records and for affording such access thereto as the Administrator may find necessary to assure the correctness and verification of such reports.

(2) The Administrator shall allocate grants or contracts under section 109 in such manner as will most nearly provide an equitable distribution of the grants or contracts throughout the United States among institutions of higher education which show promise of being able to use funds effectively for the purpose of this section.

(3)(A) Payments under this section may be used in accordance with regulations of the Administrator, and subject to the terms and conditions set forth in an application approved under paragraph (1), to pay part of the compensation of students employed in connection with the operation and maintenance of treatment works, other than as an employee in connection with the operation and maintenance of treatment works or as an employee in any branch of the Government of the United States, as part of a program for which a grant has been approved pursuant to this section.

(B) Departments and agencies of the United States are encouraged, to the extent consistent with efficient Administration, to enter into arrangements with institutions of higher education for the full-time, part-time, or temporary employment, whether in the competitive or excepted service, of students enrolled in programs set forth in applications approved under paragraph (1).

(33 U.S.C. 1260)

AWARD OF SCHOLARSHIPS

SEC. 111. (1) The Administrator is authorized to award scholarships in accordance with the provisions of this section for undergraduate study by persons who plan to enter an occupation involving the operation and maintenance of treatment works. Such scholarships shall be awarded for such periods as the Administrator may determine but not to exceed four academic years.

(2) The Administrator shall allocate scholarships under this section among institutions of higher education with programs approved under the provisions of this section for the use of individuals accepted into such programs, in such manner and accordance to such plan as will insofar as practicable—

(A) provide an equitable distribution of such scholarships throughout the United States; and

(B) attract recent graduates of secondary schools to enter an occupation involving the operation and maintenance of treatment works.

(3) The Administrator shall approve a program of any institution of higher education for the purposes of this section only upon application by the institution and only upon his finding—

(A) that such program has as a principal objective the education and training of persons in the operation and maintenance of treatment works;

(B) that such program is in effect and of high quality, or can be readily put into effect and may reasonably be expected to be of high quality;

(C) that the application describes the relation of such program to any program, activity, research, or development set forth by the applicant in an application, if any, submitted pursuant to section 110 of this Act; and

(D) that the application contains satisfactory assurances that (i) the institution will recommend to the Administrator for the award of scholarships under this section, for study in such program, only persons who have demonstrated to the satisfaction of the institution a serious intent, upon completing the program, to enter an occupation involving the operation and maintenance of treatment works, and (ii) the institution will make reasonable continuing efforts to encourage recipients of scholarships under this section, enrolled in such program, to enter occupations involving the operation and maintenance of treatment works upon completing the program.

(4)(A) The Administrator shall pay to persons awarded scholarships under this section such stipends (including such allowances for subsistence and other expenses for such persons and their dependents) as he may determine to be consistent with prevailing practices under comparable federally supported programs.

(B) The Administrator shall (in addition to the stipends paid to persons under paragraph (1)) pay to the institution of higher education at which such person is pursuing his course of study such amount as he may determine to be consistent with prevailing practices under comparable federally supported programs.

(5) A person awarded a scholarship under the provisions of this section shall continue to receive the payments provided in this section only during such periods as the Administrator finds that he is maintaining satisfactory proficiency and devoting full time to study or research in the field in which such scholarship was awarded in an institution of higher education, and is not engaging in gainful employment other than employment approved by the Administrator by or pursuant to regulation.

(6) The Administrator shall by regulation provide that any person awarded a scholarship under this section shall agree in writing to enter and remain in an occupation involving the design, operation, or maintenance of treatment works for such period after completion of his course of studies as the Administrator determines appropriate.

(33 U.S.C. 1261)

DEFINITIONS AND AUTHORIZATIONS

SEC. 112. (a) As used in sections 109 through 112 of this Act—

(1) The term "institution of higher education" means an educational institution described in the first sentence of section 101 of the Higher Education Act of 1965 (other than an institution of any agency of the United States) which is accredited by a nationally recognized accrediting agency or association approved by the Administrator for this purpose. For purposes of this subsection, the Administrator shall publish a list of nationally recognized accrediting agencies or associations which he determines to be reliable authority as to the quality of training offered.

(2) The term "academic year" means an academic year or its equivalent, as determined by the Administrator.

(b) The Administrator shall annually report his activities under sections 109 through 112 of this Act, including recommendations for needed revisions in the provisions thereof.

(c) There are authorized to be appropriated $25,000,000 per fiscal year for fiscal years ending June 30, 1973, June 30, 1974, and June 30, 1975, $6,000,000 for the fiscal year ending September 30, 1977, $7,000,000 for the fiscal year ending September 30, 1978, $7,000,000 for the fiscal year ending September 30, 1979, $7,000,000 for the fiscal year ending September 30, 1980, $7,000,000 for the fiscal year ending September 30, 1981, $7,000,000 for the fiscal year ending September 30, 1982, such sums as may be necessary for fiscal years 1983 through 1985, and $7,000,000 per fiscal year for each of the fiscal years 1986 through 1990, to carry out sections 109 through 112 of this Act.

(33 U.S.C. 1262)

ALASKA VILLAGE DEMONSTRATION PROJECTS

SEC. 113. (a) The Administrator is authorized to enter into agreements with the State of Alaska to carry out one or more projects to demonstrate methods to provide for central community facilities for safe water and elimination or control of pollution in those native villages of Alaska without such facilities. Such project shall include provisions for community safe water supply systems, toilets, bathing and laundry facilities, sewage disposal facilities, and other similar facilities, and educational and informational facilities and programs relating to health and hygiene. Such demonstration projects shall be for the further purpose of developing preliminary plans for providing such safe water and such elimination or control of pollution for all native villages in such State.

(b) In carrying out this section the Administrator shall cooperate with the Secretary of Health, Education, and Welfare for the purpose of utilizing such of the personnel and facilities of that Department as may be appropriate.

(c) The Administrator shall report to Congress not later than July 1, 1973, the results of the demonstration projects authorized by this section together with his recommendations, including and necessary legislation, relating to the establishment of a statewide program.

(d) There is authorized to be appropriated not to exceed $2,000,000 to carry out this section. In addition, there is authorized to be appropriated to carry out this section not to exceed $200,000 for the fiscal year ending September 30, 1978, and $220,000 for the fiscal year ending September 30, 1979.

(e) The Administrator is authorized to coordinate with the Secretary of the Department of Health, Education, and Welfare, the Secretary of the Department of Housing and Urban Development, the Secretary of the Department of the Interior, the Secretary of the Department of Agriculture, and the heads of any other departments or agencies he may deem appropriate to conduct a joint study with representatives of the State of Alaska and the appropriate Native organizations (as defined in Public Law 92–203) to develop a comprehensive program for achieving adequate sanitation services in Alaska villages. This study shall be coordinated with the programs and projects authorized by sections 104(q) and 105(e)(2) of this Act. The Administrator shall submit a report of the results of the study, together with appropriate supporting data and such recommendations as he deems desirable, to the Committee on Environment and Public Works of the Senate and to the Committee on Public Works and Transportation of the House of Representatives not later than December 31, 1979. The Administrator shall also submit recommended administrative actions, procedures, and any proposed legislation necessary to implement the recommendations of the study no later than June 30, 1980.

(f) The Administrator is authorized to provide technical, financial and management assistance for operation and maintenance of the demonstration projects constructed under this section, until such time as the recommendations of subsection (e) are implemented.

(g) For the purpose of this section, the term "village" shall mean an incorporated or unincorporated community with a population of ten to six hundred people living within a two-mile radius. The term "sanitation services" shall mean water supply, sewage disposal, solid waste disposal and other services necessary to maintain generally accepted standards of personal hygiene and public health.

(33 U.S.C. 1263)

LAKE TAHOE STUDY

SEC. 114. (a) The Administrator, in consultation with the Tahoe Regional Planning Agency, the Secretary of Agriculture, other Federal agencies, representatives of State and local governments, and members of the public, shall conduct a thorough and complete study on the adequacy of and need for extending Federal oversight and control in order to preserve the fragile ecology of Lake Tahoe.

(b) Such study shall include an examination of the interrelationships and responsibilities of the various agencies of the Federal Government and State and local governments with a view to establishing the necessity for redefinition of legal and other arrangements between these various governments, and making specific legislative recommendations to Congress. Such study shall consider the effect of various actions in terms of their environmental impact on the Tahoe Basin, treated as an ecosystem.

(c) The Administrator shall report on such study to Congress not later than one year after the date of enactment of this subsection.

(d) There is authorized to be appropriated to carry out this section not to exceed $500,000.

(33 U.S.C. 1264)

IN-PLACE TOXIC POLLUTANTS

SEC. 115. The Administrator is directed to identify the location of in-place pollutants with emphasis on toxic pollutants in harbors and navigable waterways and is authorized, acting through the Secretary of the Army, to make contracts for the removal and appropriate disposal of such materials from critical port and harbor areas. There is authorized to be appropriated $15,000,000 to carry out the provisions of this section, which sum shall be available until expended.

(33 U.S.C. 1265)

HUDSON RIVER PCB RECLAMATION DEMONSTRATION PROJECT

SEC. 116. (a) The Administrator is authorized to enter into contracts and other agreements with the State of New York to carry out a project to demonstrate methods for the selective removal of polychlorinated biphenyls contaminating bottom sediments of the Hudson River, treating such sediments as required, burying such sediments in secure landfills, and installing monitoring systems for such landfills. Such demonstration project shall be for the purpose of determining the feasibility of indefinite storage in secure landfills of toxic substances and of ascertaining the improvement of the rate of recovery of a toxic contaminated national waterway. No pollutants removed pursuant to this paragraph shall be placed in any landfill unless the Administrator first determines that disposal of the pollutants in such landfill would provide a higher standard of protection of the public health, safety, and welfare than disposal of such pollutants by any other method including, but not limited to, incineration or a chemical destruction process.

(b) The Administrator is authorized to make grants to the State of New York to carry out this section from funds allotted to such State under section 205(a) of this Act, except that the amount of any such grant shall be equal to 75 per centum of the cost of the project and such grant shall be made on condition that non-Federal sources provide the remainder of the cost of such project. The authority of this section shall be available until September 30, 1983. Funds allotted to the State of New York under section 205(a) shall be available under this subsection only to the extent that funds are not available, as determined by the Administrator, to the State of New York for the work authorized by this section under section 115 or 311 of this Act or a comprehensive hazardous substance response and clean up fund. Any funds used under the authority of this subsection shall be deducted from any estimate of the needs of the State of New York prepared under section 516. The Administrator may not obligate or expend more than $20,000,000 to carry out this section.

(33 U.S.C. 1266)

SEC. 117. CHESAPEAKE BAY.

(a) DEFINITIONS.—In this section, the following definitions apply:

(1) ADMINISTRATIVE COST.—The term "administrative cost" means the cost of salaries and fringe benefits incurred in administering a grant under this section.

(2) CHESAPEAKE BAY AGREEMENT.—The term "Chesapeake Bay Agreement" means the formal, voluntary agreements executed to achieve the goal of restoring and protecting the Chesapeake Bay ecosystem and the living resources of the Chesapeake Bay ecosystem and signed by the Chesapeake Executive Council.

(3) CHESAPEAKE BAY ECOSYSTEM.—The term "Chesapeake Bay ecosystem" means the ecosystem of the Chesapeake Bay and its watershed.

(4) CHESAPEAKE BAY PROGRAM.—The term "Chesapeake Bay Program" means the program directed by the Chesapeake Executive Council in accordance with the Chesapeake Bay Agreement.

(5) CHESAPEAKE EXECUTIVE COUNCIL.—The term "Chesapeake Executive Council" means the signatories to the Chesapeake Bay Agreement.

(6) SIGNATORY JURISDICTION.—The term "signatory jurisdiction" means a jurisdiction of a signatory to the Chesapeake Bay Agreement.

(b) CONTINUATION OF CHESAPEAKE BAY PROGRAM.—

(1) IN GENERAL.—In cooperation with the Chesapeake Executive Council (and as a member of the Council), the Administrator shall continue the Chesapeake Bay Program.

(2) PROGRAM OFFICE.—

(A) IN GENERAL.—The Administrator shall maintain in the Environmental Protection Agency a Chesapeake Bay Program Office.

(B) FUNCTION.—The Chesapeake Bay Program Office shall provide support to the Chesapeake Executive Council by—

(i) implementing and coordinating science, research, modeling, support services, monitoring, data collection, and other activities that support the Chesapeake Bay Program;

(ii) developing and making available, through publications, technical assistance, and other appropriate means, information pertaining to the environmental quality and living resources of the Chesapeake Bay ecosystem;

(iii) in cooperation with appropriate Federal, State, and local authorities, assisting the signatories to the Chesapeake Bay Agreement in developing and implementing specific action plans to carry out the responsibilities of the signatories to the Chesapeake Bay Agreement;

(iv) coordinating the actions of the Environmental Protection Agency with the actions of the appropriate officials of other Federal agencies and State and local authorities in developing strategies to—

(I) improve the water quality and living resources in the Chesapeake Bay ecosystem; and

(II) obtain the support of the appropriate officials of the agencies and authorities in achieving the objectives of the Chesapeake Bay Agreement; and

(v) implementing outreach programs for public information, education, and participation to foster stewardship of the resources of the Chesapeake Bay.

(c) INTERAGENCY AGREEMENTS.—The Administrator may enter into an interagency agreement with a Federal agency to carry out this section.

(d) TECHNICAL ASSISTANCE AND ASSISTANCE GRANTS.—

(1) IN GENERAL.—In cooperation with the Chesapeake Executive Council, the Administrator may provide technical assistance, and assistance grants, to nonprofit organizations, State and local governments, colleges, universities, and interstate agencies to carry out this section, subject to such terms and conditions as the Administrator considers appropriate.

(2) FEDERAL SHARE.—

(A) IN GENERAL.—Except as provided in subparagraph (B), the Federal share of an assistance grant provided under paragraph (1) shall be determined by the Administrator in accordance with guidance issued by the Administrator.

(B) SMALL WATERSHED GRANTS PROGRAM.—The Federal share of an assistance grant provided under paragraph (1) to carry out an implementing activity under subsection (g)(2) shall not exceed 75 percent of eligible project costs, as determined by the Administrator.

(3) NON-FEDERAL SHARE.—An assistance grant under paragraph (1) shall be provided on the condition that non-Federal sources provide the remainder of eligible project costs, as determined by the Administrator.

(4) ADMINISTRATIVE COSTS.—Administrative costs shall not exceed 10 percent of the annual grant award.

(e) IMPLEMENTATION AND MONITORING GRANTS.—

(1) IN GENERAL.—If a signatory jurisdiction has approved and committed to implement all or substantially all aspects of the Chesapeake Bay Agreement, on the request of the chief executive of the jurisdiction, the Administrator—

(A) shall make a grant to the jurisdiction for the purpose of implementing the management mechanisms established under the Chesapeake Bay Agreement, subject to such terms and conditions as the Administrator considers appropriate; and

(B) may make a grant to a signatory jurisdiction for the purpose of monitoring the Chesapeake Bay ecosystem.

(2) PROPOSALS.—

(A) IN GENERAL.—A signatory jurisdiction described in paragraph (1) may apply for a grant under this subsection for a fiscal year by submitting to the Administrator a comprehensive proposal to implement management mechanisms established under the Chesapeake Bay Agreement.

(B) CONTENTS.—A proposal under subparagraph (A) shall include—

(i) a description of proposed management mechanisms that the jurisdiction commits to take within a specified time period, such as reducing or preventing pollution in the Chesapeake Bay and its watershed or meeting applicable water quality standards or established goals and objectives under the Chesapeake Bay Agreement; and

(ii) the estimated cost of the actions proposed to be taken during the fiscal year.

(3) APPROVAL.—If the Administrator finds that the proposal is consistent with the Chesapeake Bay Agreement and the national goals established under section 101(a), the Administrator may approve the proposal for an award.

(4) FEDERAL SHARE.—The Federal share of a grant under this subsection shall not exceed 50 percent of the cost of implementing the management mechanisms during the fiscal year.

(5) NON-FEDERAL SHARE.—A grant under this subsection shall be made on the condition that non-Federal sources provide the remainder of the costs of implementing the management mechanisms during the fiscal year.

(6) ADMINISTRATIVE COSTS.—Administrative costs shall not exceed 10 percent of the annual grant award.

(7) REPORTING.—On or before October 1 of each fiscal year, the Administrator shall make available to the public a document that lists and describes, in the greatest practicable degree of detail—

(A) all projects and activities funded for the fiscal year;

(B) the goals and objectives of projects funded for the previous fiscal year; and

(C) the net benefits of projects funded for previous fiscal years.

(f) FEDERAL FACILITIES AND BUDGET COORDINATION.—

(1) SUBWATERSHED PLANNING AND RESTORATION.—A Federal agency that owns or operates a facility (as defined by the Administrator) within the Chesapeake Bay watershed shall participate in regional and subwatershed planning and restoration programs.

(2) COMPLIANCE WITH AGREEMENT.—The head of each Federal agency that owns or occupies real property in the Chesapeake Bay watershed shall ensure that the property, and actions taken by the agency with respect to the property, comply with the Chesapeake Bay Agreement, the Federal Agencies Chesapeake Ecosystem Unified Plan, and any subsequent agreements and plans.

(3) BUDGET COORDINATION.—

(A) IN GENERAL.—As part of the annual budget submission of each Federal agency with projects or grants related to restoration, planning, monitoring, or scientific investigation of the Chesapeake Bay ecosystem, the head of the agency shall submit to the President a report that describes plans for the expenditure of the funds under this section.

(B) DISCLOSURE TO THE COUNCIL.—The head of each agency referred to in subparagraph (A) shall disclose the report under that subparagraph with the Chesapeake Executive Council as appropriate.

(g) CHESAPEAKE BAY PROGRAM.—

(1) MANAGEMENT STRATEGIES.—The Administrator, in coordination with other members of the Chesapeake Executive Council, shall ensure that management plans are developed

and implementation is begun by signatories to the Chesapeake Bay Agreement to achieve and maintain—

(A) the nutrient goals of the Chesapeake Bay Agreement for the quantity of nitrogen and phosphorus entering the Chesapeake Bay and its watershed;

(B) the water quality requirements necessary to restore living resources in the Chesapeake Bay ecosystem;

(C) the Chesapeake Bay Basinwide Toxins Reduction and Prevention Strategy goal of reducing or eliminating the input of chemical contaminants from all controllable sources to levels that result in no toxic or bioaccumulative impact on the living resources of the Chesapeake Bay ecosystem or on human health;

(D) habitat restoration, protection, creation, and enhancement goals established by Chesapeake Bay Agreement signatories for wetlands, riparian forests, and other types of habitat associated with the Chesapeake Bay ecosystem; and

(E) the restoration, protection, creation, and enhancement goals established by the Chesapeake Bay Agreement signatories for living resources associated with the Chesapeake Bay ecosystem.

(2) SMALL WATERSHED GRANTS PROGRAM.—The Administrator, in cooperation with the Chesapeake Executive Council, shall—

(A) establish a small watershed grants program as part of the Chesapeake Bay Program; and

(B) offer technical assistance and assistance grants under subsection (d) to local governments and nonprofit organizations and individuals in the Chesapeake Bay region to implement—

(i) cooperative tributary basin strategies that address the water quality and living resource needs in the Chesapeake Bay ecosystem; and

(ii) locally based protection and restoration programs or projects within a watershed that complement the tributary basin strategies, including the creation, restoration, protection, or enhancement of habitat associated with the Chesapeake Bay ecosystem.

(h) STUDY OF CHESAPEAKE BAY PROGRAM.—

(1) IN GENERAL.—Not later than April 22, 2003, and every 5 years thereafter, the Administrator, in coordination with the Chesapeake Executive Council, shall complete a study and submit to Congress a comprehensive report on the results of the study.

(2) REQUIREMENTS.—The study and report shall—

(A) assess the state of the Chesapeake Bay ecosystem;

(B) compare the current state of the Chesapeake Bay ecosystem with its state in 1975, 1985, and 1995;

(C) assess the effectiveness of management strategies being implemented on the date of enactment of this section and the extent to which the priority needs are being met;

(D) make recommendations for the improved management of the Chesapeake Bay Program either by strengthening strategies being implemented on the date of enactment of this section or by adopting new strategies; and

(E) be presented in such a format as to be readily transferable to and usable by other watershed restoration programs.

(i) SPECIAL STUDY OF LIVING RESOURCE RESPONSE.—

(1) IN GENERAL.—Not later than 180 days after the date of enactment of this section, the Administrator shall commence a 5-year special study with full participation of the scientific community of the Chesapeake Bay to establish and expand understanding of the response of the living resources of the Chesapeake Bay ecosystem to improvements in water quality that have resulted from investments made through the Chesapeake Bay Program.

(2) REQUIREMENTS.—The study shall—

(A) determine the current status and trends of living resources, including grasses, benthos, phytoplankton, zooplankton, fish, and shellfish;

(B) establish to the extent practicable the rates of recovery of the living resources in response to improved water quality condition;

(C) evaluate and assess interactions of species, with particular attention to the impact of changes within and among trophic levels; and

(D) recommend management actions to optimize the return of a healthy and balanced ecosystem in response to improvements in the quality and character of the waters of the Chesapeake Bay.

(j) AUTHORIZATION OF APPROPRIATIONS.—There is authorized to be appropriated to carry out this section $40,000,000 for each of fiscal years 2001 through 2005. Such sums shall remain available until expended.

(33 U.S.C. 1267)

SEC. 118. GREAT LAKES.

(a) FINDINGS, PURPOSE, AND DEFINITIONS.—

(1) FINDINGS.—The Congress finds that—

(A) the Great Lakes are a valuable national resource, continuously serving the people of the United States and other nations as an important source of food, fresh water, recreation, beauty, and enjoyment;

(B) the United States should seek to attain the goals embodied in the Great Lakes Water Quality Agreement of 1978, as amended by the Water Quality Agreement of 1987 and any other agreements and amendments, with particular emphasis on goals related to toxic pollutants; and

(C) the Environmental Protection Agency should take the lead in the effort to meet those goals, working with other Federal agencies and State and local authorities.

(2) PURPOSE.—It is the purpose of this section to achieve the goals embodied in the Great Lakes Water Quality Agreement of 1978, as amended by the Water Quality Agreement of 1987 and any other agreements and amendments, through im-

proved organization and definition of mission on the part of the Agency, funding of State grants for pollution control in the Great Lakes area, and improved accountability for implementation of such agreement.

(3) DEFINITIONS.—For purposes of this section, the term—

(A) "Agency" means the Environmental Protection Agency;

(B) "Great Lakes" means Lake Ontario, Lake Erie, Lake Huron (including Lake St. Clair), Lake Michigan, and Lake Superior, and the connecting channels (Saint Mary's River, Saint Clair River, Detroit River, Niagara River, and Saint Lawrence River to the Canadian Border);

(C) "Great Lakes System" means all the streams, rivers, lakes, and other bodies of water within the drainage basin of the Great Lakes;

(D) "Program Office" means the Great Lakes National Program Office established by this section;

(E) "Research Office" means the Great Lakes Research Office established by subsection (d);

(F) "area of concern" means a geographic area located within the Great Lakes, in which beneficial uses are impaired and which has been officially designated as such under Annex 2 of the Great Lakes Water Quality Agreement;

(G) "Great Lakes States" means the States of Illinois, Indiana, Michigan, Minnesota, New York, Ohio, Pennsylvania, and Wisconsin;

(H) "Great Lakes Water Quality Agreement" means the bilateral agreement, between the United States and Canada which was signed in 1978 and amended by the Protocol of 1987;

(I) "Lakewide Management Plan" means a written document which embodies a systematic and comprehensive ecosystem approach to restoring and protecting the beneficial uses of the open waters of each of the Great Lakes, in accordance with article VI and Annex 2 of the Great Lakes Water Quality Agreement; and

(J) "Remedial Action Plan" means a written document which embodies a systematic and comprehensive ecosystem approach to restoring and protecting the beneficial uses of areas of concern, in accordance with article VI and Annex 2 of the Great Lakes Water Quality Agreement.

(b) GREAT LAKES NATIONAL PROGRAM OFFICE.—The Great Lakes National Program Office (previously established by the Administrator) is hereby established within the Agency. The Program Office shall be headed by a Director who, by reason of management experience and technical expertise relating to the Great Lakes, is highly qualified to direct the development of programs and plans on a variety of Great Lakes issues. The Great Lakes National Program Office shall be located in a Great Lakes State.

(c) GREAT LAKES MANAGEMENT.—

(1) FUNCTIONS.—The Program Office shall—

(A) in cooperation with appropriate Federal, State, tribal, and international agencies, and in accordance with section 101(e) of this Act, develop and implement specific action plans to carry out the responsibilities of the United States under the Great Lakes Water Quality Agreement of 1978, as amended by the Water Quality Agreement of 1987 and any other agreements and amendments,;[1]

(B) establish a Great Lakes system-wide surveillance network to monitor the water quality of the Great Lakes, with specific emphasis on the monitoring of toxic pollutants;

(C) serve as the liaison with, and provide information to, the Canadian members of the International Joint Commission and the Canadian counterpart to the Agency;

(D) coordinate actions of the Agency (including actions by headquarters and regional offices thereof) aimed at improving Great Lakes water quality; and

(E) coordinate actions of the Agency with the actions of other Federal agencies and State and local authorities, so as to ensure the input of those agencies and authorities in developing water quality strategies and obtain the support of those agencies and authorities in achieving the objectives of such agreement.

(2) GREAT LAKES WATER QUALITY GUIDANCE.—

(A) By June 30, 1991, the Administrator, after consultation with the Program Office, shall publish in the Federal Register for public notice and comment proposed water quality guidance for the Great Lakes System. Such guidance shall conform with the objectives and provisions of the Great Lakes Water Quality Agreement, shall be no less restrictive than the provisions of this Act and national water quality criteria and guidance, shall specify numerical limits on pollutants in ambient Great Lakes waters to protect human health, aquatic life, and wildlife, and shall provide guidance to the Great Lakes States on minimum water quality standards, antidegradation policies, and implementation procedures for the Great Lakes System.

(B) By June 30, 1992, the Administrator, in consultation with the Program Office, shall publish in the Federal Register, pursuant to this section and the Administrator's authority under this chapter, final water quality guidance for the Great Lakes System.

(C) Within two years after such Great Lakes guidance is published, the Great Lakes States shall adopt water quality standards, antidegradation policies, and implementation procedures for waters within the Great Lakes System which are consistent with such guidance. If a Great Lakes State fails to adopt such standards, policies, and procedures, the Administrator shall promulgate them not later than the end of such two-year period. When reviewing any Great Lakes State's water quality plan, the agency shall consider the extent to which the State has complied with the Great Lakes guidance issued pursuant to this section.

---

[1] See P.L. 100–688, section 1008.

(3) REMEDIAL ACTION PLANS.—

(A) For each area of concern for which the United States has agreed to draft a Remedial Action Plan, the Program Office shall ensure that the Great Lakes State in which such area of concern is located—

(i) submits a Remedial Action Plan to the Program Office by June 30, 1991;

(ii) submits such Remedial Action Plan to the International Joint Commission by January 1, 1992; and

(iii) includes such Remedial Action Plans within the State's water quality plan by January 1, 1993.

(B) For each area of concern for which Canada has agreed to draft a Remedial Action Plan, the Program Office shall, pursuant to subparagraph (c)(1)(C) of this section, work with Canada to assure the submission of such Remedial Action Plans to the International Joint Commission by June 30, 1991, and to finalize such Remedial Action Plans by January 1, 1993.

(C) For any area of concern designated as such subsequent to the enactment of this Act, the Program Office shall (i) if the United States has agreed to draft the Remedial Action Plan, ensure that the Great Lakes State in which such area of concern is located submits such Plan to the Program Office within two years of the area's designation, submits it to the International Joint Commission no later than six months after submitting it to the Program Office, and includes such Plan in the State's water quality plan no later than one year after submitting it to the Commission; and (ii) if Canada has agreed to draft the Remedial Action Plan, work with Canada, pursuant to subparagraph (c)(1)(C) of this section, to ensure the submission of such Plan to the International Joint Commission within two years of the area's designation and the finalization of such Plan no later than eighteen months after submitting it to such Commission.

(D) The Program Office shall compile formal comments on individual Remedial Action Plans made by the International Joint Commission pursuant to section 4(d) of Annex 2 of the Great Lakes Water Quality Agreement and, upon request by a member of the public, shall make such comments available for inspection and copying. The Program Office shall also make available, upon request, formal comments made by the Environmental Protection Agency on individual Remedial Action Plans.

(E) REPORT.—Not later than 1 year after the date of enactment of this subparagraph, the Administrator shall submit to Congress a report on such actions, time periods, and resources as are necessary to fulfill the duties of the Agency relating to oversight of Remedial Action Plans under—

(i) this paragraph; and

(ii) the Great Lakes Water Quality Agreement.

(4) LAKEWIDE MANAGEMENT PLANS.—The Administrator, in consultation with the Program Office shall—

(A) by January 1, 1992, publish in the Federal Register a proposed Lakewide Management Plan for Lake Michigan and solicit public comments;

(B) by January 1, 1993, submit a proposed Lakewide Management Plan for Lake Michigan to the International Joint Commission for review; and

(C) by January 1, 1994, publish in the Federal Register a final Lakewide Management Plan for Lake Michigan and begin implementation.

Nothing in this subparagraph shall preclude the simultaneous development of Lakewide Management Plans for the other Great Lakes.

(5) SPILLS OF OIL AND HAZARDOUS MATERIALS.—The Program Office, in consultation with the Coast Guard, shall identify areas within the Great Lakes which are likely to experience numerous or voluminous spills of oil or other hazardous materials from land based facilities, vessels, or other sources and, in consultation with the Great Lakes States, shall identify weaknesses in Federal and State programs and systems to prevent and respond to such spills. This information shall be included on at least a biennial basis in the report required by this section.

(6) 5-YEAR PLAN AND PROGRAM.—The Program Office shall develop, in consultation with the States, a five-year plan and program for reducing the amount of nutrients introduced into the Great Lakes. Such program shall incorporate any management program for reducing nutrient runoff from nonpoint sources established under section 319 of this Act and shall include a program for monitoring nutrient runoff into, and ambient levels in, the Great Lakes.

(7) 5-YEAR STUDY AND DEMONSTRATION PROJECTS.—(A) The Program Office shall carry out a five-year study and demonstration projects relating to the control and removal of toxic pollutants in the Great Lakes, with emphasis on the removal of toxic pollutants from bottom sediments. In selecting locations for conducting demonstration projects under this paragraph, priority consideration shall be given to projects at the following locations: Saginaw Bay, Michigan; Sheboygan Harbor, Wisconsin; Grand Calumet River, Indiana; Ashtabula River, Ohio; and Buffalo River, New York.

(B) The Program Office shall—

(i) by December 31, 1990, complete chemical, physical, and biological assessments of the contaminated sediments at the locations selected for the study and demonstration projects;

(ii) by December 31, 1990, announce the technologies that will be demonstrated at each location and the numerical standard of protection intended to be achieved at each location;

(iii) by December 31, 1992, complete full or pilot scale demonstration projects on site at each location of promising technologies to remedy contaminated sediments; and

(iv) by December 31, 1993, issue a final report to Congress on its findings.

(C) The Administrator, after providing for public review and comment, shall publish information concerning the public health and environmental consequences of contaminants in Great Lakes sediment. Information published pursuant to this subparagraph shall include specific numerical limits to protect health, aquatic life, and wildlife from the bioaccumulation of toxins. The Administrator shall, at a minimum, publish information pursuant to this subparagraph within 2 years of the date of the enactment of this title.

(8) ADMINISTRATOR'S RESPONSIBILITY.—The Administrator shall ensure that the Program Office enters into agreements with the various organizational elements of the Agency involved in Great Lakes activities and the appropriate State agencies specifically delineating—

(A) the duties and responsibilities of each such element in the Agency with respect to the Great Lakes;

(B) the time periods for carrying out such duties and responsibilities; and

(C) the resources to be committed to such duties and responsibilities.

(9) BUDGET ITEM.—The Administrator shall, in the Agency's annual budget submission to Congress, include a funding request for the Program Office as a separate budget line item.

(10) COMPREHENSIVE REPORT.—Within 90 days after the end of each fiscal year, the Administrator shall submit to Congress a comprehensive report which—

(A) describes the achievements in the preceding fiscal year in implementing the Great Lakes Water Quality Agreement of 1978 and shows by categories (including judicial enforcement, research, State cooperative efforts, and general administration) the amounts expended on Great Lakes water quality initiatives in such preceding fiscal year;

(B) describes the progress made in such preceding fiscal year in implementing the system of surveillance of the water quality in the Great Lakes System, including the monitoring of groundwater and sediment, with particular reference to toxic pollutants;

(C) describes the long-term prospects for improving the condition of the Great Lakes; and

(D) provides a comprehensive assessment of the planned efforts to be pursued in the succeeding fiscal year for implementing the Great Lakes Water Quality Agreement of 1978, which assessment shall—

(i) show by categories (including judicial enforcement, research, State cooperative efforts, and general administration) the amount anticipated to be expended on Great Lakes water quality initiatives in the fiscal year to which the assessment relates; and

(ii) include a report of current programs administered by other Federal agencies which make available resources to the Great Lakes water quality management efforts.

(11) CONFINED DISPOSAL FACILITIES.—(A) The Administrator, in consultation with the Assistant Secretary of the Army for Civil Works, shall develop and implement, within one year of the date of enactment of this paragraph, management plans for every Great Lakes confined disposal facility.

(B) The plan shall provide for monitoring of such facilities, including—

(i) water quality at the site and in the area of the site;

(ii) sediment quality at the site and in the area of the site;

(iii) the diversity, productivity, and stability of aquatic organisms at the site and in the area of the site; and

(iv) such other conditions as the Administrator deems appropriate.

(C) The plan shall identify the anticipated use and management of the site over the following twenty-year period including the expected termination of dumping at the site, the anticipated need for site management, including pollution control, following the termination of the use of the site.

(D) The plan shall identify a schedule for review and revision of the plan which shall not be less frequent than five years after adoption of the plan and every five years thereafter.

(12) REMEDIATION OF SEDIMENT CONTAMINATION IN AREAS OF CONCERN.—

(A) IN GENERAL.—In accordance with this paragraph, the Administrator, acting through the Program Office, may carry out projects that meet the requirements of subparagraph (B).

(B) ELIGIBLE PROJECTS.—A project meets the requirements of this subparagraph if the project is to be carried out in an area of concern located wholly or partially in the United States and the project—

(i) monitors or evaluates contaminated sediment;

(ii) subject to subparagraph (D), implements a plan to remediate contaminated sediment; or

(iii) prevents further or renewed contamination of sediment.

(C) PRIORITY.—In selecting projects to carry out under this paragraph, the Administrator shall give priority to a project that—

(i) constitutes remedial action for contaminated sediment;

(ii)(I) has been identified in a Remedial Action Plan submitted under paragraph (3); and

(II) is ready to be implemented;

(iii) will use an innovative approach, technology, or technique that may provide greater environmental benefits, or equivalent environmental benefits at a reduced cost; or

(iv) includes remediation to be commenced not later than 1 year after the date of receipt of funds for the project.

(D) LIMITATION.—The Administrator may not carry out a project under this paragraph for remediation of contaminated sediments located in an area of concern—

(i) if an evaluation of remedial alternatives for the area of concern has not been conducted, including a review of the short-term and long-term effects of the alternatives on human health and the environment; or

(ii) if the Administrator determines that the area of concern is likely to suffer significant further or renewed contamination from existing sources of pollutants causing sediment contamination following completion of the project.

(E) NON-FEDERAL SHARE.—

(i) IN GENERAL.—The non-Federal share of the cost of a project carried out under this paragraph shall be at least 35 percent.

(ii) IN-KIND CONTRIBUTIONS.—The non-Federal share of the cost of a project carried out under this paragraph may include the value of in-kind services contributed by a non-Federal sponsor.

(iii) NON-FEDERAL SHARE.—The non-Federal share of the cost of a project carried out under this paragraph—

(I) may include monies paid pursuant to, or the value of any in-kind service performed under, an administrative order on consent or judicial consent decree; but

(II) may not include any funds paid pursuant to, or the value of any in-kind service performed under, a unilateral administrative order or court order.

(iv) OPERATION AND MAINTENANCE.—The non-Federal share of the cost of the operation and maintenance of a project carried out under this paragraph shall be 100 percent.

(F) MAINTENANCE OF EFFORT.—The Administrator may not carry out a project under this paragraph unless the non-Federal sponsor enters into such agreements with the Administrator as the Administrator may require to ensure that the non-Federal sponsor will maintain its aggregate expenditures from all other sources for remediation programs in the area of concern in which the project is located at or above the average level of such expenditures in the 2 fiscal years preceding the date on which the project is initiated.

(G) COORDINATION.—In carrying out projects under this paragraph, the Administrator shall coordinate with the Secretary of the Army, and with the Governors of States in which the projects are located, to ensure that Federal and State assistance for remediation in areas of concern is used as efficiently as practicable.

(H) AUTHORIZATION OF APPROPRIATIONS.—

(i) IN GENERAL.—In addition to other amounts authorized under this section, there is authorized to be appropriated to carry out this paragraph $50,000,000 for each of fiscal years 2004 through 2008.

(ii) AVAILABILITY.—Funds made available under clause (i) shall remain available until expended.

(13) PUBLIC INFORMATION PROGRAM.—

(A) IN GENERAL.—The Administrator, acting through the Program Office and in coordination with States, Indian tribes, local governments, and other entities, may carry out a public information program to provide information relating to the remediation of contaminated sediment to the public in areas of concern that are located wholly or partially in the United States.

(B) AUTHORIZATION OF APPROPRIATIONS.—There is authorized to be appropriated to carry out this paragraph $1,000,000 for each of fiscal years 2004 through 2008.

(d) GREAT LAKES RESEARCH.—

(1) ESTABLISHMENT OF RESEARCH OFFICE.—There is established within the National Oceanic and Atmospheric Administration the Great Lakes Research Office.

(2) IDENTIFICATION OF ISSUES.—The Research Office shall identify issues relating to the Great Lakes resources on which research is needed. The Research Office shall submit a report to Congress on such issues before the end of each fiscal year which shall identify any changes in the Great Lakes system with respect to such issues.

(3) INVENTORY.—The Research Office shall identify and inventory, Federal, State, university, and tribal environmental research programs (and, to the extent feasible, those of private organizations and other nations) relating to the Great Lakes system, and shall update that inventory every four years.

(4) RESEARCH EXCHANGE.—The Research Office shall establish a Great Lakes research exchange for the purpose of facilitating the rapid identification, acquisition, retrieval, dissemination, and use of information concerning research projects which are ongoing or completed and which affect the Great Lakes system.

(5) RESEARCH PROGRAM.—The Research Office shall develop, in cooperation with the Coordination Office, a comprehensive environmental research program and data base for the Great Lakes system. The data base shall include, but not be limited to, data relating to water quality, fisheries, and biota.

(6) MONITORING.—The Research Office shall conduct, through the Great Lakes Environmental Research Laboratory, the National Sea Grant College program, other Federal laboratories, and the private sector, appropriate research and monitoring activities which address priority issues and current needs relating to the Great Lakes.

(7) LOCATION.—The Research Office shall be located in a Great Lakes State.

(e) RESEARCH AND MANAGEMENT COORDINATION.—

(1) JOINT PLAN.—Before October 1 of each year, the Program Office and the Research Office shall prepare a joint research plan for the fiscal year which begins in the following calendar year.

(2) CONTENTS OF PLAN.—Each plan prepared under paragraph (1) shall—

(A) identify all proposed research dedicated to activities conducted under the Great Lakes Water Quality Agreement of 1978;

(B) include the Agency's assessment of priorities for research needed to fulfill the terms of such Agreement; and

(C) identify all proposed research that may be used to develop a comprehensive environmental data base for the Great Lakes system and establish priorities for development of such data base.

(3) HEALTH RESEARCH REPORT.—(A) Not later than September 30, 1994, the Program Office, in consultation with the Research Office, the Agency for Toxic Substances and Disease Registry, and Great Lakes States shall submit to the Congress a report assessing the adverse effects of water pollutants in the Great Lakes System on the health of persons in Great Lakes States and the health of fish, shellfish, and wildlife in the Great Lakes System. In conducting research in support of this report, the Administrator may, where appropriate, provide for research to be conducted under cooperative agreements with Great Lakes States.

(B) There is authorized to be appropriated to the Administrator to carry out this section not to exceed $3,000,000 for each of fiscal years 1992, 1993, and 1994.

(f) INTERAGENCY COOPERATION.—The head of each department, agency, or other instrumentality of the Federal Government which is engaged in, is concerned with, or has authority over programs relating to research, monitoring, and planning to maintain, enhance, preserve, or rehabilitate the environmental quality and natural resources of the Great Lakes, including the Chief of Engineers of the Army, the Chief of the Soil Conservation Service, the Commandant of the Coast Guard, the Director of the Fish and Wildlife Service, and the Administrator of the National Oceanic and Atmospheric Administration, shall submit an annual report to the Administrator with respect to the activities of that agency or office affecting compliance with the Great Lakes Water Quality Agreement of 1978.

(g) RELATIONSHIP TO EXISTING FEDERAL AND STATE LAWS AND INTERNATIONAL TREATIES.—Nothing in this section shall be construed—

(1) to affect the jurisdiction, powers, or prerogatives of any department, agency, or officer of the Federal Government or of any State government, or of any tribe, nor any powers, jurisdiction, or prerogatives of any international body created by treaty with authority relating to the Great Lakes; or

(2) to affect any other Federal or State authority that is being used or may be used to facilitate the cleanup and protection of the Great Lakes.

(h) AUTHORIZATIONS OF GREAT LAKES APPROPRIATIONS.—There are authorized to be appropriated to the Administrator to carry out this section not to exceed—

(1) $11,000,000 per fiscal year for the fiscal years 1987, 1988, 1989, and 1990, and $25,000,000 for fiscal year 1991;

(2) such sums as are necessary for each of fiscal years 1992 through 2003; and

(3) $25,000,000 for each of fiscal years 2004 through 2008.

(33 U.S.C. 1268)

SEC. 119. LONG ISLAND SOUND.—(a) The Administrator shall continue the Management Conference of the Long Island Sound Study (hereinafter referred to as the "Conference") as established pursuant to section 320 of this Act, and shall establish an office (hereinafter referred to as the "Office") to be located on or near Long Island Sound.

(b) ADMINISTRATION AND STAFFING OF OFFICE.—The Office shall be headed by a Director, who shall be detailed by the Administrator, following consultation with the Administrators of EPA regions I and II, from among the employees of the Agency who are in civil service. The Administrator shall delegate to the Director such authority and detail such additional staff as may be necessary to carry out the duties of the Director under this section.

(c) DUTIES OF THE OFFICE.—The Office shall assist the Management Conference of the Long Island Sound Study in carrying out its goals. Specifically, the Office shall—

(1) assist and support the implementation of the Comprehensive Conservation and Management Plan for Long Island Sound developed pursuant to section 320 of this Act, including efforts to establish, within the process for granting watershed general permits, a system for promoting innovative methodologies and technologies that are cost-effective and consistent with the goals of the Plan;

(2) conduct or commission studies deemed necessary for strengthened implementation of the Comprehensive Conservation and Management Plan including, but not limited to—

(A) population growth and the adequacy of wastewater treatment facilities,

(B) the use of biological methods for nutrient removal in sewage treatment plants,

(C) contaminated sediments, and dredging activities,

(D) nonpoint source pollution abatement and land use activities in the Long Island Sound watershed,

(E) wetland protection and restoration,

(F) atmospheric deposition of acidic and other pollutants into Long Island Sound,

(G) water quality requirements to sustain fish, shellfish, and wildlife populations, and the use of indicator species to assess environmental quality,

(H) State water quality programs, for their adequacy pursuant to implementation of the Comprehensive Conservation and Management Plan, and

(I) options for long-term financing of wastewater treatment projects and water pollution control programs.

(3) coordinate the grant, research and planning programs authorized under this section;

(4) coordinate activities and implementation responsibilities with other Federal agencies which have jurisdiction over Long Island Sound and with national and regional marine monitoring and research programs established pursuant to the Marine Protection, Research, and Sanctuaries Act;

(5) provide administrative and technical support to the conference;

(6) collect and make available to the public publications, and other forms of information the conference determines to be appropriate, relating to the environmental quality of Long Island Sound;

(7) not more than two years after the date of the issuance of the final Comprehensive Conservation and Management Plan for Long Island Sound under section 320 of this Act, and biennially thereafter, issue a report to the Congress which—

(A) summarizes the progress made by the States in implementing the Comprehensive Conservation and Management Plan;

(B) summarizes any modifications to the Comprehensive Conservation and Management Plan in the twelve-month period immediately preceding such report; and

(C) incorporates specific recommendations concerning the implementation of the Comprehensive Conservation and Management Plan; and

(8) convene conferences and meetings for legislators from State governments and political subdivisions thereof for the purpose of making recommendations for coordinating legislative efforts to facilitate the environmental restoration of Long Island Sound and the implementation of the Comprehensive Conservation and Management Plan.

(d) GRANTS.—(1) The Administrator is authorized to make grants for projects and studies which will help implement the Long Island Sound Comprehensive Conservation and Management Plan. Special emphasis shall be given to implementation, research and planning, enforcement, and citizen involvement and education.

(2) State, interstate, and regional water pollution control agencies, and other public or nonprofit private agencies, institutions, and organizations held to be eligible for grants pursuant to this subsection.

(3) Citizen involvement and citizen education grants under this subsection shall not exceed 95 per centum of the costs of such work. All other grants under this subsection shall not exceed 50 per centum of the research, studies, or work. All grants shall be made on the condition that the non-Federal share of such costs are provided from non-Federal sources.

(e) ASSISTANCE TO DISTRESSED COMMUNITIES.—

(1) ELIGIBLE COMMUNITIES.—For the purposes of this subsection, a distressed community is any community that meets affordability criteria established by the State in which the community is located, if such criteria are developed after public review and comment.

(2) PRIORITY.—In making assistance available under this section for the upgrading of wastewater treatment facilities, the Administrator may give priority to a distressed community.

(f) AUTHORIZATIONS.—(1) There is authorized to be appropriated to the Administrator for the implementation of this section, other than subsection (d), such sums as may be necessary for each of the fiscal years 2001 through 2005.

(2) There is authorized to be appropriated to the Administrator for the implementation of subsection (d) not to exceed $40,000,000 for each of fiscal years 2001 through 2005.

(33 U.S.C. 1269)

SEC. 120. LAKE CHAMPLAIN BASIN PROGRAM.

(a) ESTABLISHMENT.—

(1) IN GENERAL.—There is established a Lake Champlain Management Conference to develop a comprehensive pollution prevention, control, and restoration plan for Lake Champlain. The Administrator shall convene the management conference within ninety days of the date of enactment of this section.

(2) IMPLEMENTATION.—The Administrator—

(A) may provide support to the State of Vermont, the State of New York, and the New England Interstate Water Pollution Control Commission for the implementation of the Lake Champlain Basin Program; and

(B) shall coordinate actions of the Environmental Protection Agency under subparagraph (A) with the actions of other appropriate Federal agencies.

(b) MEMBERSHIP.—The Members of the Management Conference shall be comprised of—

(1) the Governors of the States of Vermont and New York;

(2) each interested Federal agency, not to exceed a total of five members;

(3) the Vermont and New York Chairpersons of the Vermont, New York, Quebec Citizens Advisory Committee for the Environmental Management of Lake Champlain;

(4) four representatives of the State legislature of Vermont;

(5) four representatives of the State legislature of New York;

(6) six persons representing local governments having jurisdiction over any land or water within the Lake Champlain basin, as determined appropriate by the Governors; and

(7) eight persons representing affected industries, nongovernmental organizations, public and private educational institutions, and the general public, as determined appropriate by the trigovernmental Citizens Advisory Committee for the Environmental Management of Lake Champlain, but not to be current members of the Citizens Advisory Committee.

(c) TECHNICAL ADVISORY COMMITTEE.—(1) The Management Conference shall, not later than one hundred and twenty days after the date of enactment of this section, appoint a Technical Advisory Committee.

(2) Such Technical Advisory Committee shall consist of officials of: appropriate departments and agencies of the Federal Govern-

ment; the State governments of New York and Vermont; and governments of political subdivisions of such States; and public and private research institutions.

(d) RESEARCH PROGRAM.—The Management Conference shall establish a multi-disciplinary environmental research program for Lake Champlain. Such research program shall be planned and conducted jointly with the Lake Champlain Research Consortium.

(e) POLLUTION PREVENTION, CONTROL, AND RESTORATION PLAN.—(1) Not later than three years after the date of the enactment of this section, the Management Conference shall publish a pollution prevention, control, and restoration plan for Lake Champlain.

(2) The Plan developed pursuant to this section shall—

(A) identify corrective actions and compliance schedules addressing point and nonpoint sources of pollution necessary to restore and maintain the chemical, physical, and biological integrity of water quality, a balanced, indigenous population of shellfish, fish and wildlife, recreational, and economic activities in and on the lake;

(B) incorporate environmental management concepts and programs established in State and Federal plans and programs in effect at the time of the development of such plan;

(C) clarify the duties of Federal and State agencies in pollution prevention and control activities, and to the extent allowable by law, suggest a timetable for adoption by the appropriate Federal and State agencies to accomplish such duties within a reasonable period of time;

(D) describe the methods and schedules for funding of programs, activities, and projects identified in the Plan, including the use of Federal funds and other sources of funds;

(E) include a strategy for pollution prevention and control that includes the promotion of pollution prevention and management practices to reduce the amount of pollution generated in the Lake Champlain basin; and

(F) be reviewed and revised, as necessary, at least once every 5 years, in consultation with the Administrator and other appropriate Federal agencies.

(3) The Administrator, in cooperation with the Management Conference, shall provide for public review and comment on the draft Plan. At a minimum, the Management Conference shall conduct one public meeting to hear comments on the draft plan in the State of New York and one such meeting in the State of Vermont.

(4) Not less than one hundred and twenty days after the publication of the Plan required pursuant to this section, the Administrator shall approve such plan if the plan meets the requirements of this section and the Governors of the States of New York and Vermont concur.

(5) Upon approval of the plan, such plan shall be deemed to be an approved management program for the purposes of section 319(h) of this Act and such plan shall be deemed to be an approved comprehensive conservation and management plan pursuant to section 320 of this Act.

(f) GRANT ASSISTANCE.—(1) The Administrator may, in consultation with participants in the Lake Champlain Basin Program, make grants to State, interstate, and regional water pollution control agencies, and public or nonprofit agencies, institutions, and organizations.

(2) Grants under this subsection shall be made for assisting research, surveys, studies, and modeling and technical and supporting work necessary for the development and implementation of the Plan.

(3) The amount of grants to any person under this subsection for a fiscal year shall not exceed 75 per centum of the costs of such research, survey, study and work and shall be made available on the condition that non-Federal share of such costs are provided from non-Federal sources.

(4) The Administrator may establish such requirements for the administration of grants as he determines to be appropriate.

(g) DEFINITIONS.—In this section:

(1) LAKE CHAMPLAIN BASIN PROGRAM.—The term "Lake Champlain Basin Program" means the coordinated efforts among the Federal Government, State governments, and local governments to implement the Plan.

(2) LAKE CHAMPLAIN DRAINAGE BASIN.—The term "Lake Champlain drainage basin" means all or part of Clinton, Franklin, Warren, Essex, and Washington counties in the State of New York and all or part of Franklin, Hamilton, Grand Isle, Chittenden, Addison, Rutland, Bennington, Lamoille, Orange, Washington, Orleans, and Caledonia counties in Vermont, that contain all of the streams, rivers, lakes, and other bodies of water, including wetlands, that drain into Lake Champlain.

(3) PLAN.—The term "Plan" means the plan developed under subsection (e).

(h) NO EFFECT ON CERTAIN AUTHORITY.—Nothing in this section—

(1) affects the jurisdiction or powers of—

(A) any department or agency of the Federal Government or any State government; or

(B) any international organization or entity related to Lake Champlain created by treaty or memorandum to which the United States is a signatory;

(2) provides new regulatory authority for the Environmental Protection Agency; or

(3) affects section 304 of the Great Lakes Critical Programs Act of 1990 (Public Law 101–596; 33 U.S.C. 1270 note).

(i) AUTHORIZATION.—There are authorized to be appropriated to the Environmental Protection Agency to carry out this section—

(1) $2,000,000 for each of fiscal years 1991, 1992, 1993, 1994, and 1995;

(2) such sums as are necessary for each of fiscal years 1996 through 2003; and

(3) $11,000,000 for each of fiscal years 2004 through 2008.

(33 U.S.C. 1270)

SEC. 121. LAKE PONTCHARTRAIN BASIN.

(a) ESTABLISHMENT OF RESTORATION PROGRAM.—The Administrator shall establish within the Environmental Protection Agency the Lake Pontchartrain Basin Restoration Program.

(b) PURPOSE.—The purpose of the program shall be to restore the ecological health of the Basin by developing and funding restoration projects and related scientific and public education projects.

(c) DUTIES.—In carrying out the program, the Administrator shall—

(1) provide administrative and technical assistance to a management conference convened for the Basin under section 320;

(2) assist and support the activities of the management conference, including the implementation of recommendations of the management conference;

(3) support environmental monitoring of the Basin and research to provide necessary technical and scientific information;

(4) develop a comprehensive research plan to address the technical needs of the program;

(5) coordinate the grant, research, and planning programs authorized under this section; and

(6) collect and make available to the public publications, and other forms of information the management conference determines to be appropriate, relating to the environmental quality of the Basin.

(d) GRANTS.—The Administrator may make grants—

(1) for restoration projects and studies recommended by a management conference convened for the Basin under section 320; and

(2) for public education projects recommended by the management conference.

(e) DEFINITIONS.—In this section, the following definitions apply:

(1) BASIN.—The term "Basin" means the Lake Pontchartrain Basin, a 5,000 square mile watershed encompassing 16 parishes in the State of Louisiana and 4 counties in the State of Mississippi.

(2) PROGRAM.—The term "program" means the Lake Pontchartrain Basin Restoration Program established under subsection (a).

(f) AUTHORIZATION OF APPROPRIATIONS.—

(1) IN GENERAL.—There is authorized to be appropriated to carry out this section $20,000,000 for each of fiscal years 2001 through 2005. Such sums shall remain available until expended.

(2) PUBLIC EDUCATION PROJECTS.—Not more than 15 percent of the amount appropriated pursuant to paragraph (1) in a fiscal year may be expended on grants for public education projects under subsection (d)(2).

(33 U.S.C. 1273)

SEC. 121. WET WEATHER WATERSHED PILOT PROJECTS.[1]

(a) IN GENERAL.—The Administrator, in coordination with the States, may provide technical assistance and grants for treatment works to carry out pilot projects relating to the following areas of wet weather discharge control:

(1) WATERSHED MANAGEMENT OF WET WEATHER DISCHARGES.—The management of municipal combined sewer overflows, sanitary sewer overflows, and stormwater discharges, on an integrated watershed or subwatershed basis for the purpose of demonstrating the effectiveness of a unified wet weather approach.

(2) STORMWATER BEST MANAGEMENT PRACTICES.—The control of pollutants from municipal separate storm sewer systems for the purpose of demonstrating and determining controls that are cost-effective and that use innovative technologies in reducing such pollutants from stormwater discharges.

(b) ADMINISTRATION.—The Administrator, in coordination with the States, shall provide municipalities participating in a pilot project under this section the ability to engage in innovative practices, including the ability to unify separate wet weather control efforts under a single permit.

(c) FUNDING.—

(1) IN GENERAL.—There is authorized to be appropriated to carry out this section $10,000,000 for fiscal year 2002, $15,000,000 for fiscal year 2003, and $20,000,000 for fiscal year 2004. Such funds shall remain available until expended.

(2) STORMWATER.—The Administrator shall make available not less than 20 percent of amounts appropriated for a fiscal year pursuant to this subsection to carry out the purposes of subsection (a)(2).

(3) ADMINISTRATIVE EXPENSES.—The Administrator may retain not to exceed 4 percent of any amounts appropriated for a fiscal year pursuant to this subsection for the reasonable and necessary costs of administering this section.

(d) REPORT TO CONGRESS.—Not later than 5 years after the date of enactment of this section, the Administrator shall transmit to Congress a report on the results of the pilot projects conducted under this section and their possible application nationwide.

(33 U.S.C. 1274)

TITLE II—GRANTS FOR CONSTRUCTION OF TREATMENT WORKS

PURPOSE

SEC. 201. (a) It is the purpose of this title to require and to assist the development and implementation of waste treatment management plans and practices which will achieve the goals of this Act.

(b) Waste treatment management plans and practices shall provide for the application of the best practicable waste treatment

[1] The second section 121 was added by section 112(b) of the Miscellaneous Appropriations Act, 2001 (114 Stat. 2763A–225), as enacted into law by section 1(a)(6) of Public Law 106–554 (114 Stat. 2763).

technology before any discharge into receiving waters, including reclaiming and recycling of water, and confined disposal of pollutants so they will not migrate to cause water or other environmental pollution and shall provide for consideration of advanced waste treatment techniques.

(c) To the extent practicable, waste treatment management shall be on an areawide basis and provide control or treatment of all point and nonpoint sources of pollution, including in place or accumulated pollution sources.

(d) The Administrator shall encourage waste treatment management which results in the construction of revenue producing facilities providing for—

(1) the recycling of potential sewage pollutants through the production of agriculture, silviculture, or aquaculture products, or any combination thereof;

(2) the confined and contained disposal of pollutants not recycled;

(3) the reclamation of wastewater; and

(4) the ultimate disposal of sludge in a manner that will not result in environmental hazards.

(e) The Administrator shall encourage waste treatment management which results in integrating facilities for sewage treatment and recycling with facilities to treat, dispose of, or utilize other industrial and municipal wastes, including but not limited to solid waste and waste heat and thermal discharges. Such integrated facilities shall be designed and operated to produce revenues in excess of capital and operation and maintenance costs and such revenues shall be used by the designated regional management agency to aid in financing other environmental improvement programs.

(f) The Administrator shall encourage waste treatment management which combines "open space" and recreational considerations with such management.

(g)(1) The Administrator is authorized to make grants to any State, municipality, or intermunicipal or interstate agency for the construction of publicly owned treatment works. On and after October 1, 1984, grants under this title shall be made only for projects for secondary treatment or more stringent treatment, or any cost effective alternative thereto, new interceptors and appurtenances, and infiltration-in-flow correction. Notwithstanding the preceding sentences, the Administrator may make grants on and after October 1, 1984, for (A) any project within the definition set forth in section 212(2) of this Act, other than for a project referred to in the preceding sentence, and (B) any purpose for which a grant may be made under sections[1] 319 (h) and (i) of this Act (including any innovative and alternative approaches for the control of nonpoint sources of pollution), except that not more than 20 per centum (as determined by the Governor of the State) of the amount allotted to a State under section 205 of this Act for any fiscal year shall be obligated in such State under authority of this sentence.

(2) The Administrator shall not make grants from funds authorized for any fiscal year beginning after June 30, 1974, to any

[1] So in original. Probably should be "section".

State, municipality, or intermunicipal or interstate agency for the erection, building, acquisition, alteration, remodeling, improvement, or extension of treatment works unless the grant applicant has satisfactorily demonstrated to the Administrator that—

(A) alternative waste management techniques have been studied and evaluated and the works proposed for grant assistance will provide for the application of the best practicable waste treatment technology over the life of the works consistent with the purposes of this title; and

(B) as appropriate, the works proposed for grant assistance will take into account and allow to the extent practicable the application of technology at a later date which will provide for the reclaiming or recycling of water or otherwise eliminate the discharge of pollutants.

(3) The Administrator shall not approve any grant after July 1, 1973, for treatment works under this section unless the applicant shows to the satisfaction of the Administrator that each sewer collection system discharging into such treatment works is not subject to excessive infiltration.

(4) The Administrator is authorized to make grants to applicants for treatment works grants under this section for such sewer system evaluation studies as may be necessary to carry out the requirements of paragraph (3) of this subsection. Such grants shall be made in accordance with rules and regulations promulgated by the Administrator. Initial rules and regulations shall be promulgated under this paragraph not later than 120 days after the date of enactment of the Federal Water Pollution Control Act Amendments of 1972.

(5) The Administrator shall not make grants from funds authorized for any fiscal year beginning after September 30, 1978, to any State, municipality, or intermunicipal or interstate agency for the erection, building, acquisition, alteration, remodeling, improvement, or extension of treatment works unless the grant applicant has satisfactorily demonstrated to the Administrator that innovative and alternative wastewater treatment processes and techniques which provide for the reclaiming and reuse of water, otherwise eliminate the discharge of pollutants, and utilize recycling techniques, land treatment, new or improved methods of waste treatment management for municipal and industrial waste (discharged into municipal systems) and the confined disposal of pollutants, so that pollutants will not migrate to cause water or other environmental pollution, have been fully studied and evaluated by the applicant taking into account section 201(d) of this Act and taking into account and allowing to the extent practicable the more efficient use of energy and resources.

(6) The Administrator shall not make grants from funds authorized for any fiscal year beginning after September 30, 1978, to any State, municipality, or intermunicipal or interstate agency for the erection, building, acquisition, alteration, remodeling, improvement, or extension of treatment works unless the grant applicant has satisfactorily demonstrated to the Administrator that the applicant has analyzed the potential recreation and open space opportunities in the planning of the proposed treatment works.

(h) A grant may be made under this section to construct a privately owned treatment works serving one or more principal residences or small commercial establishments constructed prior to, and inhabited on the date of enactment of this subsection where the Administrator finds that—

(1) a public body otherwise eligible for a grant under subsection (g) of this section has applied on behalf of a number of such units and certified that public ownership of such works is not feasible;

(2) such public body has entered into an agreement with the Administrator which guarantees that such treatment works will be properly operated and maintained and will comply with all other requirements of section 204 of this Act and includes a system of charges to assure that each recipient of waste treatment services under such a grant will pay its proportionate share of the cost of operation and maintenance (including replacement); and

(3) the total cost and environmental impact of providing waste treatment services to such residences or commercial establishments will be less than the cost of providing a system of collection and central treatment of such wastes.

(i) The Administrator shall encourage waste treatment management methods, processes, and techniques which will reduce total energy requirements.

(j) The Administrator is authorized to make a grant for any treatment works utilizing processes and techniques meeting the guidelines promulgated under section 304(d)(3) of this Act, if the Administrator determines it is in the public interest and if in the cost effectiveness study made of the construction grant application for the purpose of evaluating alternative treatment works, the life cycle cost of the treatment works for which the grant is to be made does not exceed the life cycle cost of the most effective alternative by more than 15 per centum.

(k) No grant made after November 15, 1981, for a publicly owned treatment works, other than for facility planning and the preparation of construction plans and specifications, shall be used to treat, store, or convey the flow of any industrial user into such treatment works in excess of a flow per day equivalent to fifty thousand gallons per day of sanitary waste. This subsection shall not apply to any project proposed by a grantee which is carrying out an approved project to prepare construction plans and specifications for a facility to treat wastewater, which received its grant approval before May 15, 1980. This subsection shall not be in effect after November 15, 1981.

(l)(1) After the date of enactment of this subsection, Federal grants shall not be made for the purpose of providing assistance solely for facility plans, or plans, specifications, and estimates for any proposed project for the construction of treatment works. In the event that the proposed project receives a grant under this section for construction, the Administrator shall make an allowance in such grant for non-Federal funds expended during the facility planning and advanced engineering and design phase at the prevailing Federal share under section 202(a) of this Act, based on the percentage of total project costs which the Administrator determines is the general experience for such projects.

(2)(A) Each State shall use a portion of the funds allotted to such State each fiscal year, but not to exceed 10 per centum of such funds, to advance to potential grant applicants under this title the costs of facility planning or the preparation of plans, specifications, and estimates.

(B) Such an advance shall be limited to the allowance for such costs which the Administrator establishes under paragraph (1) of this subsection, and shall be provided only to a potential grant applicant which is a small community and which in the judgment of the State would otherwise be unable to prepare a request for a grant for construction costs under this section.

(C) In the event a grant for construction costs is made under this section for a project for which an advance has been made under this paragraph, the Administrator shall reduce the amount of such grant by the allowance established under paragraph (1) of this subsection. In the event no such grant is made, the State is authorized to seek repayment of such advance on such terms and conditions as it may determine.

(m)(1) Notwithstanding any other provisions of this title, the Administrator is authorized to make a grant from any funds otherwise allotted to the State of California under section 205 of this Act to the project (and in the amount) specified in Order WQG 81–1 of the California State Water Resources Control Board.

(2) Notwithstanding any other provision of this Act, the Administrator shall make a grant from any funds otherwise allotted to the State of California to the city of Eureka, California, in connection with project numbered C–06–2772, for the purchase of one hundred and thirty-nine acres of property as environmental mitigation for siting of the proposed treatment plant.

(3) Notwithstanding any other provision of this Act, the Administrator shall make a grant from any funds otherwise allotted to the State of California to the city of San Diego, California, in connection with that city's aquaculture sewage process (total resources recovery system) as an innovative and alternative waste treatment process.

(n)(1) On and after October 1, 1984, upon the request of the Governor of an affected State, the Administrator is authorized to use funds available to such State under section 205 to address water quality problems due to the impacts of discharges from combined storm water and sanitary sewer overflows, which are not otherwise eligible under this subsection, where correction of such discharges is a major priority for such State.

(2) Beginning fiscal year 1983, the Administrator shall have available $200,000,000 per fiscal year in addition to those funds authorized in section 207 of this Act to be utilized to address water quality problems of marine bays and estuaries subject to lower levels of water quality due to the impacts of discharges from combined storm water and sanitary sewer overflows from adjacent urban complexes, not otherwise eligible under this subsection. Such sums may be used as deemed appropriate by the Administrator as provided in paragraphs (1) and (2) of this subsection, upon the request

of and demonstration of water quality benefits by the Governor of an affected State.

(o) The Administrator shall encourage and assist applicants for grant assistance under this title to develop and file with the Administrator a capital financing plan which, at a minimum—

(1) projects the future requirements for waste treatment services within the applicant's jurisdiction for a period of no less than ten years;

(2) projects the nature, extent, timing, and costs of future expansion and reconstruction of treatment works which will be necessary to satisfy the applicant's projected future requirements for waste treatment services; and

(3) sets forth with specificity the manner in which the applicant intends to finance such future expansion and reconstruction.

(p) TIME LIMIT ON RESOLVING CERTAIN DISPUTES.—In any case in which a dispute arises with respect to the awarding of a contract for construction of treatment works by a grantee of funds under this title and a party to such dispute files an appeal with the Administrator under this title for resolution of such dispute, the Administrator shall make a final decision on such appeal within 90 days of the filing of such appeal.

(33 U.S.C. 1281)

FEDERAL SHARE

SEC. 202. (a)(1) The amount of any grant for treatment works made under this Act from funds authorized for any fiscal year beginning after June 30, 1971, and ending before October 1, 1984, shall be 75 per centum of the cost of construction thereof (as approved by the Administrator), and for any fiscal year beginning on or after October 1, 1984, shall be 55 per centum of the cost of construction thereof (as approved by the Administrator), unless modified to a lower percentage rate uniform throughout a State by the Governor of that State with the concurrence of the Administrator. Within ninety days after the enactment of this sentence the Administrator, shall issue guidelines for concurrence in any such modification, which shall provide for the consideration of the unobligated balance of sums allocated to the State under section 205 of this Act, the need for assistance under this title in such State, and the availability of State grant assistance to replace the Federal share reduced by such modification. The payment of any such reduced Federal share shall not constitute an obligation on the part of the United States or a claim on the part of any State or grantee to reimbursement for the portion of the Federal share reduced in any such State. Any grant (other than for reimbursement) made prior to the date of enactment of the Federal Water Pollution Control Act Amendments of 1972 from any funds authorized for any fiscal year beginning after June 30, 1971, shall, upon the request of the applicant, be increased to the applicable percentage under this section. Notwithstanding the first sentence of this paragraph, in any case where a primary, secondary, or advanced waste treatment facility or its related interceptors or a project for infiltration-in-flow correction has received a grant for erection, building, acquisition, alteration, remodeling, improvement, extension, or correction before October 1, 1984, all segments and phases of such facility, interceptors, and project for infiltration-in-flow correction shall be eligible for grants at 75 per centum of the cost of construction thereof for any grant made pursuant to a State obligation which obligation occurred before October 1, 1990. Notwithstanding the first sentence of this paragraph, in the case of a project for which an application for a grant under this title has been made to the Administrator before October 1, 1984, and which project is under judicial injunction on such date prohibiting its construction, such project shall be eligible for grants at 75 percent of the cost of construction thereof. Notwithstanding the first sentence of this paragraph, in the case of the Wyoming Valley Sanitary Authority project mandated by judicial order under a proceeding begun prior to October 1, 1984, and a project for wastewater treatment for Altoona, Pennsylvania, such projects shall be eligible for grants at 75 percent of the cost of construction thereof.

(2) The amount of any grant made after September 30, 1978, and before October 1, 1981, for any eligible treatment works or significant portion thereof utilizing innovative or alternative wastewater treatment processes and techniques referred to in section 201(g)(5) shall be 85 per centum of the cost of construction thereof, unless modified by the Governor of the State with the concurrence of the Administrator to a percentage rate no less than 15 per centum greater than the modified uniform percentage rate in which the Administrator has concurred pursuant to paragraph (1) of this subsection. The amount of any grant made after September 30, 1981, for any eligible treatment works or unit processes and techinques thereof utilizing innovative or alternative wastewater treatment processes and techniques referred to in section 201(g)(5) shall be a percentage of the cost of construction thereof equal to 20 per centum greater than the percentage in effect under paragraph (1) of this subsection for such works or unit processes and techniques, but in no event greater than 85 per centum of the cost of construction thereof. No grant shall be made under this paragraph for construction of a treatment works in any State unless the proportion of the State contribution to the non-Federal share of construction costs for all treatment works in such State receiving a grant under this paragraph is the same as or greater than the proportion of the State contribution (if any) to the non-Federal share of construction costs for all treatment works receiving grants in such State under paragraph (1) of this subsection.

(3) In addition to any grant made pursuant to paragraph (2) of this subsection, the Administrator is authorized to make a grant to fund all of the costs of the modification or replacement of any facilities constructed with a grant made pursuant to paragraph (2) if the Administrator finds that such facilities have not met design performance specifications unless such failure is attributable to negligence on the part of any person and if such failure has significantly increased capital or operating and maintenance expenditures. In addition, the Administrator is authorized to make a grant to fund all of the costs of the modification or replacement of biodisc equipment (rotating biological contractors) in any publicly owned treatment works if the Administrator finds that such equipment

has failed to meet design performance specifications, unless such failure is attributable to negligence on the part of any person, and if such failure has significantly increased capital or operating and maintenance expenditures.

(4) For the purposes of this section, the term "eligible treatment works" means those treatment works in each State which meet the requirements of section 201(g)(5) of this Act and which can be fully funded from funds available for such purpose in such State.

(b) The amount of the grant for any project approved by the Administrator after January 1, 1971, and before July 1, 1971, for the construction of treatment works, the actual erection, building or acquisition of which was not commenced prior to July 1, 1971, shall, upon the request of the applicant, be increased to the applicable percentage under subsection (a) of this section for grants for treatment works from funds for fiscal years beginning after June 30, 1971, with respect to the cost of such actual erection, building, or acquisition. Such increased amount shall be paid from any funds allocated to the State in which the treatment works is located without regard to the fiscal year for which such funds were authorized. Such increased amount shall be paid for such project only if—

(1) a sewage collection system that is a part of the same total waste treatment system as the treatment works for which such grant was approved is under construction or is to be constructed for use in conjunction with such treatment works, and if the cost of such sewage collection system exceeds the cost of such treatment works, and

(2) the State water pollution control agency or other appropriate State authority certifies that the quantity of available ground water will be insufficient, inadequate, or unsuitable for public use, including the ecological preservation and recreational use of surface water bodies, unless effluents from publicly-owned treatment works after adequate treatment are returned to the ground water consistent with acceptable technological standards.

(c) Notwithstanding any other provision of law, sums allotted to the Commonwealth of Puerto Rico under section 205 of this Act for fiscal year 1981 shall remain available for obligation for the fiscal year for which authorized and for the period of the next succeeding twenty-four months. Such sums and any unobligated funds available to Puerto Rico from allotments for fiscal years ending prior to October 1, 1981, shall be available for obligation by the Administrator of the Environmental Protection Agency only to fund the following systems: Aguadilla, Arecibo, Mayaguez, Carolina, and Camuy Hatillo. These funds may be used by the Commonwealth of Puerto Rico to fund the non-Federal share of the costs of such projects. To the extent that these funds are used to pay the non-Federal share, the Commonwealth of Puerto Rico shall repay to the Environmental Protection Agency such amounts on terms and conditions developed and approved by the Administrator in consultation with the Governor of the Commonwealth of Puerto Rico. Agreement on such terms and conditions including the payment of interest to be determined by the Secretary of the Treasury, shall be reached prior to the use of these funds for the Commonwealth's non-Federal share. No Federal funds awarded under this provision shall be used to replace local governments funds previously expended on these projects.

(33 U.S.C. 1282)

PLANS, SPECIFICATIONS, ESTIMATES, AND PAYMENTS

SEC. 203. (a)(1) Each applicant for a grant shall submit to the Administrator for his approval, plans, specifications, and estimates for each proposed project for the construction of treatment works for which a grant is applied for under section 201(g)(1) from funds allotted to the State under section 205 and which otherwise meets the requirements of this Act. The Administrator shall act upon such plans, specifications, and estimates as soon as practicable after the same have been submitted, and his approval of any such plans, specifications, and estimates shall be deemed a contractual obligation of the United States for the payment of its proportional contribution to such project.

(2) AGREEMENT ON ELIGIBLE COSTS.—

(A) LIMITATION ON MODIFICATIONS.—Before taking final action on any plans, specifications, and estimates submitted under this subsection after the 60th day following the date of the enactment of the Water Quality Act of 1987, the Administrator shall enter into a written agreement with the applicant which establishes and specifies which items of the proposed project are eligible for Federal payments under this section. The Administrator may not later modify such eligibility determinations unless they are found to have been made in violation of applicable Federal statutes and regulations.

(B) LIMITATION ON EFFECT.—Eligibility determinations under this paragraph shall not preclude the Administrator from auditing a project pursuant to section 501 of this Act, or other authority, or from withholding or recovering Federal funds for costs which are found to be unreasonable, unsupported by adequate documentation, or otherwise unallowable under applicable Federal costs principles, or which are incurred on a project which fails to meet the design specifications or effluent limitations contained in the grant agreement and permit pursuant to section 402 of this Act for such project.

(3) In the case of a treatment works that has an estimated total cost of $8,000,000 of less (as determined by the Administrator), and the population of the applicant municipality is twenty-five thousand or less (according to the most recent United States census), upon completion of an approved facility plan, a single grant may be awarded for the combined Federal share of the cost of preparing construction plans and specifications, and the building and erection of the treatment works.

(b) The Administrator shall, from time to time as the work progresses, make payments to the recipient of a grant for costs of construction incurred on a project. These payments shall at no time exceed the Federal share of the cost of construction incurred to the date of the voucher covering such payment plus the Federal share of the value of the materials which have been stockpiled in the vicinity of such construction in conformity to plans and specifications for the project.

(c) After completion of a project and approval of the final voucher by the Administrator, he shall pay out of the appropriate sums the unpaid balance of the Federal share payable on account of such project.

(d) Nothing in this Act shall be construed to require, or to authorize the Administrator to require, that grants under this Act for construction of treatment works be made only for projects which are operable units usable for sewage collection, transportation, storage, waste treatment, or for similar purposes without additional construction.

(e) At the request of a grantee under this title, the Administrator is authorized to provide technical and legal assistance in the administration and enforcement of any contract in connection with treatment works assisted under this title, and to intervene any civil action involving the enforcement of such a contract.

(f) DESIGN/BUILD PROJECTS.—

(1) AGREEMENT.—Consistent with State law, an applicant who proposes to construct waste water treatment works may enter into an agreement with the Administrator under this subsection providing for the preparation of construction plans and specifications and the erection of such treatment works, in lieu of proceeding under the other provisions of this section.

(2) LIMITATION ON PROJECTS.—Agreements under this subsection shall be limited to projects under an approved facility plan which projects are—

(A) treatment works that have an estimated total cost of $8,000,000 or less; and

(B) any of the following types of waste water treatment systems: aerated lagoons, trickling filters, stabilization ponds, land application systems, sand filters, and subsurface disposal systems.

(3) REQUIRED TERMS.—An agreement entered into under this subsection shall—

(A) set forth an amount agreed to as the maximum Federal contribution to the project, based upon a competitively bid document of basic design data and applicable standard construction specifications and a determination of the federally eligible costs of the project at the applicable Federal share under section 202 of this Act;

(B) set forth dates for the start and completion of construction of the treatment works by the applicant and a schedule of payments of the Federal contribution to the project;

(C) contain assurances by the applicant that (i) engineering and management assistance will be provided to manage the project; (ii) the proposed treatment works will be an operable unit and will meet all the requirements of this title; and (iii) not later than 1 year after the date specified as the date of completion of construction of the treatment works, the treatment works will be operating so as to meet the requirements of any applicable permit for such treatment works under section 402 of this Act;

(D) require the applicant to obtain a bond from the contractor in an amount determined necessary by the Administrator to protect the Federal interest in the project; and

(E) contain such other terms and conditions as are necessary to assure compliance with this title (except as provided in paragraph (4) of this subsection).

(4) LIMITATION ON APPLICATION.—Subsections (a), (b), and (c) of this section shall not apply to grants made pursuant to this subsection.

(5) RESERVATION TO ASSURE COMPLIANCE.—The Administrator shall reserve a portion of the grant to assure contract compliance until final project approval as defined by the Administrator. If the amount agreed to under paragraph (3)(A) exceeds the cost of designing and constructing the treatment works, the Administrator shall reallot the amount of the excess to the State in which such treatment works are located for the fiscal year in which such audit is completed.

(6) LIMITATION ON OBLIGATIONS.—The Administrator shall not obligate more than 20 percent of the amount allotted to a State for a fiscal year under section 205 of this Act for grants pursuant to this subsection.

(7) ALLOWANCE.—The Administrator shall determine an allowance for facilities planning for projects constructed under this subsection in accordance with section 201(l).

(8) LIMITATION ON FEDERAL CONTRIBUTIONS.—In no event shall the Federal contribution for the cost of preparing construction plans and specifications and the building and erection of treatment works pursuant to this subsection exceed the amount agreed upon under paragraph (3).

(9) RECOVERY ACTION.—In any case in which the recipient of a grant made pursuant to this subsection does not comply with the terms of the agreement entered into under paragraph (3), the Administrator is authorized to take such action as may be necessary to recover the amount of the Federal contribution to the project.

(10) PREVENTION OF DOUBLE BENEFITS.—A recipient of a grant made pursuant to this subsection shall not be eligible for any other grants under this title for the same project.

(33 U.S.C. 1283)

LIMITATIONS AND CONDITIONS

SEC. 204. (a) Before approving grants for any project for any treatment works under section 201(g)(1) the Administrator shall determine—

(1) that any required areawide waste treatment management plan under section 208 of this Act (A) is being implemented for such area and the proposed treatment works are included in such plan, or (B) is being developed for such area and reasonable progress is being made toward its implementation

and the proposed treatment works will be included in such plan;

(2) that (A) the State in which the project is to be located (i) is implementing any required plan under section 303(e) of this Act and the proposed treatment works are in conformity with such plan, or (ii) is developing such a plan and the proposed treatment works will be in conformity with such plan, and (B) such State is in compliance with section 305(b) of this Act;

(3) that such works have been certified by the appropriate State water pollution control agency as entitled to priority over such other works in the State in accordance with any applicable State plan under section 303(e) of this Act, except that any priority list developed pursuant to section 303(e)(3)(H) may be modified by such State in accordance with regulations promulgated by the Administrator to give higher priority for grants for the Federal share of the cost of preparing construction drawings and specifications for any treatment works utilizing processes and techniques meeting the guidelines promulgated under section 304(d)(3) of this Act for grants for the combined Federal share of the cost of preparing construction drawings and specifications and the building and erection of any treatment works meeting the requirements of the next to the last sentence of section 203(a) of this Act which utilizes processes and techniques meeting the guidelines promulgated under section 304(d)(3) of this Act.[1]

(4) that the applicant proposing to construct such works agrees to pay the non-Federal costs of such works and has made adequate provisions satisfactory to the Administrator for assuring proper and efficient operation, including the employment of trained management and operations personnel, and the maintenance of such works in accordance with a plan of operation approved by the state water pollution control agency or, as appropriate, the interstate agency, after construction thereof;

(5) that the size and capacity of such works relate directly to the needs to be served by such works, including sufficient reserve capacity. The amount of reserve capacity provided shall be approved by the Administrator on the basis of a comparison of the cost of constructing such reserves as a part of the works to be funded and the anticipated cost of providing expanded capacity at a date when such capacity will be required, after taking into account, in accordance with regulations promulgated by the Administrator, efforts to reduce total flow of sewage and unnecessary water consumption. The amount of reserve capacity eligible for a grant under this title shall be determined by the Administrator taking into account the projected population and associated commercial and industrial establishments within the jurisdiction of the applicant to be served by such treatment works as identified in an approved facilities plan, an areawide plan under section 208, or an applicable municipal master plan of development. For the purpose of this paragraph,

[1] So in law. The period should be a semicolon.

section 208, and any such plan, projected population shall be determined on the basis of the latest information available from the United States Department of Commerce or from the States as the Administrator, by regulation, determines appropriate. Beginning October 1, 1984, no grants shall be made under this title to construct that portion of any treatment works providing reserve capacity in excess of existing needs (including existing needs of residential, commercial, industrial, and other users) on the date of approval of a grant for the erection, building, acquisition, alteration, remodeling, improvement, or extension of a project for secondary treatment or more stringent treatment or new interceptors and appurtenances, except that in no event shall reserve capacity of a facility and its related interceptors to which this subsection applies be in excess of existing needs on October 1, 1990. In any case in which an applicant proposes to provide reserve capacity greater than that eligible for Federal financial assistance under this title, the incremental costs of the additional reserve capacity shall be paid by the applicant;

(6) that no specification for bids in connection with such works shall be written in such a manner as to contain proprietary, exclusionary, or discriminatory requirements other than those based upon performance, unless such requirements are necessary to test or demonstrate a specific thing or to provide for necessary interchangeability of parts and equipment. When in the judgment of the grantee, it is impractical or uneconomical to make a clear and accurate description of the technical requirements, a "brand name or equal" description may be used as a means to define the performance or other salient requirements of a procurement, and in doing so the grantee need not establish the existence of any source other than the brand or source so named.

(b)(1) Notwithstanding any other provision of this title, the Administrator shall not approve any grant for any treatment works under section 201(g)(1) after March 1, 1973, unless he shall first have determined that the applicant (A) has adopted or will adopt a system of charges to assure that each recipient of waste treatment services within the applicant's jurisdiction, as determined by the Administrator, will pay its proportionate share (except as otherwise provided in this paragraph) of the costs of operation and maintenance (including replacement) of any waste treatment services provided by the applicant; and (B) has legal, institutional, managerial, and financial capability to insure adequate construction, operation, and maintenance of treatment works throughout the applicant's jurisdiction, as determined by the Administrator. In any case where an applicant which, as of the date of enactment of this sentence, uses a system of dedication ad valorem taxes and the Administrator determines that the applicant has a system of charges which results in the distribution of operation and maintenance costs for treatment works within the applicant's jurisdiction, to each user class, in proportion to the contribution to the total cost of operation and maintenance of such works by each user class (taking into account total waste water loading of such works, the constituent elements of the waste, and other appropriate factors), and such applicant is otherwise in compliance with clause (A) of this paragraph with respect to each industrial user, then such dedication ad valorem tax system shall be deemed to be the user charge system meeting the requirements of clause (A) of this paragraph for the residential user class and such small non-residential user classes as defined by the Administrator. In defining small non-residential users, the Administrator shall consider the volume of wastes discharged into the treatment works by such users and the constituent elements of such wastes as well as such other factors as he deems appropriate. A system of user charges which imposes a lower charge for low-income residential users (as defined by the Administrator) shall be deemed to be a user charge system meeting the requirements of clause (A) of this paragraph if the Administrator determines that such system was adopted after public notice and hearing.

(2) The Administrator shall, within one hundred and eighty days after the date of enactment of the Federal Water Pollution Control Act Amendments of 1972, and after consultation with appropriate State, interstate, municipal and intermunicipal agencies, issue guidelines applicable to payment of waste treatment costs by industrial and nonindustrial receipts of waste treatment services which shall establish (A) classes of users of such services, including categories of industrial users; (B) criteria against which to determine the adequacy of charges imposed on classes and categories of users reflecting all factors that influence the cost of waste treatment, including strength, volume, and delivery flow rate characteristics of waste; and (C) model systems and rates of user charges typical of various treatment works serving municipal-industrial communities.

(3) Approval by the Administrator of a grant to an interstate agency established by interstate compact for any treatment works shall satisfy any other requirement that such works be authorized by Act of Congress.

(4) A system of charges which meets the requirement of clause (A) of paragraph (1) of this subsection may be based on something other than metering the sewage or water supply flow of residential recipients of waste treatment services, including ad valorem taxes. If the system of charges is based on something other than metering the Administrator shall require (A) the applicant to establish a system by which the necessary funds will be available for the proper operation and maintenance of the treatment works; and (B) the applicant to establish a procedure under which the residential user will be notified as to that portion of his total payment which will be allocated to the costs of the waste treatment services.

(c) The next to the last sentence of paragraph (5) of subsection (a) of this section shall not apply in any case where a primary, secondary, or advanced waste treatment facility or its related interceptors has received a grant for erection, building, acquisition, alteration, remodeling, improvement, or extension before October 1, 1984, and all segments and phases of such facility and interceptors shall be funded based on a 20-year reserve capacity in the case of such facility and a 20-year reserve capacity in the case of such interceptors, except that, if a grant for such interceptors has been approved prior to the date of enactment of the Municipal Wastewater Treatment Construction Grant Amendments of 1981, such interceptors shall be funded based on the approved reserve capacity not to exceed 40 years.

(d)(1) A grant for the construction of treatment works under this title shall provide that the engineer or engineering firm supervising construction or providing architect engineering services during construction shall continue its relationship to the grant applicant for a period of one year after the completion of construction and initial operation of such treatment works. During such period such engineer or engineering firm shall supervise operation of the treatment works, train operating personnel, and prepare curricula and training material for operating personnel. Costs associated with the implementation of this paragraph shall be eligible for Federal assistance in accordance with this title.

(2) On the date one year after the completion of construction and initial operation of such treatment works, the owner and operator of such treatment works shall certify to the Administrator whether or not such treatment works meet the design specifications and effluent limitations contained in the grant agreement and permit pursuant to section 402 of the Act for such works. If the owner and operator of such treatment works cannot certify that such treatment works meet such design specifications and effluent limitations, any failure to meet such design specifications and effluent limitations shall be corrected in a timely manner, to allow such affirmative certification, at other than Federal expense.

(3) Nothing in this section shall be construed to prohibit a grantee under this title from requiring more assurances, guarantees, or indemnity or other contractual requirements from any party to a contract pertaining to a project assisted under this title, than those provided under this subsection.

(33 U.S.C. 1284)

ALLOTMENT

SEC. 205. (a) Sums authorized to be appropriated pursuant to section 207 for each fiscal year beginning after June 30, 1972, before September 30, 1977, shall be allotted by the Administrator not later than the January 1st immediately preceding the beginning of the fiscal year for which authorized, except that the allotment for fiscal year 1973 shall be made not later than 30 days after the date of enactment of the Federal Water Pollution Control Act Amendments of 1972. Such sums shall be allotted among the States by the Administrator in accordance with regulations promulgated by him, in the ratio that the estimated cost of constructing all needed publicly owned treatment works in each State bears to the estimated cost of construction of all needed publicly owned treatment works in all of the States. For the fiscal years ending June 30, 1973, and June 30, 1974, such ratio shall be determined on the basis of table III of House Public Works Committee Print No. 92–50. For the fiscal year ending June 30, 1975, such ratio shall be determined one-half on the basis of table I of House Public Works Committee Print Numbered 93–28 and one-half on the basis of table II of such print, except that no State shall receive an allotment less than that which it received for the fiscal year ending

June 30, 1972, as set forth in table III of such print. Allotments for fiscal years which begin after the fiscal year ending June 30, 1975 shall be made only in accordance with a revised cost estimate made and submitted to Congress in accordance with section 516 of this Act and only after such revised cost estimate shall have been approved by law specifically enacted hereafter.

(b)(1) Any sums allotted to a State under subsection (a) shall be available for obligation under section 203 on and after the date of such allotment. Such sums shall continue available for obligation in such State for a period of one year after the close of the fiscal year for which such sums are authorized. Any amounts so allotted which are not obligated by the end of such one-year period shall be immediately reallotted by the Administrator, in accordance with regulations promulgated by him, generally on the basis of the ratio used in making the last allotment of sums under this section. Such reallotted sums shall be added to the last allotments made to the States. Any sum made available to a State by reallotment under this subsection shall be in addition to any funds otherwise allotted to such State for grants under this title during any fiscal year.

(2) Any sums which have been obligated under section 203 and which are released by the payment of the final voucher for the project shall be immediately credited to the State to which such sums were last allotted. Such released sums shall be added to the amounts last allotted to such State and shall be immediately available for obligation in the same manner and to the same extent as such last allotment.

(c)(1) Sums authorized to be appropriated pursuant to section 207 for the fiscal years during the period beginning October 1, 1977, and ending September 30, 1981, shall be allotted for each such year by the Administrator not later than the tenth day which begins after the date of enactment of the Clean Water Act of 1977. Not-withstanding any other provision of law, sums authorized for the fiscal years ending September 30, 1978, September 30, 1979, September 30, 1980, and September 30, 1981, shall be allotted in accordance with table 3 of Committee Print Numbered 95–30 of the Committee on Public Works and Transportation of the House of Representatives.

(2) Sums authorized to be appropriated pursuant to section 207 for the fiscal years 1982, 1983, 1984, and 1985 shall be allotted for each such year by the Administrator not later than the tenth day which begins after the date of enactment of the Municipal Wastewater Treatment Construction Grant Amendments of 1981. Notwithstanding any other provision of law, sums authorized for the fiscal year ending September 30, 1982, shall be allotted in accordance with table 3 of Committee Print Numbered 95–30 of the Committee on Public Works and Transportation of the House of Representatives. Sums authorized for the fiscal years ending September 30, 1983, September 30, 1984, September 30, 1985, and September 30, 1986, shall be allotted in accordance with the following table:

	Fiscal years 1983 through 1985 [1]
States:	
Alabama	.011398
Alaska	.006101
Arizona	.006885
Arkansas	.006668
California	.072901
Colorado	.008154
Connecticut	.012487
Delaware	.004965
District of Columbia	.004965
Florida	.034407
Georgia	.017234
Hawaii	.007895
Idaho	.004965
Illinois	.046101
Indiana	.024566
Iowa	.013796
Kansas	.009201
Kentucky	.012973
Louisiana	.011205
Maine	.007788
Maryland	.024653
Massachusetts	.034608
Michigan	.043829
Minnesota	.018735
Mississippi	.009184
Missouri	.028257
Montana	.004965
Nebraska	.005214
Nevada	.004965
New Hampshire	.010186
New Jersey	.041654
New Mexico	.004965
New York	.113097
North Carolina	.018396
North Dakota	.004965
Ohio	.057383
Oklahoma	.008235
Oregon	.011515
Pennsylvania	.040377
Rhode Island	.006750
South Carolina	.010442
South Dakota	.004965
Tennessee	.014807
Texas	.038726
Utah	.005371
Vermont	.004965
Virginia	.020861
Washington	.017726
West Virginia	.015890
Wisconsin	.027557
Wyoming	.004965
Samoa	.000915
Guam	.000662
Northern Marianas	.000425
Puerto Rico	.013295
Pacific Trust Territories	.001305
Virgin Islands	.000531
United States totals	.999996

[1] So in original. Probably should be “1986”.

(3) FISCAL YEARS 1987–1990.—Sums authorized to be appropriated pursuant to section 207 for the fiscal years 1987, 1988, 1989, and 1990 shall be allotted for each such year by the Administrator not later than the 10th day which begins after the

date of the enactment of this paragraph. Sums authorized for such fiscal years shall be allotted in accordance with the following table:

States:	
Alabama	.011309
Alaska	.006053
Arizona	.006831
Arkansas	.006616
California	.072333
Colorado	.008090
Connecticut	.012390
Delaware	.004965
District of Columbia	.004965
Florida	.034139
Georgia	.017100
Hawaii	.007833
Idaho	.004965
Illinois	.045741
Indiana	.024374
Iowa	.013688
Kansas	.009129
Kentucky	.012872
Louisiana	.011118
Maine	.007829
Maryland	.024461
Massachusetts	.034338
Michigan	.043487
Minnesota	.018589
Mississippi	.009112
Missouri	.028037
Montana	.004965
Nebraska	.005173
Nevada	.004965
New Hampshire	.010107
New Jersey	.041329
New Mexico	.004965
New York	.111632
North Carolina	.018253
North Dakota	.004965
Ohio	.056936
Oklahoma	.008171
Oregon	.011425
Pennsylvania	.040062
Rhode Island	.006791
South Carolina	.010361
South Dakota	.004965
Tennessee	.014692
Texas	.046226
Utah	.005329
Vermont	.004965
Virginia	.020698
Washington	.017588
West Virginia	.015766
Wisconsin	.027342
Wyoming	.004965
American Samoa	.000908
Guam	.000657
Northern Marianas	.000422
Puerto Rico	.013191
Pacific Trust Territories	.001295
Virgin Islands	.000527

(d) Sums allotted to the States for a fiscal year shall remain available for obligation for the fiscal year for which authorized and for the period of the next succeeding twelve months. The amount of any allotment not obligated by the end of such twenty-four-month period shall be immediately reallotted by the Administrator on the basis of the same ratio as applicable to sums allotted for the then current fiscal year, except that none of the funds reallotted by the Administrator for fiscal year 1978 and for fiscal years thereafter shall be allotted to any State which failed to obligate any of the funds being reallotted. Any sum made available to a State by reallotment under this subsection shall be in addition to any funds otherwise allotted to such State for grants under this title during any fiscal year.

(e) For the fiscal years 1978, 1979, 1980, 1981, 1982, 1983, 1984, 1985, 1986, 1987, 1988, 1989, and 1990, no State shall receive less than one-half of 1 per centum of the total allotment under subsection (c) of this section, except that in the case of Guam, Virgin Islands, American Samoa, and the Trust Territories not more than thirty-three one-hundredths of 1 per centum in the aggregate shall be allotted to all four for these jurisdictions. For the purpose of carrying out this subsection there are authorized to be appropriated, subject to such amounts as are provided in appropriation Acts, not to exceed $75,000,000 for each fiscal years 1978, 1979, 1980, 1981, 1982, 1983, 1984, 1985, 1986, 1987, 1988, 1989, and 1990. If for any fiscal year the amount appropriated under authority of this subsection is less than the amount necessary to carry out this subsection, the amount each State receives under this subsection for such year shall be the same ratio for the amount such State would have received under this subsection in such year if the amount necessary to carry it out had been appropriated as the amount appropriated for such year bears to the amount necessary to carry out this subsection for such year.

(f) Notwithstanding any other provision of this section, sums made available between January 1, 1975, and March 1, 1975, by the Administrator for obligation shall be available for obligation until September 30, 1978.

(g)(1) The Administrator is authorized to reserve each fiscal year not to exceed 2 per centum of the amount authorized under section 207 of this title for purposes of the allotment made to each State under this section on or after October 1, 1977, except in the case of any fiscal year beginning on or after October 1, 1981, and ending before October 1, 1994, in which case the percentage authorized to be reserved shall not exceed 4 per centum.[1] or $400,000 whichever amount is the greater. Sums so reserved shall be available for making grants to such State under paragraph (2) of this subsection for the same period as sums are available from such allotment under subsection (d) of this section, and any such grant shall be available for obligation only during such period. Any grant made from sums reserved under this subsection which has not been obligated by the end of the period for which available shall be added to the amount last allotted to such State under this section and shall be immediately available for obligation in the same manner and to the same extent as such last allotment. Sums authorized to be reserved by this paragraph shall be in addition to and

[1] P.L. 97–117 added this phrase with a period at the end; probably should be a comma.

not in lieu of any other funds which may be authorized to carry out this subsection.

(2) The Administrator is authorized to grant to any State from amounts reserved to such State under this subsection, the reasonable costs of administering any aspects of sections 201, 203, 204, and 212 of this Act the responsibility for administration of which the Administrator has delegated to such State. The Administrator may increase such grant to take into account the reasonable costs of administering an approved program under section 402 or 404, administering a statewide waste treatment management planning program under section 208(b)(4), and managing waste treatment construction grants for small communities.

(h) The Administrator shall set aside from funds authorized for each fiscal year beginning on or after October 1, 1978, a total (as determined by the Governor of the State) of not less than 4 percent nor more than 7½ percent of the sums allotted to any State with a rural population of 25 per centum or more of the total population of such State, as determined by the Bureau of the Census. The Administrator may set aside no more than 7½ percent of the sums allotted to any other State for which the Governor requests such action. Such sums shall be available only for alternatives to conventional sewage treatment works for municipalities having a population of three thousand five hundred or less, or for the highly dispersed sections of larger municipalities, as defined by the Administrator.

(i) SET-ASIDE FOR INNOVATIVE AND ALTERNATIVE PROJECTS.—Not less than ½ of 1 percent of funds allotted to a State for each of the fiscal years ending September 30, 1979, through September 30, 1990, under subsection (c) of this section shall be expended only for increasing the Federal share of grants for construction of treatment works utilizing innovative processes and techniques pursuant to section 202(a)(2) of this Act. Including the expenditures authorized by the preceding sentence, a total of 2 percent of the funds allotted to a State for each of the fiscal years ending September 30, 1979, and September 30, 1980, and 3 percent of the funds allotted to a State for the fiscal year ending September 30, 1981, under subsection (c) of this section shall be expended only for increasing grants for construction of treatment works pursuant to section 202(a)(2) of this Act. Including the expenditures authorized by the first sentence of this subsection, a total (as determined by the Governor of the State) of not less than 4 percent nor more than 7½ percent of the funds allotted to such State under subsection (c) of this section for each of the fiscal years ending September 30, 1982, through September 30, 1990, shall be expended only for increasing the Federal share of grants for construction of treatment works pursuant to section 202(a)(2) of this Act.

(j)(1) The Administrator shall reserve each fiscal year not to exceed 1 per centum of the sums allotted and available for obligation to each State under this section for each fiscal year beginning on or after October 1, 1981, or $100,000, whichever amount is the greater.

(2) Such sums shall be used by the Administrator to make grants to the States to carry out water quality management planning, including, but not limited to—

(A) identifying most cost effective and locally acceptable facility and non-point measures to meet and maintain water quality standards;

(B) developing an implementation plan to obtain State and local financial and regulatory commitments to implement measures developed under subparagraph (A);

(C) determining the nature, extent, and causes of water quality problems in various areas of the State and interstate region, and reporting on these annually; and

(D) determining those publicly owned treatment works which should be constructed with assistance under this title, in which areas and in what sequence, taking into account the relative degree of effluent reduction attained, the relative contributions to water quality of other point or nonpoint sources, and the consideration of alternatives to such construction, and implementing section 303(e) of this Act.

(3) In carrying out planning with grants made under paragraph (2) of this subsection, a State shall develop jointly with local, regional, and interstate entities, a plan for carrying out the program and give funding priority to such entities and designated or undesignated public comprehensive planning organizations to carry out the purposes of this subsection. In giving such priority, the State shall allocate at least 40 percent of the amount granted to such State for a fiscal year under paragraph (2) of this subsection to regional public comprehensive planning organizations in such State and appropriate interstate organizations for the development and implementation of the plan described in this paragraph. In any fiscal year for which the Governor, in consultation with such organizations and with the approval of the Administrator, determines that allocation of at least 40 percent of such amount to such organizations will not result in significant participation by such organizations in water quality management planning and not significantly assist in development and implementation of the plan described in this paragraph and achieving the goals of this Act, the allocation to such organization may be less than 40 percent of such amount.

(4) All activities undertaken under this subsection shall be in coordination with other related provisions of this Act.

(5) NONPOINT SOURCE RESERVATION.—In addition to the sums reserved under paragraph (1), the Administrator shall reserve each fiscal year for each State 1 percent of the sums allotted and available for obligation to such State under this section for each fiscal year beginning on or after October 1, 1986, or $100,000, whichever is greater, for the purpose of carrying out section 319 of this Act. Sums so reserved in a State in any fiscal year for which such State does not request the use of such sums, to the extent such sums exceed $100,000, may be used by such State for other purposes under this title.

(k) The Administrator shall allot to the State of New York from sums authorized to be appropriated for the fiscal year ending September 30, 1982, an amount necessary to pay the entire cost of conveying sewage from the Convention Center of the City of New York to the Newtown sewage treatment plant, Brooklyn-Queens area, New York. The amount allotted under this subsection shall be in addition to and not in lieu of any other amounts authorized to be allotted to such State under this Act.

(l) MARINE ESTUARY RESERVATION.—

(1) RESERVATION OF FUNDS.—

(A) GENERAL RULE.—Prior to making allotments among the States under subsection (c) of this section, the Administrator shall reserve funds from sums appropriated pursuant to section 207 for each fiscal year beginning after September 30, 1986.

(B) FISCAL YEARS 1987 AND 1988.—For each of fiscal years 1987 and 1988 the reservation shall be 1 percent of the sums appropriated pursuant to section 207 for such fiscal year.

(C) FISCAL YEARS 1989 AND 1990.—For each of fiscal years 1989 and 1990 the reservation shall be 1½ percent of the funds appropriated pursuant to section 207 for such fiscal year.

(2) USE OF FUNDS.—Of the sums reserved under this subsection, two-thirds shall be available to address water quality problems of marine bays and estuaries subject to lower levels of water quality due to the impacts of discharges from combined storm water and sanitary sewer overflows from adjacent urban complexes, and one-third shall be available for the implementation of section 320 of this Act, relating to the national estuary program.

(3) PERIOD OF AVAILABILITY.—Sums reserved under this subsection shall be subject to the period of availability for obligation established by subsection (d) of this section.

(4) TREATMENT OF CERTAIN BODY OF WATER.—For purposes of this section and section 201(n), Newark Bay, New Jersey, and the portion of the Passaic River up to Little Falls, in the vicinity of Beatties Dam, shall be treated as a marine bay and estuary.

(m) DISCRETIONARY DEPOSITS INTO STATE WATER POLLUTION CONTROL REVOLVING FUNDS.—

(1) FROM CONSTRUCTION GRANT ALLOTMENTS.—In addition to any amounts deposited in a water pollution control revolving fund established by a State under title VI, upon request of the Governor of such State, the Administrator shall make available to the State for deposit, as capitalization grants, in such fund in any fiscal year beginning after September 30, 1986, such portion of the amounts allotted to such State under this section for such fiscal year as the Governor considers appropriate; except that (A) in fiscal year 1987 such deposit may not exceed 50 percent of the amounts allotted to such State under this section for such fiscal year, and (B) in fiscal year 1988, such deposit may not exceed 75 percent of the amounts allotted to such State under this section for this [1] fiscal year.

(2) NOTICE REQUIREMENT.—The Governor of a State may make a request under paragraph (1) for a deposit into the water pollution control revolving fund of such State—

[1] So in original. Probably should be "such".

(A) in fiscal year 1987 only if no later than 90 days after the date of the enactment of this subsection, and

(B) in each fiscal year thereafter only if 90 days before the first day of such fiscal year,

the State provides notice of its intent to make such deposit.

(3) EXCEPTION.—Sums reserved under section 205(j) of this Act shall not be available for obligation under this subsection.

(33 U.S.C. 1285)

REIMBURSEMENT AND ADVANCED CONSTRUCTION

SEC. 206. (a) Any publicly owned treatment works in a State on which construction was initiated after June 30, 1966, but before July 1, 1973, which was approved by the appropriate State water pollution control agency and which the Administrator finds meets the requirements of section 8 of this Act in effect at the time of the initiation of construction shall be reimbursed a total amount equal to the difference between the amount of Federal financial assistance, if any, received under such section 8 for such project and 50 per centum of the cost of such project, or 55 per centum of the project cost where the Administrator also determines that such treatment works was constructed in conformity with a comprehensive metropolitan treatment plan as described in section 8(f) of the Federal Water Pollution Control Act as in effect immediately prior to the date of enactment of the Federal Water Pollution Control Act Amendments of 1972. Nothing in this subsection shall result in any such works receiving Federal grants from all sources in excess of 80 per centum of the cost of such project.

(b) Any publicly owned treatment works constructed with or eligible for Federal financial assistance under this Act in a State between June 30, 1956, and June 30, 1966, which was approved by the State water pollution control agency and which the Administrator finds meets the requirements of section 8 of this Act prior to the date of enactment of the Federal Water Pollution Control Act Amendments of 1972 but which was constructed without assistance under such section 8 or which received such assistance in an amount less than 30 per centum of the cost of such project shall qualify for payments and reimbursement of State or local funds used for such project from sums allocated to such State under this section in an amount which shall not exceed the difference between the amount of such assistance, if any, received for such project and 30 per centum of the cost of such project.

(c) No publicly owned treatment works shall receive any payment or reimbursement under subsection (a) or (b) of this section unless an application for such assistance is filed with the Administrator within the one year period which begins on the date of enactment of the Federal Water Pollution Control Act Amendments of 1972. Any application filed within such one year period may be revised from time to time, as may be necessary.

(d) The Administrator shall allocate to each qualified project under subsection (a) of this section each fiscal year for which funds are appropriated under subsection (e) of this section an amount which bears the same ratio to the unpaid balance of the reimbursement due such project as the total of such funds for such year bears to the total unpaid balance of reimbursement due all such approved projects on the date of enactment of such appropriation. The Administrator shall allocate to each qualified project under subsection (b) of this section each fiscal year for which funds are appropriated under subsection (e) of this section an amount which bears the same ratio to the unpaid balance of the reimbursement due such project as the total of such funds for such year bears to the total unpaid balance of reimbursement due all such approved projects on the date of enactment of such appropriation.

(e) There is authorized to be appropriated to carry out subsection (a) of this section not to exceed $2,600,000,000 and, to carry out subsection (b) of this section, not to exceed $750,000,000. The authorizations contained in this subsection shall be the sole source of funds for reimbursements authorized by this section.

(f)(1) In any case where a substantial portion of the funds allotted to a State for the current fiscal year under this title have been obligated under section 201(g), or will be so obligated in a timely manner (as determined by the Administrator), and there is construction of any treatment work project without the aid of Federal funds and in accordance with all procedures and all requirements applicable to treatment works projects, except those procedures and requirements which limit construction of projects to those constructed with the aid of previously allotted Federal funds, the Administrator, upon his approval of an application made under this subsection therefore, is authorized to pay the Federal share of the cost of construction of such project when additional funds are allotted to the State under this title if prior to the construction of the project the Administrator approves plans, specifications, and estimates therefor in the same manner as other treatment works projects. The Administrator may not approve an application under this subsection unless an authorization is in effect for the first fiscal year in the period for which the application requests payment and such requested payment for that fiscal year does not exceed the State's expected allotment from such authorization. The Administrator shall not be required to make such requested payment for any fiscal year—

(A) to the extent that such payment would exceed such State's allotment of the amount appropriated for such fiscal year; and

(B) unless such payment is for a project which, on the basis of an approved funding priority list of such State, is eligible to receive such payment based on the allotment and appropriation for such fiscal year.

To the extent that sufficient funds are not appropriated to pay the full Federal share with respect to a project for which obligations under the provisions of this subsection have been made, the Administrator shall reduce the Federal share to such amount less than 75 per centum as such appropriations do provide.

(2) In determining the allotment for any fiscal year under this title, any treatment works project constructed in accordance with this section and without the aid of Federal funds shall not be considered completed until an application under the provisions of this subsection with respect to such project has been approved by the

Administrator, or the availability of funds from which this project is eligible for reimbursement has expired, whichever first occurs.

(33 U.S.C. 1286)

AUTHORIZATION

SEC. 207. There is authorized to be appropriated to carry out this title, other than sections 206(e), 208 and 209, for the fiscal year ending June 30, 1973, not to exceed $5,000,000,000, for the fiscal year ending June 30, 1974, not to exceed $6,000,000,000, and for the fiscal year ending June 30, 1975, not to exceed $7,000,000,000, and, subject to such amounts as are provided in appropriation Acts for the fiscal year ending September 30, 1977, $1,000,000,000 for the fiscal year ending September 30, 1978, $4,500,000,000 and for the fiscal years ending September 30, 1979, September 30, 1980, not to exceed $5,000,000,000; for the fiscal year ending September 30, 1981, not to exceed $2,548,837,000; and for the fiscal years ending September 30, 1982, September 30, 1983, September 30, 1984, and September 30, 1985, not to exceed $2,400,000,000 per fiscal year; and for each of the fiscal years ending September 30, 1986, September 30, 1987, and September 30, 1988, not to exceed $2,400,000,000; and for each of the fiscal years ending September 30, 1989, and September 30, 1990, not to exceed $1,200,000,000.

(33 U.S.C. 1287)

AREAWIDE WASTE TREATMENT MANAGEMENT

SEC. 208. (a) For the purpose of encouraging and facilitating the development and implementation of areawide waste treatment management plans—

(1) The Administrator, within ninety days after the date of enactment of this Act and after consultation with appropriate Federal, State, and local authorities, shall by regulation publish guidelines for the identification of those areas which, as a result of urban-industrial concentrations or other factors, have substantial water quality control problems.

(2) The Governor of each State, within sixty days after publication of the guidelines issued pursuant to paragraph (1) of this subsection, shall identify each area within the State which, as a result of urban-industrial concentrations or other factors, has substantial water quality control problems. Not later than one hundred and twenty days following such identification and after consultation with appropriate elected and other officials of local governments having jurisdiction in such areas, the Governor shall designate (A) the boundaries of each such area, and (B) a single representative organization, including elected officials from local governments or their designees, capable of developing effective areawide waste treatment management plans for such an area. The Governor may in the same manner at any later time identify any additional area (or modify an existing area) for which he determines areawide waste treatment management to be appropriate, designate the boundaries of such area, and designate an organization capable

of developing effective areawide waste treatment management plans for such area.

(3) With respect to any area which, pursuant to the guidelines published under paragraph (1) of this subsection, is located in two or more States, the Governors of the respective States shall consult and cooperate in carrying out the provisions of paragraph (2), with a view toward designating the boundaries of the interstate area having common water quality control problems and for which areawide waste treatment management plans would be most effective, and toward designating, within one hundred and eighty days after publication of guidelines issued pursuant to paragraph (1) of this subsection, of a single representative organization capable of developing effective areawide waste treatment management plans for such area.

(4) If a Governor does not act, either by designating or determining not to make a designation under paragraph (2) of this subsection, within the time required by such paragraph, or if, in the case of an interstate area, the Governors of the States involved do not designate a planning organization within the time required by paragraph (3) of this subsection, the chief elected officials of local governments within an area may by agreement designate (A) the boundaries for such an area, and (B) a single representative organization including elected officials from such local governments, or their designees, capable of developing an areawide waste treatment management plan for such area.

(5) Existing regional agencies may be designated under paragraphs (2), (3), and (4) of this subsection.

(6) The State shall act as a planning agency for all portions of such State which are not designated under paragraphs (2), (3), or (4) of this subsection.

(7) Designations under this subsection shall be subject to the approval of the Administrator.

(b)(1)(A) Not later than one year after the date of designation of any organization under subsection (a) of this section such organization shall have in operation a continuing areawide waste treatment management planning process consistent with section 201 of this Act. Plans prepared in accordance with this process shall contain alternatives for waste treatment management, and be applicable to all wastes generated within the area involved. The initial plan prepared in accordance with such process shall be certified by the Governor and submitted to the Administrator not later than two years after the planning process is in operation.

(B) For any agency designated after 1975 under subsection (a) of this section and for all portions of a State for which the State is required to act as the planning agency in accordance with subsection (a)(6), the initial plan prepared in accordance with such process shall be certified by the Governor and submitted to the Administrator not later than three years after the receipt of the initial grant award authorized under subsection (f) of this section.

(2) Any plan prepared under such process shall include, but not be limited to—

(A) the identification of treatment works necessary to meet the anticipated municipal and industrial waste treatment needs of the area over a twenty-year period, annually updated (including an analysis of alternative waste treatment systems), including any requirements for the acquisition of land for treatment purposes; the necessary waste water collection and urban storm water runoff systems; and a program to provide the necessary financial arrangements for the development of such treatment works, and an identification of open space and recreation opportunities that can be expected to result from improved water quality, including consideration of potential use of lands associated with treatment works and increased access to water-based recreation;

(B) the establishment of construction priorities for such treatment works and time schedules for the initiation and completion of all treatment works;

(C) the establishment of a regulatory program to—

(i) implement the waste treatment management requirements of section 201(c),

(ii) regulate the location, modification, and construction of any facilities within such area which may result in any discharge in such area, and

(iii) assure that any industrial or commercial waste discharged into any treatment works in such area meet applicable pretreatment requirements;

(D) the identification of those agencies necessary to construct, operate, and maintain all facilities required by the plan and otherwise to carry out the plan;

(E) the identification of the measures necessary to carry out the plan (including financing), the period of time necessary to carry out the plan, the costs of carrying out the plan within such time, and the economic, social, and environmental impact of carrying out the plan within such time;

(F) a process to (i) identify, if appropriate, agriculturally and silviculturally related nonpoint sources of pollution, including return flows from irrigated agriculture, and their cumulative effects, runoff from manure disposal areas, and from land used for livestock and crop production, and (ii) set forth procedures and methods (including land use requirements) to control to the extent feasible such sources;

(G) a process of (i) identify, if appropriate, mine-related sources of pollution including new, current, and abandoned surface and underground mine runoff, and (ii) set forth procedures and methods (including land use requirements) to control to the extent feasible such sources;

(H) a process to (i) identify construction activity related sources of pollution, and (ii) set forth procedures and methods (including land use requirements) to control to the extent feasible such sources;

(I) a process to (i) identify, if appropriate, salt water intrusion into rivers, lakes, and estuaries resulting from reduction of fresh water flow from any cause, including irrigation, obstruction, ground water extraction, and diversion, and (ii) set forth procedures and methods to control such intrusion to the extent feasible where such procedures and methods are otherwise a part of the waste treatment management plan;

(J) a process to control the disposition of all residual waste generated in such area which could affect water quality; and

(K) a process to control the disposal of pollutants on land or in subsurface excavations within such area to protect ground and surface water quality.

(3) Areawide waste treatment management plans shall be certified annually by the Governor or his designee (or Governors or their designees, where more than one State is involved) as being consistent with applicable basin plans and such areawide waste treatment management plans shall be submitted to the Administrator for his approval.

(4)(A) Whenever the Governor of any State determines (and notifies the Administrator) that consistency with a statewide regulatory program under section 303 so requires, the requirements of clauses (F) through (K) of paragraph (2) of this subsection shall be developed and submitted by the Governor to the Administrator for approval for application to a class or category of activity throughout such State.

(B) Any program submitted under subparagraph (A) of this paragraph which, in whole or in part, is to control the discharge or other placement of dredged or fill material into the navigable waters shall include the following:

(i) A consultation process which includes the State agency with primary jurisdiction over fish and wildlife resources.

(ii) A process to identify and manage the discharge or other placement of dredged or fill material which adversely affects navigable waters, which shall complement and be coordinated with a State program under section 404 conducted pursuant to this Act.

(iii) A process to assure that any activity conducted pursuant to a best management practice will comply with the guidelines established under section 404(b)(1), and sections 307 and 403 of this Act.

(iv) A process to assure that any activity conducted pursuant to a best management practice can be terminated or modified for cause including, but not limited to, the following:

(I) violation of any condition of the best management practice;

(II) change in any activity that requires either a temporary or permanent reduction or elimination of the discharge pursuant to the best management practice.

(v) A process to assure continued coordination with Federal and Federal-State water-related planning and reviewing processes, including the National Wetlands Inventory.

(C) If the Governor of a State obtains approval from the Administrator of a statewide regulatory program which meets the requirements of subparagraph (B) of this paragraph and if such State is administering a permit program under section 404 of this Act, no person shall be required to obtain an individual permit pursuant to such section, or to comply with a general permit issued pursuant to such section, with respect to any appropriate activity within such State for which a best management practice has been approved by the Administrator under the program approved by the Administrator pursuant to this paragraph.

(D)(i) Whenever the Administrator determines after public hearing that a State is not administering a program approved under this section in accordance with the requirements of this section, the Administrator shall so notify the State, and if appropriate corrective action is not taken within a reasonable time, not to exceed ninety days, the Administrator shall withdraw approval of such program. The Administrator shall not withdraw approval of any such program unless he shall first have notified the State, and made public, in writing, the reasons for such withdrawal.

(ii) In the case of a State with a program submitted and approved under this paragraph, the Administrator shall withdraw approval of such program under this subparagraph only for a substantial failure of the State to administer its program in accordance with the requirements of this paragraph.

(c)(1) The Governor of each State, in consultation with the planning agency designated under subsection (a) of this section, at the time a plan is submitted to the Administrator, shall designate one or more waste treatment management agencies (which may be an existing or newly created local, regional or State agency or potential subdivision) for each area designated under subsection (a) of this section and submit such designations to the Administrator.

(2) The Administrator shall accept any such designation, unless, within 120 days of such designation, he finds that the designated management agency (or agencies) does not have adequate authority—

(A) to carry out appropriate portions of an areawide waste treatment management plan developed under subsection (b) of this section;

(B) to manage effectively waste treatment works and related facilities serving such area in conformance with any plan required by subsection (b) of this section;

(C) directly or by contract, to design and construct new works, and to operate and maintain new and existing works as required by any plan developed pursuant to subsection (b) of this section;

(D) to accept and utilize grants, or other funds from any source, for waste treatment management purposes;

(E) to raise revenues, including the assessment of waste treatment charges;

(F) to incur short- and long-term indebtedness;

(G) to assure in implementation of an areawide waste treatment management plan that each participating community pays its proportionate share of treatment costs;

(H) to refuse to receive any wastes from any municipality or subdivision thereof, which does not comply with any provisions of an approved plan under this section applicable to such area; and

(I) to accept for treatment industrial wastes.

(d) After a waste treatment management agency having the authority required by subsection (c) has been designated under such subsection for an area and a plan for such area has been approved under subsection (b) of this section, the Administrator shall not make any grant for construction of a publicy owned treatment works under section 201(g)(1) within such area except to such designated agency and for works in conformity with such plan.

(e) No permit under section 402 of this Act shall be issued for any point source which is in conflict with a plan approved pursuant to subsection (b) of this section.

(f)(1) The Administrator shall make grants to any agency designated under subsection (a) of this section for payment of the reasonable costs of developing and operating a continuing areawide waste treatment management planning process under subsection (b) of this section.

(2) For the two-year period beginning on the date of the first grant is made under paragraph (1) of this subsection to an agency, if such first grant is made before October 1, 1977, the amount of each such grant to such agency shall be 100 per centum of the costs of developing and operating a continuing areawide waste treatment management planning process under subsection (b) of this section, and thereafter the amount granted to such agency shall not exceed 75 per centum of such costs in each succeeding one-year period. In the case of any other grant made to an agency under such paragraph (1) of this subsection, the amount of such grant shall not exceed 75 per centum of the costs of developing and operating a continuing areawide waste treatment management planning process in any year.

(3) Each applicant for a grant under this subsection shall submit to the Administrator for his approval each proposal for which a grant is applied for under this subsection. The Administrator shall act upon such proposal as soon as practicable after it has been submitted, and his approval of that proposal shall be deemed a contractual obligation of the United States for the payment of its contribution to such proposal, subject to such amounts as are provided in appropriation Acts. There is authorized to be appropriated to carry out this subsection not to exceed $50,000,000 for the fiscal year ending June 30, 1973, not to exceed $100,000,000 for the fiscal year ending June 30, 1974, not to exceed $150,000,000 per fiscal year for the fiscal years ending June 30, 1975, September 30, 1977, September 30, 1978, September 30, 1979, and September 30, 1980, not to exceed $100,000,000 per fiscal year for the fiscal years ending September 30, 1981, and September 30, 1982, and such sums as may be necessary for fiscal years 1983 through 1990.

(g) The Administrator is authorized, upon request of the Governor or the designated planning agency, and without reimbursement, to consult with, and provide technical assistance to, any agency designated under subsection (a) of this section in the development of areawide waste treatment management plans under subsection (b) of this section.

(h)(1) The Secretary of the Army, acting through the Chief of Engineers, in cooperation with the Administrator is authorized and directed, upon request of the Governor or the designated planning organization, to consult with, and provide technical assistance to, any agency designed[1] under subsection (a) of this section in devel-

[1] So in original. Probably should be "designated".

oping and operating a continuing areawide waste treatment management planning process under subsection (b) of this section.

(2) There is authorized to be appropriated to the Secretary of the Army, to carry out this subsection, not to exceed $50,000,000 per fiscal year for the fiscal years ending June 30, 1973, and June 30, 1974.

(i)(1) The Secretary of the Interior, acting through the Director of the United States Fish and Wildlife Service, shall, upon request of the Governor of a State, and without reimbursement, provide technical assistance to such State in developing a statewide program for submission to the Administrator under subsection (b)(4)(B) of this section and in implementing such program after its approval.

(2) There is authorized to be appropriated to the Secretary of the Interior $6,000,000 to complete the National Wetlands Inventory of the United States, by December 31, 1981, and to provide information from such Inventory to States as it becomes available to assist such States in the development and operation of programs under this Act.

(j)(1) The Secretary of Agriculture, with the concurrence of the Administrator, and acting through the Soil Conservation Service and such other agencies of the Department of Agriculture as the Secretary may designate, is authorized and directed to establish and administer a program to enter into contracts, subject to such amounts as are provided in advance by appropriation acts, of not less than five years nor more than ten years with owners and operators having control of rural land for the purpose of installing and maintaining measures incorporating best management practices to control nonpoint source pollution for improved water quality in those States or areas for which the Administrator has approved a plan under subsection (b) of this section where the practices to which the contracts apply are certified by the management agency designated under subsection (c)(1) of this section to be consistent with such plans and will result in improved water quality. Such contracts may be entered into during the period ending not later than September 31, 1988. Under such contracts the land owners or operator shall agree—

(i) to effectuate a plan approved by a soil conservation district, where one exists, under this section for his farm, ranch, or other land substantially in accordance with the schedule outlined therein unless any requirement thereof is waived or modified by the Secretary;

(ii) to forfeit all rights to further payments or grants under the contract and refund to the United States all payments and grants received thereunder, with interest, upon his violation of the contract at any stage during the time he has control of the land if the Secretary, after considering the recommendations of the soil conservation district, where one exists, and the Administrator, determines that such violation is of such a nature as to warrant termination of the contract, or to make refunds or accept such payment adjustments as the Secretary may deem appropriate if he determines that the violation by the owner or operator does not warrant termination of the contract;

(iii) upon transfer of his right and interest in the farm, ranch, or other land during the contract period to forfeit all rights to further payments or grants under the contract and refund to the United States all payments or grants received thereunder, with interest, unless the transferee of any such land agrees with the Secretary to assume all obligations of the contract;

(iv) not to adopt any practice specified by the Secretary on the advice of the Administrator in the contract as a practice which would tend to defeat the purposes of the contract;

(v) to such additional provisions as the Secretary determines are desirable and includes in the contract to effectuate the purposes of the program or to facilitate the practical administration of the program.

(2) In return for such agreement by the landowner or operator the Secretary shall agree to provide technical assistance and share the cost of carrying out those conservation practices and measures set forth in the contract for which he determines that cost sharing is appropriate and in the public interest and which are approved for cost sharing by the agency designated to implement the plan developed under subsection (b) of this section. The portion of such cost (including labor) to be shared shall be that part which the Secretary determines is necessary and appropriate to effectuate the installation of the water quality management practices and measures under the contract, but not to exceed 50 per centum of the total cost of the measures set forth in the contract; except the Secretary may increase the matching cost share where he determines that (1) the main benefits to be derived from the measures are related to improving offsite water quality, and (2) the matching share requirement would place a burden on the landowner which would probably prevent him from participating in the program.

(3) The Secretary may terminate any contract with a landowner or operator by mutual agreement with the owner or operator if the Secretary determines that such termination would be in the public interest, and may agree to such modification of contracts previously entered into as he may determine to be desirable to carry out the purposes of the program or facilitate the practical administration thereof or to accomplish equitable treatment with respect to other conservation, land use, or water quality programs.

(4) In providing assistance under this subsection the Secretary will give priority to those areas and sources that have the most significant effect upon water quality. Additional investigations or plans may be made, where necessary, to supplement approved water quality management plans, in order to determine priorities.

(5) The Secretary shall, where practicable, enter into agreements with soil conservation districts, State soil and water conservation agencies, or State water quality agencies to administer all or part of the program established in this subsection under regulations developed by the Secretary. Such agreements shall provide for the submission of such reports as the Secretary deems necessary, and for payment by the United States of such portion of the costs incurred in the administration of the program as the Secretary may deem appropriate.

(6) The contracts under this subsection shall be entered into only in areas where the management agency designated under subsection (c)(1) of this section assures an adequate level of participation by owners and operators having control of rural land in such areas. Within such areas the local soil conservation district, where one exists, together with the Secretary of Agriculture, will determine the priority of assistance among individual land owners and operators to assure that the most critical water quality problems are addressed.

(7) The Secretary, in consultation with the Administrator and subject to section 304(k) of this Act, shall, not later than September 30, 1978, promulgate regulations for carrying out this subsection and for support and cooperation with other Federal and non-Federal agencies for implementation of this subsection.

(8) This program shall not be used to authorize or finance projects that would otherwise be eligible for assistance under the terms of Public Law 83–566.

(9) There are hereby authorized to be appropriated to the Secretary of Agriculture $200,000,000 for fiscal year 1979, $400,000,000 for fiscal year 1980, $100,000,000 for fiscal year 1981, $100,000,000 for fiscal year 1982, and such sums as may be necessary for fiscal years 1983 through 1990, to carry out this subsection. The program authorized under this subsection shall be in addition to, and not in substitution of, other programs in such area authorized by this or any other public law.

(33 U.S.C. 1288)

BASIN PLANNING

SEC. 209. (a) The President, acting through the Water Resources Council, shall, as soon as practicable, prepare a Level B plan under the Water Resource Planning Act for all basins in the United States. All such plans shall be completed not later than January 1, 1980, except that priority in the preparation of such plans shall be given to those basins and portions thereof which are within those areas designated under paragraphs (2), (3), and (4) of subsection (a) of section 208 of this Act.

(b) The President, acting through the Water Resources Council, shall report annually to Congress on progress being made in carrying out this section. The first such report shall be submitted not later than January 31, 1973.

(c) There is authorized to be appropriated to carry out this section not to exceed $200,000,000.

(33 U.S.C. 1289)

ANNUAL SURVEY

SEC. 210. The Administrator shall annually make a survey to determine the efficiency of the operation and maintenance of treatment works constructed with grants made under this Act, as compared to the efficiency planned at the time the grant was made. The results of such annual survey shall be reported to Congress not later than 90 days after the date of convening of each session of Congress.

(33 U.S.C. 1290)

SEWAGE COLLECTION SYSTEMS

SEC. 211. (a) No grant shall be made for a sewage collection system under this title unless such grant (1) is for replacement or major rehabilitation of an existing collection system and is necessary to the total integrity and performance of the waste treatment works serving such community, or (2) is for a new collection system in an existing community with sufficient existing or planned capacity adequately to treat such collected sewage and is consistent with section 201 of this Act.

(b) If the Administrator uses population density as a test for determining the eligibility of a collector sewer for assistance it shall be only for the purpose of evaluating alternatives and determining the needs for such system in relation to ground or surface water quality impact.

(c) No grant shall be made under this title from funds authorized for any fiscal year during the period beginning October 1, 1977, and ending September 30, 1990, for treatment works for control of pollutant discharges from separate storm sewer systems.

(33 U.S.C. 1291)

DEFINITIONS

SEC. 212. As used in this title—

(1) The term "construction" means any one or more of the following: preliminary planning to determine the feasibility of treatment works, engineering, architectural, legal, fiscal, or economic investigations or studies, surveys, designs, plans, working drawings, specifications, procedures, field testing of innovative or alternative waste water treatment processes and techniques meeting guidelines promulgated under section 304(d)(3) of this Act, or other necessary actions, erection, building, acquisition, alteration, remodeling, improvement, or extension of treatment works, or the inspection or supervision of any of the foregoing items.

(2)(A) The term "treatment works" means any devices and systems used in the storage, treatment, recycling, and reclamation of municipal sewage or industrial wastes of a liquid nature to implement section 201 of this act, or necessary to recycle or reuse water at the most economical cost over the estimated life of the works, including intercepting sewers, outfall sewers, sewage collection systems, pumping, power, and other equipment, and their appurtenances; extensions, improvements, remodeling, additions, and alterations thereof; elements essential to provide a reliable recycled supply such as standby treatment units and clear well facilities; and any works, including site acquisition of the land that will be an integral part of the treatment process (including land use for the storage of treated wastewater in land treatment systems prior to land application) or is used for ultimate disposal of residues resulting from such treatment.

(B) In addition to the definition contained in subparagraph (A) of this paragraph, "treatment works" means any other method or system for preventing, abating, reducing, storing, treating, separating, or disposing of municipal waste, including storm water runoff, or industrial waste, including waste in combined storm water and sanitary sewer systems. Any application for construction grants which includes wholly or in part such methods or systems shall, in accordance with guidelines published by the Administrator pursuant to subparagraph (C) of this paragraph, contain adequate data and analysis demonstrating such proposal to be, over the life of such works, the most cost efficient alternative to comply with sections 301 or 302 of this Act, or the requirements of section 201 of this Act.

(C) For the purposes of subparagraph (B) of this paragraph, the Administrator shall, within one hundred and eighty days after the date of enactment of this title, publish and thereafter revise no less often than annually, guidelines for the evaluation of methods, including cost-effective analysis, described in subparagraph (B) of this paragraph.

(3) The term "replacement" as used in this title means those expenditures for obtaining and installing equipment, accessories, or appurtenances during the useful life of the treatment works necessary to maintain the capacity and performance for which such works are designed and constructed.

(33 U.S.C. 1292)

LOAN GUARANTEES FOR CONSTRUCTION OF TREATMENT WORKS

SEC. 213. (a) Subject to the conditions of this section and to such terms and conditions as the Administrator determines to be necessary to carry out the purposes of this title, the Administrator is authorized to guarantee, and to make commitments to guarantee, the principal and interest (including interest accruing between the date of default and the date of the payment in full of the guarantee) of any loan, obligation, or participation therein of any State, municipality, or intermunicipal or interstate agency issued directly and exclusively to the Federal Financing Bank to finance that part of the cost of any grant-eligible project for the construction of publicly owned treatment works not paid for with Federal financial assistance under this title (other than this section), which project the Administrator has determined to be eligible for such financial assistance under this title, including, but not limited to, projects eligible for reimbursement under section 206 of this title.

(b) No guarantee, or commitment to make a guarantee, may be made pursuant to this section—

(1) unless the Administrator certifies that the issuing body is unable to obtain on reasonable terms sufficient credit to finance its actual needs without such guarantee; and

(2) unless the Administrator determines that there is a reasonable assurance or repayment of the loan, obligation, or participation therein.

A determination of whether financing is available at reasonable rates shall be made by the Secretary of the Treasury with relationship to the current average yield on outstanding marketable obligations of municipalities of comparable maturity.

(c) The Administrator is authorized to charge reasonable fees for the investigation of an application for a guarantee and for the issuance of a commitment to make a guarantee.

(d) The Administrator, in determining whether there is a reasonable assurance of repayment, may require a commitment which would apply to such repayment. Such commitment may include, but not be limited to, any funds received by such grantee from the amounts appropriated under section 206 of this Act.

(33 U.S.C. 1293)

PUBLIC INFORMATION

SEC. 214. The Administrator shall develop and operate within one year of the date of enactment of this section, a continuing program of public information and education on recycling and reuse of wastewater (including sludge), the use of land treatment, and methods for the reduction of wastewater volume.

(33 U.S.C. 1294)

REQUIREMENTS FOR AMERICAN MATERIALS

SEC. 215. Notwithstanding any other provision of law, no grant for which application is made after February 1, 1978, shall be made under this title for any treatment works unless only such unmanufactured articles, materials, and supplies as have been mined or produced in the United States, and only such manufactured articles, materials, and supplies as have been manufactured in the United States, substantially all from articles, materials, or supplies mined, produced, or manufactured, as the case may be, in the United States will be used in such treatment works. This section shall not apply in any case where the Administrator determines, based upon those factors the Administrator deems relevant, including the available resources of the agency, it to be inconsistent with the public interest (including multilateral government procurement agreements) or the cost to be unreasonable, or if articles, materials, or supplies of the class or kind to be used or the articles, materials, or supplies from which they are manufactured are not mined, produced, or manufactured, as the case may be, in the United States in sufficient and reasonably available commercial quantities and of a satisfactory quality.

(33 U.S.C. 1295)

DETERMINATION OF PRIORITY

SEC. 216. Notwithstanding any other provision of this Act, the determination of the priority to be given each category of projects for construction of publicly owned treatment works within each State shall be made solely by that State, except that if the Administrator, after a public hearing, determines that a specific project will not result in compliance with the enforceable requirements of this Act, such project shall be removed from the State's priority list and such State shall submit a revised priority list. These categories shall include, but not be limited to (A) secondary treatment, (B) more stringent treatment, (C) infiltration-in-flow correction, (D) major sewer system rehabilitation, (E) new collector sewers and appurtenances, (F) new interceptors and appurtenances, and (G) correction of combined sewer overflows. Not less than 25 per centum of funds allocated to a State in any fiscal year under this title for construction of publicly owned treatment works in such State shall be obligated for those types of projects referred to in clauses (D), (E), (F), and (G) of this section, if such projects are on such State's priority list for that year and are otherwise eligible for funding in that fiscal year. It is the policy of Congress that projects for wastewater treatment and management undertaken with Federal financial assistance under this Act by any State, municipality, or intermunicipal or interstate agency shall be projects which, in the estimation of the State, are designed to achieve optimum water quality management, consistent with the public health and water quality goals and requirements of the Act.

(33 U.S.C. 1296)

COST-EFFECTIVENESS GUIDELINES

SEC. 217. Any guidelines for cost-effectiveness analysis published by the Administrator under this title shall provide for the identification and selection of cost effective alternatives to comply with the objective and goals of this Act and sections 201(b), 201(d), 201(g)(2)(A), and 301(b)(2)(B) of this Act.

(33 U.S.C. 1297)

COST EFFECTIVENESS

SEC. 218. (a) It is the policy of Congress that a project for waste treatment and management undertaken with Federal financial assistance under this Act by any State, municipality, or intermunicipal or interstate agency shall be considered as an overall waste treatment system for waste treatment and management, and shall be that system which constitutes the most economical and cost-effective combination of devices and systems used in the storage, treatment, recycling, and reclamation of municipal sewage or industrial wastes of a liquid nature to implement section 201 of this Act, or necessary to recycle or reuse water at the most economical cost over the estimated life of the works, including intercepting sewers, outfall sewers, sewage collection systems, pumping power, and other equipment, and their appurtenances; extension, improvements, remodeling, additions, and alterations thereof; elements essential to provide a reliable recycled supply such as standby treatment units and clear well facilities; and any works, including site acquisition of the land that will be an integral part of the treatment process (including land use for the storage of treated wastewater in land treatment systems prior to land application) or which is used for ultimate disposal of residues resulting from such treatment; water efficiency measures and devices; and any other method or system for preventing, abating, reducing, storing, treating, separating, or disposing of municipal waste, including storm water runoff, or industrial waste, including waste in combined storm water and sanitary sewer systems; to meet the requirements of this Act.

(b) In accordance with the policy set forth in subsection (a) of this section, before the Administrator approves any grant to any State, municipality, or intermunicipal or interstate agency for the erection, building, acquisition, alteration, remodeling, improvement, or extension of any treatment works the Administrator shall determine that the facilities plan of which such treatment works are a part constitutes the most economical and cost-effective combination of treatment works over the life of the project to meet the requirements of this Act, including, but not limited to, consideration of construction costs, operation, maintenance, and replacement costs.

(c) In furtherance of the policy set forth in subsection (a) of this section, the Administrator shall require value engineering review in connection with any treatment works, prior to approval of any grant for the erection, building, acquisition, alteration, remodeling, improvement, or extension of such treatment works, in any case in which the cost of such erection, building, acquisition, alteration, remodeling, improvement, or extension is projected to be in excess of $10,000,000. For purposes of this subsection, the term "value engineering review" means a specialized cost control technique which uses a systematic and creative approach to identify and to focus on unnecessarily high cost in a project in order to arrive at a cost saving without sacrificing the reliability or efficiency of the project.

(d) This section applies to projects for waste treatment and management for which no treatment works including a facilities plan for such project have received Federal financial assistance for the preparation of construction plans and specifications under this Act before the date of enactment of this section.

(33 U.S.C. 1298)

STATE CERTIFICATION OF PROJECTS

SEC. 219. Whenever the Governor of a State which has been delegated sufficient authority to administer the construction grant program under this title in that State certifies to the Administrator that a grant application meets applicable requirements of Federal and State law for assistance under this title, the Administrator shall approve or disapprove such application within 45 days of the date of receipt of such application. If the Administrator does not approve or disapprove such application within 45 days of receipt, the application shall be deemed approved. If the Administrator disapproves such application the Administrator shall state in writing the reasons for such disapproval. Any grant approved or deemed approved under this section shall be subject to amounts provided in appropriation Acts.

(33 U.S.C. 1299)

SEC. 220. PILOT PROGRAM FOR ALTERNATIVE WATER SOURCE PROJECTS.

(a) POLICY.—Nothing in this section shall be construed to affect the application of section 101(g) of this Act and all of the provisions of this section shall be carried out in accordance with the provisions of section 101(g).

(b) IN GENERAL.—The Administrator may establish a pilot program to make grants to State, interstate, and intrastate water resource development agencies (including water management districts and water supply authorities), local government agencies, private utilities, and nonprofit entities for alternative water source projects to meet critical water supply needs.

(c) ELIGIBLE ENTITY.—The Administrator may make grants under this section to an entity only if the entity has authority under State law to develop or provide water for municipal, industrial, and agricultural uses in an area of the State that is experiencing critical water supply needs.

(d) SELECTION OF PROJECTS.—

(1) LIMITATION.—A project that has received funds under the reclamation and reuse program conducted under the Reclamation Projects Authorization and Adjustment Act of 1992 (43 U.S.C. 390h et seq.) shall not be eligible for grant assistance under this section.

(2) ADDITIONAL CONSIDERATION.—In making grants under this section, the Administrator shall consider whether the project is located within the boundaries of a State or area referred to in section 1 of the Reclamation Act of June 17, 1902 (32 Stat. 385), and within the geographic scope of the reclamation and reuse program conducted under the Reclamation Projects Authorization and Adjustment Act of 1992 (43 U.S.C. 390h et seq.).

(3) GEOGRAPHICAL DISTRIBUTION.—Alternative water source projects selected by the Administrator under this section shall reflect a variety of geographical and environmental conditions.

(e) COMMITTEE RESOLUTION PROCEDURE.—

(1) IN GENERAL.—No appropriation shall be made for any alternative water source project under this section, the total Federal cost of which exceeds $3,000,000, if such project has not been approved by a resolution adopted by the Committee on Transportation and Infrastructure of the House of Representatives or the Committee on Environment and Public Works of the Senate.

(2) REQUIREMENTS FOR SECURING CONSIDERATION.—For purposes of securing consideration of approval under paragraph (1), the Administrator shall provide to a committee referred to in paragraph (1) such information as the committee requests and the non-Federal sponsor shall provide to the committee information on the costs and relative needs for the alternative water source project.

(f) USES OF GRANTS.—Amounts from grants received under this section may be used for engineering, design, construction, and final testing of alternative water source projects designed to meet critical water supply needs. Such amounts may not be used for planning, feasibility studies or for operation, maintenance, replacement, repair, or rehabilitation.

(g) COST SHARING.—The Federal share of the eligible costs of an alternative water source project carried out using assistance made available under this section shall not exceed 50 percent.

(h) REPORTS.—On or before September 30, 2004, the Administrator shall transmit to Congress a report on the results of the pilot program established under this section, including progress made toward meeting the critical water supply needs of the participants in the pilot program.

(i) DEFINITIONS.—In this section, the following definitions apply:

(1) ALTERNATIVE WATER SOURCE PROJECT.—The term "alternative water source project" means a project designed to

provide municipal, industrial, and agricultural water supplies in an environmentally sustainable manner by conserving, managing, reclaiming, or reusing water or wastewater or by treating wastewater. Such term does not include water treatment or distribution facilities.

(2) CRITICAL WATER SUPPLY NEEDS.—The term "critical water supply needs" means existing or reasonably anticipated future water supply needs that cannot be met by existing water supplies, as identified in a comprehensive statewide or regional water supply plan or assessment projected over a planning period of at least 20 years.

(j) AUTHORIZATION OF APPROPRIATIONS.—There is authorized to be appropriated to carry out this section a total of $75,000,000 for fiscal years 2002 through 2004. Such sums shall remain available until expended.

(33 U.S.C. 1300)

SEC. 221. SEWER OVERFLOW CONTROL GRANTS.

(a) IN GENERAL.—In any fiscal year in which the Administrator has available for obligation at least $1,350,000,000 for the purposes of section 601—

(1) the Administrator may make grants to States for the purpose of providing grants to a municipality or municipal entity for planning, design, and construction of treatment works to intercept, transport, control, or treat municipal combined sewer overflows and sanitary sewer overflows; and

(2) subject to subsection (g), the Administrator may make a direct grant to a municipality or municipal entity for the purposes described in paragraph (1).

(b) PRIORITIZATION.—In selecting from among municipalities applying for grants under subsection (a), a State or the Administrator shall give priority to an applicant that—

(1) is a municipality that is a financially distressed community under subsection (c);

(2) has implemented or is complying with an implementation schedule for the nine minimum controls specified in the CSO control policy referred to in section 402(q)(1) and has begun implementing a long-term municipal combined sewer overflow control plan or a separate sanitary sewer overflow control plan;

(3) is requesting a grant for a project that is on a State's intended use plan pursuant to section 606(c); or

(4) is an Alaska Native Village.

(c) FINANCIALLY DISTRESSED COMMUNITY.—

(1) DEFINITION.—In subsection (b), the term "financially distressed community" means a community that meets affordability criteria established by the State in which the community is located, if such criteria are developed after public review and comment.

(2) CONSIDERATION OF IMPACT ON WATER AND SEWER RATES.—In determining if a community is a distressed community for the purposes of subsection (b), the State shall consider, among other factors, the extent to which the rate of growth of a community's tax base has been historically slow such that

implementing a plan described in subsection (b)(2) would result in a significant increase in any water or sewer rate charged by the community's publicly owned wastewater treatment facility.

(3) INFORMATION TO ASSIST STATES.—The Administrator may publish information to assist States in establishing affordability criteria under paragraph (1).

(d) COST-SHARING.—The Federal share of the cost of activities carried out using amounts from a grant made under subsection (a) shall be not less than 55 percent of the cost. The non-Federal share of the cost may include, in any amount, public and private funds and in-kind services, and may include, notwithstanding section 603(h), financial assistance, including loans, from a State water pollution control revolving fund.

(e) ADMINISTRATIVE REPORTING REQUIREMENTS.—If a project receives grant assistance under subsection (a) and loan assistance from a State water pollution control revolving fund and the loan assistance is for 15 percent or more of the cost of the project, the project may be administered in accordance with State water pollution control revolving fund administrative reporting requirements for the purposes of streamlining such requirements.

(f) AUTHORIZATION OF APPROPRIATIONS.—There is authorized to be appropriated to carry out this section $750,000,000 for each of fiscal years 2002 and 2003. Such sums shall remain available until expended.

(g) ALLOCATION OF FUNDS.—

(1) FISCAL YEAR 2002.—Subject to subsection (h), the Administrator shall use the amounts appropriated to carry out this section for fiscal year 2002 for making grants to municipalities and municipal entities under subsection (a)(2), in accordance with the criteria set forth in subsection (b).

(2) FISCAL YEAR 2003.—Subject to subsection (h), the Administrator shall use the amounts appropriated to carry out this section for fiscal year 2003 as follows:

(A) Not to exceed $250,000,000 for making grants to municipalities and municipal entities under subsection (a)(2), in accordance with the criteria set forth in subsection (b).

(B) All remaining amounts for making grants to States under subsection (a)(1), in accordance with a formula to be established by the Administrator, after providing notice and an opportunity for public comment, that allocates to each State a proportional share of such amounts based on the total needs of the State for municipal combined sewer overflow controls and sanitary sewer overflow controls identified in the most recent survey conducted pursuant to section 516(b)(1).

(h) ADMINISTRATIVE EXPENSES.—Of the amounts appropriated to carry out this section for each fiscal year—

(1) the Administrator may retain an amount not to exceed 1 percent for the reasonable and necessary costs of administering this section; and

(2) the Administrator, or a State, may retain an amount not to exceed 4 percent of any grant made to a municipality or

municipal entity under subsection (a), for the reasonable and necessary costs of administering the grant.

(i) REPORTS.—Not later than December 31, 2003, and periodically thereafter, the Administrator shall transmit to Congress a report containing recommended funding levels for grants under this section. The recommended funding levels shall be sufficient to ensure the continued expeditious implementation of municipal combined sewer overflow and sanitary sewer overflow controls nationwide.

(33 U.S.C. 1301)

TITLE III—STANDARDS AND ENFORCEMENT

EFFLUENT LIMITATIONS

SEC. 301. (a) Except as in compliance with this section and sections 302, 306, 307, 318, 402, and 404 of this Act, the discharge of any pollutant by any person shall be unlawful.

(b) In order to carry out the objective of this Act there shall be achieved—

(1)(A) not later than July 1, 1977, effluent limitations for point sources, other than publicly owned treatment works, (i) which shall require the application of the best practicable control technology currently available as defined by the Administrator pursuant to section 304(b) of this Act, or (ii) in the case of a discharge into a publicly owned treatment works which meets the requirements of subparagraph (B) of this paragraph, which shall require compliance with any applicable pretreatment requirements and any requirements under section 307 of this Act; and

(B) for publicly owned treatment works in existence on July 1, 1977, or approved pursuant to section 203 of this Act prior to June 30, 1974 (for which construction must be completed within four years of approval), effluent limitations based upon secondary treatment as defined by the Administrator pursuant to section 304(d)(1) of this Act; or,

(C) not later than July 1, 1977, any more stringent limitation, including those necessary to meet water quality standards, treatment standards, or schedule of compliance, established pursuant to any State law or regulations, (under authority preserved by section 510) or any other Federal law or regulation, or required to implement any applicable water quality standard established pursuant to this Act.

(2)(A) for pollutants identified in subparagraphs (C), (D), and (F) of this paragraph, effluent limitations for categories and classes of point sources, other than publicly owned treatment works, which (i) shall require application of the best available technology economically achievable for such category or class, which will result in reasonable further progress toward the national goal of eliminating the discharge of all pollutants, as determined in accordance with regulations issued by the Administrator pursuant to section 304(b)(2) of this Act, which such effluent limitations shall require the elimination of discharges of all pollutants if the Administrator finds, on the basis of information available to him (including information developed pursuant to section 315), that such elimination is technologically and economically achievable for category or class of point sources as determined in accordance with regulations issued by the Administrator pursuant to section 304(b)(2) of this Act, or (ii) in the case of the introduction of a pollutant into a publicly owned treatment works which meets the requirements of subparagraph (B) of this paragraph, shall require compliance with any applicable pretreatment requirements and any other requirement under section 307 of this Act;

[(B) subparagraph (B) repealed by section 21(b) of P.L. 97–117.]

(C) with respect to all toxic pollutants referred to in table 1 of Committee Print Numbered 95–30 of the Committee on Public Works and Transportation of the House of Representatives compliance with effluent limitations in accordance with subparagraph (A) of this paragraph as expeditiously as practicable but in no case later than three years after the date such limitations are promulgated under section 304(b), and in no case later than March 31, 1989;

(D) for all toxic pollutants listed under paragraph (1) of subsection (a) of section 307 of this Act which are not referred to in subparagraph (C) of this paragraph compliance with effluent limitation in accordance with subparagraph (A) of this paragraph as expeditiously as practicable, but in no case later than three years after the date such limitations are promulgated under section 304(b), and in no case later than March 31, 1989;

(E) as expeditiously as practicable but in no case later than three years after the date such limitations are promulgated under section 304(b), and in no case later than March 31, 1989, compliance with effluent limitations for categories and classes of point sources, other than publicly owned treatment works, which in the case of pollutants identified pursuant to section 304(a)(4) of this Act shall require application of the best conventional pollutant control technology as determined in accordance with regulations issued by the Administrator pursuant to section 304(b)(4) of this Act; and

(F) for all pollutants (other than those subject to subparagraphs (C), (D), or (E) of this paragraph) compliance with effluent limitations in accordance with subparagraph (A) of this paragraph as expeditiously as practicable but in no case later than 3 years after the date such limitations are established, and in no case later than March 31, 1989.

(3)(A) for effluent limitations under paragraph (1)(A)(i) of this subsection promulgated after January 1, 1982, and requiring a level of control substantially greater or based on fundamentally different control technology than under permits for an industrial category issued before such date, compliance as expeditiously as practicable but in no case later than three years after the date such limitations are promulgated under section 304(b), and in no case later than March 31, 1989; and

(B) for any effluent limitation in accordance with paragraph (1)(A)(i), (2)(A)(i), or (2)(E) of this subsection established only on the basis of section 402(a)(1) in a permit issued after enactment of the Water Quality Act of 1987, compliance as expeditiously as practicable but in no case later than three years after the date such limitations are established, and in no case later than March 31, 1989.

(c) The Administrator may modify the requirements of subsection (b)(2)(A) of this section with respect to any point source for which a permit application is filed after July 1, 1977, upon a showing by the owner or operator of such point source satisfactory to the Administrator that such modified requirements (1) will represent the maximum use of technology within the economic capability of the owner or operator; and (2) will result in reasonable further progress toward the elimination of the discharge of pollutants.

(d) Any effluent limitation required by paragraph (2) of subsection (b) of this section shall be reviewed at least every five years and, if appropriate, revised pursuant to the procedure established under such paragraph.

(e) Effluent limitations established pursuant to this section or section 302 of this Act shall be applied to all point sources of discharge of pollutants in accordance with the provisions of this Act.

(f) Notwithstanding any other provisions of this Act it shall be unlawful to discharge any radiological, chemical, or biological warfare agent, any high-level radioactive waste, or any medical waste, into the navigable waters.

(g) MODIFICATIONS FOR CERTAIN NONCONVENTIONAL POLLUTANTS.—

(1) GENERAL AUTHORITY.—The Administrator, with the concurrence of the State, may modify the requirements of subsection (b)(2)(A) of this section with respect to the discharge from any point source of ammonia, chlorine, color, iron, and total phenols (4AAP) (when determined by the Administrator to be a pollutant covered by subsection (b)(2)(F)) and any other pollutant which the Administrator lists under paragraph (4) of this subsection.

(2) REQUIREMENTS FOR GRANTING MODIFICATIONS.—A modification under this subsection shall be granted only upon a showing by the owner or operator of a point source satisfactory to the Administrator that—

(A) such modified requirements will result at a minimum in compliance with the requirements of subsection (b)(1)(A) or (C) of this section, whichever is applicable;

(B) such modified requirements will not result in any additional requirements on any other point or nonpoint source; and

(C) such modification will not interfere with the attainment or maintenance of that water quality which shall assure protection of public water supplies, and the protection and propagation of a balanced population of shellfish, fish, and wildlife, and allow recreational activities, in and on the water and such modification will not result in the discharge of pollutants in quantities which may reasonably be anticipated to pose an unacceptable risk to human health or the environment because of bioaccumulation, persistency in the environment, acute toxicity, chronic toxicity (including carcinogenicity, mutagenicity or teratogenicity), or synergistic propensities.

(3) LIMITATION ON AUTHORITY TO APPLY FOR SUBSECTION (c) MODIFICATION.—If an owner or operator of a point source applies for a modification under this subsection with respect to the discharge of any pollutant, such owner or operator shall be eligible to apply for modification under subsection (c) of this section with respect to such pollutant only during the same time-period as he is eligible to apply for a modification under this subsection.

(4) PROCEDURES FOR LISTING ADDITIONAL POLLUTANTS.—

(A) GENERAL AUTHORITY.—Upon petition of any person, the Administrator may add any pollutant to the list of pollutants for which modification under this section is authorized (except for pollutants identified pursuant to section 304(a)(4) of this Act, toxic pollutants subject to section 307(a) of this Act, and the thermal component of discharges) in accordance with the provisions of this paragraph.

(B) REQUIREMENTS FOR LISTING.—

(i) SUFFICIENT INFORMATION.—The person petitioning for listing of an additional pollutant under this subsection shall submit to the Administrator sufficient information to make the determinations required by this subparagraph.

(ii) TOXIC CRITERIA DETERMINATION.—The Administrator shall determine whether or not the pollutant meets the criteria for listing as a toxic pollutant under section 307(a) of this Act.

(iii) LISTING AS TOXIC POLLUTANT.—If the Administrator determines that the pollutant meets the criteria for listing as a toxic pollutant under section 307(a), the Administrator shall list the pollutant as a toxic pollutant under section 307(a).

(iv) NONCONVENTIONAL CRITERIA DETERMINATION.—If the Administrator determines that the pollutant does not meet the criteria for lising as a toxic pollutant under such section and determines that adequate test methods and sufficient data are available to make the determinations required by paragraph (2) of this subsection with respect to the pollutant, the Administrator shall add the pollutant to the list of pollutants specified in paragraph (1) of this subsection for which modifications are authorized under this subsection.

(C) REQUIREMENTS FOR FILING OF PETITIONS.—A petition for lising of a pollutant under this paragraph—

(i) must be filed not later than 270 days after the date of promulgation of an applicable effluent guideline under section 304;

(ii) may be filed before promulgation of such guideline; and

(iii) may be filed with an application for a modification under paragraph (1) with respect to the discharge of such pollutant.

(D) DEADLINE FOR APPROVAL OF PETITION.—A decision to add a pollutant to the list of pollutants for which modifications under this subsection are authorized must be made within 270 days after the date of promulgation of an applicable effluent guideline under section 304.

(E) BURDEN OF PROOF.—The burden of proof for making the determinations under subparagraph (B) shall be on the petitioner.

(5) REMOVAL OF POLLUTANTS.—The Administrator may remove any pollutant from the list of pollutants for which modifications are authorized under this subsection if the Administrator determines that adequate test methods and sufficient data are no longer available for determining whether or not modifications may be granted with respect to such pollutant under paragraph (2) of this subsection.

(h) The Administrator, with the concurrence of the State, may issue a permit under section 402 which modifies the requirements of subsection (b)(1)(B) of this section with respect to the discharge of any pollutant from a publicly owned treatment works into marine waters, if the applicant demonstrates to the satisfaction of the Administrator that—

(1) there is an applicable water quality standard specific to the pollutant for which the modification is requested, which has been identified under section 304(a)(6) of this Act;

(2) the discharge of pollutants in accordance with such modified requirements will not interfere, alone or in combination with pollutants from other sources, with the attainment or maintenance of that water quality which assures protection of public water supplies and the protection and propagation of a balanced, indigenous population of shellfish, fish and wildlife, and allows recreational activities, in and on the water;

(3) the applicant has established a system for monitoring the impact of such discharge on a representative sample of aquatic biota, to the extent practicable, and the scope of such monitoring is limited to include only those scientific investigations which are necessary to study the effects of the proposed discharge;

(4) such modified requirements will not result in any additional requirements on any other point or nonpoint source;

(5) all applicable pretreatment requirements for sources introducing waste into such treatment works will be enforced;

(6) in the case of any treatment works serving a population of 50,000 or more, with respect to any toxic pollutant introduced into such works by an industrial discharger for which pollutant there is no applicable pretreatment requirement in effect, sources introducing waste into such works are in compliance with all applicable pretreatment requirements, the applicant will enforce such requirements, and the applicant has in effect a pretreatment program which, in combination with the treatment of discharges from such works, removes the same amount of such pollutant as would be removed if such works

were to apply secondary treatment to discharges and if such works had no pretreatment program with respect to such pollutant;

(7) to the extent practicable the applicant has established a schedule of activities designed to eliminate the entrance of toxic pollutants from nonindustrial sources into such treatment works;

(8) there will be no new or substantially increased discharges from the point source of the pollutant to which the modification applies above that volume of discharge specified in the permit;

(9) the applicant at the time such modification becomes effective will be discharging effluent which has received at least primary or equivalent treatment and which meets the criteria established under section 304(a)(1) of this Act after initial mixing in the waters surrounding or adjacent to the point at which such effluent is discharged.

For the purposes of this subsection the phrase "the discharge of any pollutant into marine waters" refers to a discharge into deep waters of the territorial sea or the waters of the contiguous zone, or into saline estuarine waters where there is strong tidal movement and other hydrological and geological characteristics which the Administrator determines necessary to allow compliance with paragraph (2) of this subsection, and section 101(a)(2) of this Act. For the purposes of paragraph (9), "primary or equivalent treatment" means treatment by screening, sedimentation, and skimming adequate to remove at least 30 percent of the biological oxygen demanding material and of the suspended solids in the treatment works influent, and disinfection, where appropriate. A municipality which applies secondary treatment shall be eligible to receive a permit pursuant to this subsection which modifies the requirements of subsection (b)(1)(B) of this section with respect to the discharge of any pollutant from any treatment works owned by such municipality into marine waters. No permit issued under this subsection shall authorize the discharge of sewage sludge into marine waters. In order for a permit to be issued under this subsection for the discharge of a pollutant into marine waters, such marine waters must exhibit characteristics assuring that water providing dilution does not contain significant amounts of previously discharged effluent from such treatment works. No permit issued under this subsection shall authorize the discharge of any pollutant into saline estuarine waters which at the time of application do not support a balanced indigenous population of shellfish, fish and wildlife, or allow recreation in and on the waters or which exhibit ambient water quality below applicable water quality standards adopted for the protection of public water supplies, shellfish, fish and wildlife or recreational activities or such other standards necessary to assure support and protection of such uses. The prohibition contained in the preceding sentence shall apply without regard to the presence or absence of a causal relationship between such characteristics and the applicant's current or proposed discharge. Notwithstanding any other provisions of this subsection, no permit may be issued under this subsection for discharge of a pollutant into the New York Bight Apex consisting of the ocean waters of the Atlantic Ocean westward of 73 degrees 30 minutes west longitude and northward of 40 degrees 10 minutes north latitude.

(i)(1) Where construction is required in order for a planned or extisting publicly owned treatment works to achieve limitations under subsection (b)(1)(B) or (b)(1)(C) of this section, but (A) construction cannot be completed with the time required in such subsection, or (B) the United States has failed to make financial assistance under this Act available in time to achieve such limitations by the time specified in such subsection, the owner or operator of such treatment works may request the Administrator (or if appropriate the State) to issue a permit pursuant to section 402 of this Act or to modify a permit issued pursuant to that section to extend such time for compliance. Any such request shall be filed with the Administrator (or if appropriate the State) within 180 days after the date of enactment of the Water Quality Act of 1987. The Administrator (or if appropriate the State) may grant such request and issue or modify such a permit, which shall contain a schedule of compliance for the publicly owned treatment works based on the earliest date by which such financial assistance will be available from the United States and construction can be completed, but in no event later than July 1, 1988, and shall contain such other terms and conditions, including those necessary to carry out subsections (b) through (g) of section 201 of this Act, section 307 of this Act, and such interim effluent limitations applicable to that treatment works as the Administrator determines are necessary to carry out the provisions of this Act.

(2)(A) Where a point source (other than a publicly owned treatment works) will not achieve the requirements of subsections (b)(1)(A) and (b)(1)(C) of this section and—

(i) if a permit issued prior to July 1, 1977, to such point source is based upon a discharge into a publicly owned treatment works; or

(ii) if such point source (other than a publicly owned treatment works) had before July 1, 1977, a contract (enforceable against such point source) to discharge into a publicly owned treatment works; or

(iii) if either an application made before July 1, 1977, for a construction grant under this Act for a publicly owned treatment works, or engineering or architectural plans or working drawings made before July 1, 1977, for a publicly owned treatment works, show that such point source was to discharge into such publicly owned treatment works,

and such publicly owned treatment works is presently unable to accept such discharge without construction, and in the case of a discharge to an existing publicly owned treatment works, such treatment works has an extension pursuant to paragraph (1) of this subsection, the owner or operator of such point source may request the Administrator (or if appropriate the State) to issue or modify such a permit pursuant to such section 402 to extend such time for compliance. Any such request shall be filed with the Administrator (or if appropriate the State) within 180 days after the date of enactment of this subsection or the filing of a request by the appropriate publicly owned treatment works under paragraph (1) of this subsection, whichever is later. If the Administrator (or if appropriate the State) finds that the owner or operator of such point source has acted in good faith, he may grant such request and issue or modify such a permit, which shall contain a schedule of compliance for the point source to achieve the requirements of subsections (b)(1)(A) and (C) of this section and shall contain such other terms and conditions, including pretreatment and interim effluent limitations and water conservation requirements applicable to that point source, as the Administrator determines are necessary to carry out the provisions of this Act.

(B) No time modification granted by the Administrator (or if appropriate the State) pursuant to paragraph (2)(A) of this subsection shall extend beyond the earliest date practicable for compliance or beyond the date of any extension granted to the appropriate publicly owned treatment works pursuant to paragraph (1) of this subsection, but in no event shall it extend beyond July 1, 1988, and no such time modification shall be granted unless (i) the publicly owned treatment works will be in operation and available to the point source before July 1, 1988, and will meet the requirements to subsections (b)(1) (B) and (C) of this section after receiving the discharge from that point source; and (ii) the point source and the publicly owned treatment works have entered into an enforceable contract requiring the point source to discharge into the publicly owned treatment works, the owner or operator of such point source to pay the costs required under section 204 of this Act, and the publicly owned treatment works to accept the discharge from the point source; and (iii) the permit for such point source requires point source to meet all requirements under section 307 (a) and (b) during the period of such time modification.

(j)(1) Any application filed under this section for a modification of the provisions of—

(A) subsection (b)(1)(B) under subsection (h) of this section shall be filed not later that[1] the 365th day which begins after the date of enactment of the Municipal Wastewater Treatment Construction Grant Amendments of 1981, except that a publicly owned treatment works which prior to December 31, 1982, had a contractual arrangement to use a portion of the capacity of an ocean outfall operated by another publicly owned treatment works which has applied for or received modification under subsection (h), may apply for a modification of subsection (h) in its own right not later than 30 days after the date of the enactment of the Water Quality Act of 1987, and except as provided in paragraph (5);

(B) subsection (b)(2)(A) as it applies to pollutants identified in subsection (b)(2)(F) shall be filed not later than 270 days after the date of promulgation of an applicable effluent guideline under section 304 or not later than 270 days after the date of enactment of the Clean Water Act of 1977, whichever is later.

(2) Subject to paragraph (3) of this section, any application for a modification filed under subsection (g) of this section shall not operate to stay any requirement under this Act, unless in the judgment of the Administrator such a stay or the modification sought

[1] So in law. Probably should be "than".

will not result in the discharge of pollutants in quantities which may reasonably be anticipated to pose an unacceptable risk to human health or the environment because of bioaccumulation, persistency in the environment, acute toxicity, chronic toxicity (including carcinogenicity, mutagenicity or teratogenicity), or synergistic propensities, and that there is a substantial likelihood that the applicant will succeed on the merits of such application. In the case of an application filed under subsection (g) of this section, the Administrator may condition any stay granted under this paragraph on requiring the filing of a bond or other appropriate security to assure timely compliance with the requirements from which a modification is sought.

(3) COMPLIANCE REQUIREMENTS UNDER SUBSECTION (g).—

(A) EFFECT OF FILING.—An application for a modification under subsection (g) and a petition for listing of a pollutant as a pollutant for which modifications are authorized under such subsection shall not stay the requirement that the person seeking such modification or listing comply with effluent limitations under this Act for all pollutants not the subject of such application or petition.

(B) EFFECT OF DISAPPROVAL.—Disapproval of an application for a modification under subsection (g) shall not stay the requirement that the person seeking such modification comply with all applicable effluent limitations under this Act.

(4) DEADLINE FOR SUBSECTION (g) DECISION.—An application for a modification with respect to a pollutant filed under subsection (g) must be approved or disapproved not later than 365 days after the date of such filing; except that in any case in which a petition for listing such pollutant as a pollutant for which modifications are authorized under such subsection is approved, such application must be approved or disapproved not later than 365 days after the date of approval of such petition.

(5) EXTENSION OF APPLICATION DEADLINE.—

(A) IN GENERAL.—In the 180-day period beginning on the date of the enactment of this paragraph, the city of San Diego, California, may apply for a modification pursuant to subsection (h) of the requirements of subsection (b)(1)(B) with respect to biological oxygen demand and total suspended solids in the effluent discharged into marine waters.

(B) APPLICATION.—An application under this paragraph shall include a commitment by the applicant to implement a waste water reclamation program that, at a minimum, will—

(i) achieve a system capacity of 45,000,000 gallons of reclaimed waste water per day by January 1, 2010; and

(ii) result in a reduction in the quantity of suspended solids discharged by the applicant into the marine environment during the period of the modification.

(C) ADDITIONAL CONDITIONS.—The Administrator may not grant a modification pursuant to an application submitted under this paragraph unless the Administrator determines that such modification will result in removal of not less than 58 percent of the biological oxygen demand (on an annual average) and not less than 80 percent of total suspended solids (on a monthly average) in the discharge to which the application applies.

(D) PRELIMINARY DECISION DEADLINE.—The Administrator shall announce a preliminary decision on an application submitted under this paragraph not later than 1 year after the date the application is submitted.

(k) In the case of any facility subject to a permit under section 402 which proposes to comply with the requirements of subsection (b)(2)(A) or (b)(2)(E) of this section by replacing existing production capacity with an innovative production process which will result in an effluent reduction significantly greater than that required by the limitation otherwise applicable to such facility and moves toward the national goal of eliminating the discharge of all pollutants, or with the installation of an innovative control technique that has a substantial likelihood for enabling the facility to comply with the applicable effluent limitation by achieving a significantly greater effluent reduction than that required by the applicable effluent limitation and moves toward the national goal of eliminating the discharge of all pollutants, or by achieving the required reduction with an innovative system that has the potential for significantly lower costs than the systems which have been determined by the Administrator to be economically achievable, the Administrator (or the State with an approved program under section 402, in consultation with the Administrator) may establish a date for compliance under subsection (b)(2)(A) or (b)(2)(E) of this section no later than two years after the date for compliance with such effluent limitation which would otherwise be applicable under such subsection, if it is also determined that such innovative system has the potential for industrywide application.

(l) Other than as provided in subsection (n) of this section, the Administrator may not modify any requirement of this section as it applies to any specific pollutant which is on the toxic pollutant list under section 307(a)(1) of this Act.

(m)(1) The Administrator, with the concurrence of the State, may issue a permit under section 402 which modifies the requirements of subsections (b)(1)(A) and (b)(2)(E) of this section, and of section 403, with respect to effluent limitations to the extent such limitations relate to biochemical oxygen demand and pH from discharges by an industrial discharger in such State into deep waters of the territorial seas, if the applicant demonstrates and the Administrator finds that—

(A) the facility for which modification is sought is covered at the time of the enactment of this subsection by National Pollutant Discharge Elimination System permit number CA0005894 or CA0005282;

(B) the energy and environmental costs of meeting such requirements of subsections (b)(1)(A) and (b)(2)(E) and section

403 exceed by an unreasonable amount the benefits to be obtained, including the objectives of this Act;

(C) the applicant has established a system for monitoring the impact of such discharges on a representative sample of aquatic biota;

(D) such modified requirements will not result in any additional requirements on any other point or nonpoint source;

(E) there will be no new or substantially increased discharges from the point source of the pollutant to which the modification applies above that volume of discharge specified in the permit;

(F) the discharge is into waters where there is strong tidal movement and other hydrological and geological characteristics which are necessary to allow compliance with this subsection and section 101(a)(2) of this Act;

(G) the applicant accepts as a condition to the permit a contractural obligation to use funds in the amount required (but not less than $250,000 per year for ten years) for research and development of water pollution control technology, including but not limited to closed cycle technology;

(H) the facts and circumstances present a unique situation which, if relief is granted, will not establish a precedent or the relaxation of the requirements of this Act applicable to similarly situated discharges; and

(I) no owner or operator of a facility comparable to that of the applicant situated in the United States has demonstrated that it would be put at a competitive disadvantage to the applicant (or the parent company or any subsidiary thereof) as a result of the issuance of a permit under this subsection.

(2) The effluent limitations established under a permit issued under paragraph (1) shall be sufficient to implement the applicable State water quality standards, to assure the protection of public water supplies and protection and propagation of a balanced, indigenous population of shellfish, fish, fauna, wildlife, and other aquatic organisms, and to allow recreational activities in and on the water. In setting such limitations, the Administrator shall take into account any seasonal variations and the need for an adequate margin of safety, considering the lack of essential knowledge concerning the relationship between effluent limitations and water quality and the lack of essential knowledge of the effects of discharges on beneficial uses of the receiving waters.

(3) A permit under this subsection may be issued for a period not to exceed five years, and such a permit may be renewed for one additional period not to exceed five years upon a demonstration by the applicant and a finding by the Administrator at the time of application for any such renewal that the provisions of this subsection are met.

(4) The Administrator may terminate a permit issued under this subsection if the Administrator determines that there has been a decline in ambient water quality of the receiving waters during the period of the permit even if a direct cause and effect relationship cannot be shown: *Provided,* That if the effluent from a source with a permit issued under this subsection is contributing to a de-

cline in ambient water quality of the receiving waters, the Administrator shall terminate such permit.

(n) FUNDAMENTALLY DIFFERENT FACTORS.—

(1) GENERAL RULE.—The Administrator, with the concurrance of the State, may establish an alternative requirement under subsection (b)(2) or section 307(b) for a facility that modifies the requirements of national effluent limitation guidelines or categorical pretreatment standards that would otherwise be applicable to such facility, if the owner or operator of such facility demonstrates to the satisfaction of the Administrator that—

(A) the facility is fundamentally different with respect to the factors (other than cost) specified in section 304(b) or 304(g) and considered by the Administrator in establishing such national effluent limitation guidelines or categorical pretreatment standards;

(B) the application—

(i) is based solely on information and supporting data submitted to the Administrator during the rule making for establishment of the applicable national effluent limitation guidelines or categorical pretreatment standard specifically raising the factors that are fundamentally different for such facility; or

(ii) is based on information and supporting data referred to in clause (i) and information and supporting data the applicant did not have a reasonable opportunity to submit during such rulemaking;

(C) the alternative requirement is no less stringent than justified by the fundamental difference; and

(D) the alternative requirement will not result in a non-water quality environmental impact which is markedly more adverse than the impact considered by the Administrator in establishing such national affluent limitation guideline or categorical pretreatment standard.

(2) TIME LIMIT FOR APPLICATIONS.—An application for an alternative requirement which modifies the requirements of an effluent limitation or pretreatment standard under this subsection must be submitted to the Administrator within 180 days after the date on which such limitation or standard is established or revised, as the case may be.

(3) TIME LIMIT FOR DECISION.—The Administrator shall approve or deny by final agency action an application submitted under this subsection within 180 days after the date such application is filed with the Administrator.

(4) SUBMISSION OF INFORMATION.—The Administrator may allow an applicant under this subsection to submit information and supporting data until the earlier of the date the application is approved or denied or the last day that the Administrator has to approve or deny such application.

(5) TREATMENT OF PENDING APPLICATIONS.—For the purposes of this subsection, an application for an alternative requirement based on fundamentally different factors which is pending on the date of the enactment of this subsection shall be treated as having been submitted to the Administrator on

the 180th day following such date of enactment. The applicant may amend the application to take into account the provisions of this subsection.

(6) EFFECT OF SUBMISSION OF APPLICATION.—An application for an alternative requirement under this subsection shall not stay the applicant's obligation to comply with the effluent limitation guideline or categorical pretreatment standard which is the subject of the application.

(7) EFFECT OF DENIAL.—If an application for an alternative requirement which modifies the requirements of an effluent limitation or pretreatment standard under this subsection is denied by the Administrator, the applicant must comply with such limitation or standard as established or revised, as the case may be.

(8) REPORTS.—By January 1, 1997, and January 1 of every odd-numbered year thereafter, the Administrator shall submit to the Committee on Environment and Public Works of the Senate and the Committee on Transportation and Infrastructure of Representatives a report on the status of applications for alternative requirements which modify the requirements of effluent limitations under section 301 or 304 of this Act or any national categorical pretreatment standard under section 307(b) of this Act filed before, on, or after such date of enactment.

(o) APPLICATION FEES.—The Administrator shall prescribe and collect from each applicant fees reflecting the reasonable administrative costs incurred in reviewing and processing applications for modifications submitted to the Administrator pursuant to subsections (c), (g), (i), (k), (m), and (n) of section 301, section 304(d)(4), and section 316(a) of this Act. All amounts collected by the Administrator under this subsection shall be deposited into a special fund of the Treasury entitled "Water Permits and Related Services" which shall thereafter be available for appropriation to carry out activities of the Environmental Protection Agency for which such fees were collected.

(p) MODIFIED PERMIT FOR COAL REMINING OPERATIONS.—

(1) IN GENERAL.—Subject to paragraphs (2) through (4) of this subsection, the Administrator, or the State in any case which the State has an approved permit program under section 402(b), may issue a permit under section 402 which modifies the requirements of subsection (b)(2)(A) of this section with respect to the pH level of any pre-existing discharge, and with respect to pre-existing discharges of iron and manganese from the remined area of any coal remining operation or with respect to the pH level or level of iron or manganese in any pre-existing discharge affected by the remining operation. Such modified requirements shall apply the best available technology economically achievable on a case-by-case basis, using best professional judgment, to set specific numerical effluent limitations in each permit.

(2) LIMITATIONS.—The Administrator or the State may only issue a permit pursuant to paragraph (1) if the applicant demonstrates to the satisfaction of the Administrator or the State, as the case may be, that the coal remining operation will

result in the potential for improved water quality from the remining operation but in no event shall such a permit allow the pH level of any discharge, and in no event shall such a permit allow the discharges of iron and manganese, to exceed the levels being discharged from the remined area before the coal remining operation begins. No discharge from, or affected by, the remining operation shall exceed State water quality standards established under section 303 of this Act.

(3) DEFINITIONS.—For purposes of this subsection—

(A) COAL REMINING OPERATION.—The term "coal remining operation" means a coal mining operation which begins after the date of the enactment of this subsection at a site on which coal mining was conducted before the effective date of the Surface Mining Control and Reclamation Act of 1977.

(B) REMINED AREA.—The term "remined area" means only that area of any coal remining operation on which coal mining was conducted before the effective date of the Surface Mining Control and Reclamation Act of 1977.

(C) PRE-EXISTING DISCHARGE.—The term "pre-existing discharge" means any discharge at the time of permit application under this subsection.

(4) APPLICABILITY OF STRIP MINING LAWS.—Nothing in this subsection shall affect the application of the Surface Mining Control and Reclamation Act of 1977 to any coal remining operation, including the application of such Act to suspended solids.

(33 U.S.C. 1311)

WATER QUALITY RELATED EFFLUENT LIMITATIONS

SEC. 302. (a) Whenever, in the judgment of the Administrator or as identified under section 304(l), discharges of pollutants from a point source or group of point sources, with the application of effluent limitations required under section 301(b)(2) of this Act, would interfere with the attainment or maintenance of that water quality in a specific portion of the navigable waters which shall assure protection of public health, public water supplies, agricultural and industrial uses, and the protection and propagation of a balanced population of shellfish, fish and wildlife, and allow recreational activities in and on the water, effluent limitations (including alternative effluent control strategies) for such point source or sources shall be established which can reasonably be expected to contribute to the attainment or maintenance of such water quality.

(b) MODIFICATIONS OF EFFLUENT LIMITATIONS.—

(1) NOTICE AND HEARING.—Prior to establishment of any effluent limitation pursuant to subsection (a) of this section, the Administrator shall publish such proposed limitation and within 90 days of such publication hold a public hearing.

(2) PERMITS.—

(A) NO REASONABLE RELATIONSHIP.—The Administrator, with the concurrence of the State, may issue a permit which modifies the effluent limitations required by subsection (a) of this section for pollutants other than toxic pollutants if the applicant demonstrates at such hearing that (whether or not technology or other alternative control strategies are available) there is no reasonable relationship between the economic and social costs and the benefits to be obtained (including attainment of the objective of this Act) from achieving such limitation.

(B) REASONABLE PROGRESS.—The Administrator, with the concurrence of the State, may issue a permit which modifies the effluent limitations required by subsection (a) of this section for toxic pollutants for a single period not to exceed 5 years if the applicant demonstrates to the satisfaction of the Administrator that such modified requirements (i) will represent the maximum degree of control within the economic capability of the owner and operator of the source, and (ii) will result in reasonable further progress beyond the requirements of section 301(b)(2) toward the requirements of subsection (a) of this section.

(c) The establishment of effluent limitations under this section shall not operate to delay the application of any effluent limitation established under section 301 of this Act.

(33 U.S.C. 1312)

WATER QUALITY STANDARDS AND IMPLEMENTATION PLANS

SEC. 303. (a)(1) In order to carry out the purpose of this Act, any water quality standard applicable to interstate waters which was adopted by any State and submitted to, and approved by, or is awaiting approval by, the Administrator pursuant to this Act as in effect immediately prior to the date of enactment of the Federal Water Pollution Control Act Amendments of 1972, shall remain in effect unless the Administrator determined that such standard is not consistent with the applicable requirements of this Act as in effect immediately prior to the date of enactment of the Federal Water Pollution Control Act Amendments of 1972. If the Administrator makes such a determination he shall, within three months after the date of enactment of the Federal Water Pollution Control Act Amendments of 1972, notify the State and specify the changes needed to meet such requirements. If such changes are not adopted by the State within ninety days after the date of such notification, the Administrator shall promulgate such changes in accordance with subsection (b) of this section.

(2) Any State which, before the date of enactment of the Federal Water Pollution Control Act Amendments of 1972, has adopted, pursuant to its own law, water quality standards applicable to intrastate waters shall submit such standards to the Administrator within thirty days after the date of enactment of the Federal Water Pollution Control Act Amendments of 1972. Each such standard shall remain in effect, in the same manner and to the same extent as any other water quality standard established under this Act unless the Administrator determines that such standard is inconsistent with the applicable requirements of this Act as in effect immediately prior to the date of enactment of the Federal Water Pollution Control Act Amendments of 1972. If the Administrator makes such a determination he shall not later than the one hun-

dred and twentieth day after the date of submission of such standards, notify the State and specify the changes needed to meet such requirements. If such changes are not adopted by the State within ninety days after such notification, the Administrator shall promulgate such changes in accordance with subsection (b) of this section.

(3)(A) Any State which prior to the date of enactment of the Federal Water Pollution Control Act Amendments of 1972 has not adopted pursuant to its own laws water quality standards applicable to intrastate waters shall, not later than one hundred and eighty days after the date of enactment of the Federal Water Pollution Control Act Amendments of 1972, adopt and submit such standards to the Administrator.

(B) If the Administrator determines that any such standards are consistent with the applicable requirements of this Act as in effect immediately prior to the date of enactment of the Federal Water Pollution Control Act Amendments of 1972, he shall approve such standards.

(C) If the Administrator determines that any such standards are not consistent with the applicable requirements of this Act as in effect immediately prior to the date of enactment of the Federal Water Pollution Control Act Amendments of 1972, he shall, not later than the ninetieth day after the date of submission of such standards, notify the State and specify the changes to meet such requirements. If such changes are not adopted by the State within ninety days after the date of notification, the Administrator shall promulgate such standards pursuant to subsection (b) of this section.

(b)(1) The Administrator shall promptly prepare and publish proposed regulations setting forth water quality standards for a State in accordance with the applicable requirements of this Act as in effect immediately prior to the date of enactment of the Federal Water Pollution Control Act Amendments of 1972, if—

(A) the State fails to submit water quality standards within the times prescribed in subsection (a) of this section,

(B) a water quality standard submitted by such State under subsection (a) of this section is determined by the Administrator not to be consistent with the applicable requirements of subsection (a) of this section.

(2) The Administrator shall promulgate any water quality standard published in a proposed regulation not later than one hundred and ninety days after the date he publishes any such proposed standard, unless prior to such promulgation, such State has adopted a water quality standard which the Administrator determines to be in accordance with subsection (a) of this section.

(c)(1) The Governor of a State or the State water pollution control agency of such State shall from time to time (but at least once each three year period beginning with the date of enactment of the Federal Water Pollution Control Act Amendments of 1972) hold public hearings for the purpose of reviewing applicable water quality standards and, as appropriate, modifying and adopting standards. Results of such review shall be made available to the Administrator.

(2)(A) Whenever the State revises or adopts a new standard, such revised or new standard shall be submitted to the Administrator. Such revised or new water quality standard shall consist of the designated uses of the navigable waters involved and the water quality criteria for such waters based upon such uses. Such standards shall be such as to protect the public health or welfare, enhance the quality of water and serve the purposes of this Act. Such standards shall be established taking into consideration their use and value for public water supplies, propagation of fish and wildlife, recreational purposes, and agricultural, industrial, and other purposes, and also taking into consideration their use and value for navigation.

(B) Whenever a State reviews water quality standards pursuant to paragraph (1) of this subsection, or revises or adopts new standards pursuant to this paragraph, such State shall adopt criteria for all toxic pollutants listed pursuant to section 307(a)(1) of this Act for which criteria have been published under section 304(a), the discharge or presence of which in the affected waters could reasonably be expected to interfere with those designated uses adopted by the State, as necessary to support such designated uses. Such criteria shall be specific numerical criteria for such toxic pollutants. Where such numerical criteria are not available, whenever a State reviews water quality standards pursuant to paragraph (1), or revises or adopts new standards pursuant to this paragraph, such State shall adopt criteria based on biological monitoring or assessment methods consistent with information published pursuant to section 304(a)(8). Nothing in this section shall be construed to limit or delay the use of effluent limitations or other permit conditions based on or involving biological monitoring or assessment methods or previously adopted numerical criteria.

(3) If the Administrator, within sixty days after the date of submission of the revised or new standard, determines that such standard meets the requirements of this Act, such standard shall thereafter be the water quality standard for the applicable waters of that State. If the Administrator determines that any such revised or new standard is not consistent with the applicable requirements of this Act, he shall not later than the ninetieth day after the date of submission of such standard notify the State and specify the changes to meet such requirements. If such changes are not adopted by the State within ninety days after the date of notification, the Administrator shall promulgate such standard pursuant to paragraph (4) of this subsection.

(4) The Administrator shall promptly prepare and publish proposed regulations setting forth a revised or new water quality standard for the navigable waters involved—

(A) if a revised or new water quality standard submitted by such State under paragraph (3) of this subsection for such waters is determined by the Administrator not to be consistent with the applicable requirements of this Act, or

(B) in any case where the Administrator determines that a revised or new standard is necessary to meet the requirements of this Act.

The Administrator shall promulgate any revised or new standard under this paragraph not later than ninety days after he publishes such proposed standards, unless prior to such promulgation, such

State has adopted a revised or new water quality standard which the Administrator determines to be in accordance with this Act.

(d)(1)(A) Each State shall identify those waters within its boundaries for which the effluent limitations required by section 301(b)(1)(A) and section 301(b)(1)(B) are not stringent enough to implement any water quality standard applicable to such waters. The State shall establish a priority ranking for such waters, taking into account the severity of the pollution and the uses to be made of such waters.

(B) Each State shall identify those waters or parts thereof within its boundaries for which controls on thermal discharges under section 301 are not stringent enough to assure protection and propagation of a balanced indigenous population of shellfish, fish, and wildlife.

(C) Each State shall establish for the waters identified in paragraph (1)(A) of this subsection, and in accordance with the priority ranking, the total maximum daily load, for those pollutants which the Administrator identifies under section 304(a)(2) as suitable for such calculation. Such load shall be established at a level necessary to implement the applicable water quality standards with seasonal variations and a margin of safety which takes into account any lack of knowledge concerning the relationship between effluent limitations and water quality.

(D) Each State shall estimate for the waters identified in paragraph (1)(D) of this subsection the total maximum daily thermal load required to assure protection and propagation of a balanced, indigenous population of shellfish, fish and wildlife. Such estimates shall take into account the normal water temperatures, flow rates, seasonal variations, existing sources of heat input, and the dissipative capacity of the identified waters or parts thereof. Such estimates shall include a calculation of the maximum heat input that can be made into each such part and shall include a margin of safety which takes into account any lack of knowledge concerning the development of thermal water quality criteria for such protection and propagation in the identified waters or parts thereof.

(2) Each State shall submit to the Administrator from time to time, with the first such submission not later than one hundred and eighty days after the date of publication of the first identification of pollutants under section 304(a)(2)(D), for his approval the waters identified and the loads established under paragraphs (1)(A), (1)(B), (1)(C), and (1)(D) of this subsection. The Administrator shall either approve or disapprove such identification and load not later than thirty days after the date of submission. If the Administrator approves such identification and load, such State shall incorporate them into its current plan under subsection (e) of this section. If the Administrator disapproves such identification and load, he shall not later than thirty days after the date of such disapproval identify such waters in such State and establish such loads for such waters as he determines necessary to implement the water quality standards applicable to such waters and upon such identification and establishment the State shall incorporate them into its current plan under subsection (e) of this section.

(3) For the specific purpose of developing information, each State shall identify all waters within its boundaries which it has

not identified under paragraph (1)(A) and (1)(B) of this subsection and estimate for such waters the total maximum daily load with seasonal variations and margins of safety, for those pollutants which the Administrator identifies under section 304(a)(2) as suitable for such calculation and for thermal discharges, at a level that would assure protection and propagation of a balanced indigenous population of fish, shellfish and wildlife.

(4) LIMITATIONS ON REVISION OF CERTAIN EFFLUENT LIMITATIONS.—

(A) STANDARD NOT ATTAINED.—For waters identified under paragraph (1)(A) where the applicable water quality standard has not yet been attained, any effluent limitation based on a total maximum daily load or other waste load allocation established under this section may be revised only if (i) the cumulative effect of all such revised effluent limitations based on such total maximum daily load or waste load allocation will assure the attainment of such water quality standard, or (ii) the designated use which is not being attained is removed in accordance with regulations established under this section.

(B) STANDARD ATTAINED.—For waters identified under paragraph (1)(A) where the quality of such waters equals or exceeds levels necessary to protect the designated use for such waters or otherwise required by applicable water quality standard, any effluent limitation based on a total maximum daily load or other waste load allocation established under this section, or any water quality standard established under this section, or any other permitting standard may be revised only if such revision is subject to and consistent with the antidegradation policy established under this section.

(e)(1) Each State shall have a continuing planning process approved under paragraph (2) of this subsection which is consistent with this Act.

(2) Each State shall submit not later than 120 days after the date of the enactment of the Water Pollution Control Amendments of 1972 to the Administrator for his approval a proposed continuing planning process which is consistent with this Act. Not later than thirty days after the date of submission of such a process the Administrator shall either approve or disapprove such process. The Administrator shall from time to time review each State's approved planning process for the purpose of insuring that such planning process is at all times consistent with this Act. The Administrator shall not approve any State permit program under title IV of this Act for any State which does not have an approved continuing planning process under this section.

(3) The Administrator shall approve any continuing planning process submitted to him under this section which will result in plans for all navigable waters within such State, which include, but are not limited to, the following:

(A) effluent limitations and schedules of compliance at least as stringent as those required by section 301(b)(1), section 301(b)(2), section 306, and section 307, and at least as stringent as any requirements contained in any applicable water quality standard in effect under authority of this section;

(B) the incorporation of all elements of any applicable areawide waste management plans under section 208, and applicable basin plans under section 209 of this Act;

(C) total maximum daily load for pollutants in accordance with subsection (d) of this section;

(D) procedures for revision;

(E) adequate authority for intergovernmental cooperation;

(F) adequate implementation, including schedules of compliance, for revised or new water quality standards, under subsection (c) of this section;

(G) controls over the disposition of all residual waste from any water treatment processing;

(H) an inventory and ranking, in order of priority, of needs for construction of waste treatment works required to meet the applicable requirements of sections 301 and 302.

(f) Nothing in this section shall be construed to affect any effluent limitation, or schedule of compliance required by any State to be implemented prior to the dates set forth in sections 301(b)(1) and 301(b)(2) nor to preclude any State from requiring compliance with any effluent limitation or schedule of compliance at dates earlier than such dates.

(g) Water quality standards relating to heat shall be consistent with the requirements of section 316 of this Act.

(h) For the purposes of this Act the term "water quality standards" includes thermal water quality standards.

(i) COASTAL RECREATION WATER QUALITY CRITERIA.—

(1) ADOPTION BY STATES.—

(A) INITIAL CRITERIA AND STANDARDS.—Not later than 42 months after the date of the enactment of this subsection, each State having coastal recreation waters shall adopt and submit to the Administrator water quality criteria and standards for the coastal recreation waters of the State for those pathogens and pathogen indicators for which the Administrator has published criteria under section 304(a).

(B) NEW OR REVISED CRITERIA AND STANDARDS.—Not later than 36 months after the date of publication by the Administrator of new or revised water quality criteria under section 304(a)(9), each State having coastal recreation waters shall adopt and submit to the Administrator new or revised water quality standards for the coastal recreation waters of the State for all pathogens and pathogen indicators to which the new or revised water quality criteria are applicable.

(2) FAILURE OF STATES TO ADOPT.—

(A) IN GENERAL.—If a State fails to adopt water quality criteria and standards in accordance with paragraph (1)(A) that are as protective of human health as the criteria for pathogens and pathogen indicators for coastal recreation waters published by the Administrator, the Administrator shall promptly propose regulations for the State setting forth revised or new water quality standards for pathogens and pathogen indicators described in paragraph (1)(A) for coastal recreation waters of the State.

(B) EXCEPTION.—If the Administrator proposes regulations for a State described in subparagraph (A) under subsection (c)(4)(B), the Administrator shall publish any revised or new standard under this subsection not later than 42 months after the date of the enactment of this subsection.

(3) APPLICABILITY.—Except as expressly provided by this subsection, the requirements and procedures of subsection (c) apply to this subsection, including the requirement in subsection (c)(2)(A) that the criteria protect public health and welfare.

(33 U.S.C. 1313)

INFORMATION AND GUIDELINES

SEC. 304. (a)(1) The Administrator, after consultation with appropriate Federal and State agencies and other interested persons, shall develop and publish, within one year after the date of enactment of this title (and from time to time thereafter revise) criteria for water quality accurately reflecting the latest scientific knowledge (A) on the kind and extent of all identifiable effects on health and welfare including, but not limited to, plankton, fish, shellfish, wildlife, plant life, shorelines, beaches, esthetics, and recreation which may be expected from the presence of pollutants in any body of water, including ground water; (B) on the concentration and dispersal of pollutants, or their byproducts, through biological, physical, and chemical processes; and (C) on the effects of pollutants on biological community diversity, productivity, and stability, including information on the factors affecting rates of eutrophication and rates of organic and inorganic sedimentation for varying types of receiving waters.

(2) The Administrator, after consultation with appropriate Federal and State agencies and other interested persons, shall develop and publish, within one year after the date of enactment of this title (and from time to time thereafter revise) information (A) on the factors necessary to restore and maintain the chemical, physical, and biological integrity of all navigable waters, ground waters, waters of the contiguous zone, and the oceans; (B) on the factors necessary for the protection and propagation of shellfish, fish, and wildlife for classes and categories of receiving waters and to allow recreational activities in and on the water; and (C) on the measurement and classification of water quality; and (D) for the purpose of section 303, on and the identification of pollutants suitable for maximum daily load measurement correlated with the achievement of water quality objectives.

(3) Such criteria and information and revisions thereof shall be issued to the States and shall be published in the Federal Register and otherwise made available to the public.

(4) The Administrator shall, within 90 days after the date of enactment of the Clean Water Act of 1977 and from time to time thereafter, publish and revise as appropriate information identifying conventional pollutants, including but not limited to, pollut-

ants classified as biological oxygen demanding, suspended solids, fecal coliform, and pH. The thermal component of any discharge shall not be identified as a conventional pollutant under this paragraph.

(5)(A) The Administrator, to the extent practicable before consideration of any request under section 301(g) of this Act and within six months after the date of enactment of the Clean Water Act of 1977, shall develop and publish information on the factors necessary for the protection of public water supplies, and the protection and propagation of a balanced population of shellfish, fish and wildlife, and to allow recreational activities, in and on the water.

(B) The Administrator, to the extent practicable before consideration of any application under section 301(h) of this Act and within six months after the date of enactment of Clean Water Act of 1977, shall develop and publish information on the factors necessary for the protection of public water supplies, and the protection and propagation of a balanced indigenous population of shellfish, fish and wildlife, and to allow recreational activities, in and on the water.

(6) The Administrator shall, within three months after enactment of the Clean Water Act of 1977 and annually thereafter, for purposes of section 301(h) of this Act publish and revise as appropriate information identifying each water quality standard in effect under this Act of State law, the specific pollutants associated with such water quality standard, and the particular waters to which such water quality standard applies.

(7) GUIDANCE TO STATES.—The Administrator, after consultation with appropriate State agencies and on the basis of criteria and information published under paragraphs (1) and (2) of this subsection, shall develop and publish, within 9 months after the date of the enactment of the Water Quality Act of 1987, guidance to the States on performing the identification required by section 304(l)(1) of this Act.

(8) INFORMATION ON WATER QUALITY CRITERIA.—The Administrator, after consultation with appropriate State agencies and within 2 years after the date of the enactment of the Water Quality Act of 1987, shall develop and publish information on methods for establishing and measuring water quality criteria for toxic pollutants on other bases than pollutant-by-pollutant criteria, including biological monitoring and assessment methods.

(9) REVISED CRITERIA FOR COASTAL RECREATION WATERS.—

(A) IN GENERAL.—Not later than 5 years after the date of the enactment of this paragraph, after consultation and in cooperation with appropriate Federal, State, tribal, and local officials (including local health officials), the Administrator shall publish new or revised water quality criteria for pathogens and pathogen indicators (including a revised list of testing methods, as appropriate), based on the results of the studies conducted under section 104(v), for the purpose of protecting human health in coastal recreation waters.

(B) REVIEWS.—Not later than the date that is 5 years after the date of publication of water quality criteria under this paragraph, and at least once every 5 years thereafter, the Administrator shall review and, as necessary, revise the water quality criteria.

(b) For the purposes of adopting or revising effluent limitations under this Act the Administrator shall, after consultation with appropriate Federal and State agencies and other interested persons, publish within one year of enactment of this title, regulations, providing guidelines for effluent limitations, and, at least annually thereafter, revise, if appropriate, such regulations. Such regulations shall—

(1)(A) identify, in terms of amounts of constituents and chemical, physical, and biological characteristics of pullutants, the degree of effluent reduction attainable through the application of the best practicable control technology currently available for classes and categories to point sources (other than publicly owned treatment works); and

(B) specify factors to be taken into account in determining the control measures and practices to be applicable to point sources (other than publicly owned treatment works) within such categories of classes. Factors relating to the assessment of best practical control technology currently available to comply with subsection (b)(1) of section 301 of this Act shall include consideration of the total cost of application of technology in relation to the effluent reduction benefits to be achieved from such application, and shall also take into account the age of equipment and facilities involved, the process employed, the engineering aspects of the application of various types of control techniques, process changes, non-water quality environmental impact (including energy requirements), and such other factors as the Administrator deems appropriate;

(2)(A) identify, in terms of amounts of constituents and chemical, physical, and biological characteristics of pollutants, the degree of effluent reduction attainable through the application of the best control measures and practices achievable including treatment techniques, process and procedure innovations, operating methods, and other alternatives for classes and categories of point sources (other than publicly owned treatment works); and

(B) specify factors to be taken into account in determining the best measures and practices available to comply with subsection (b)(2) of section 301 of this Act to be applicable to any point source (other than publicly owned treatment works) within such categories of classes. Factors relating to the assessment of best available technology shall take into account the age of equipment and facilities involved, the process employed, the engineering aspects of the application of various types of control techniques, process changes, the cost of achieving such effluent reduction, non-water quality environmental impact (including energy requirements), and such other factors as the Administrator deems appropriate;

(3) identify control measures and practices available to eliminate the discharge of pollutants from categories and classes of point sources, taking into account the cost of achieving such elimination of the discharge of pollutants; and

(4)(A) identify, in terms of amounts of constituents and chemical, physical, and biological characteristics of pollutants, the degree of effluent reduction attainable through the application of the best conventional pollutant control technology (including measures and practices) for classes and categories of point sources (other than publicly owned treatment works); and

(B) specify factors to be taken into account in determining the best conventional pollutant control technology measures and practices to comply with section 301(b)(2)(E) of this Act to be applicable to any point source (other than publicly owned treatment works) within such categories or classes. Factors relating to the assessment of best conventional pollutant control technology (including measures and practices) shall include consideration of the reasonableness of the relationship between the costs of attaining a reduction in effluents and the effluent reduction benefits derived, and the comparison of the cost and level of reduction of such pollutants from the discharge from publicly owned treatment works to the cost and level of reduction of such pollutants from a class or category of industrial sources, and shall take into account the age of equipment and facilities involved, the process employed, the engineering aspects of the application of various types of control techniques, process changes, non-water quality environmental impact (including energy requirements), and such other factors as the Administrator deems appropriate.

(c) The Administrator, after consultation, with appropriate Federal and State agencies and other interested persons, shall issue to the States and appropriate water pollution control agencies within 270 days after enactment of this title (and from time to time thereafter) information on the processes, procedures, or operating methods which result in the elimination or reduction of the discharge of pollutants to implement standards of performance under section 306 of this Act. Such information shall include technical and other data, including costs, as are available on alternative methods of elimination or reduction of the discharge of pollutants. Such information, and revisions thereof, shall be published in the Federal Register and otherwise shall be made available to the public.

(d)(1) The Administrator, after consultation with appropriate Federal and State agencies and other interested persons, shall publish within sixty days after enactment of this title (and from time to time thereafter) information, in terms of amounts of constituents and chemical, physical, and biological characteristics of pollutants, on the degree of effluent reduction attainable through the application of secondary treatment.

(2) The Administrator, after consultation with appropriate Federal and State agencies and other interested persons, shall publish within nine months after the date of enactment of this title (and from time to time thereafter) information on alternative waste treatment management techniques and systems available to implement section 201 of this Act.

(3) The Administrator, after consultation with appropriate Federal and State agencies and other interested persons, shall promul-

gate witin one hundred and eighty days after the date of enactment of this subsection guidelines for identifying and evaluating innovative and alternative wastewater treatment process and techniques referred to in section 201(g)(5) of this Act.

(4) For the purposes of this subsection, such biological treatment facilities as oxidation ponds, lagoons, and ditches and trickling filters shall be deemed the equivalent of secondary treatment. The Administrator shall provide guidance under paragraph (1) of this subsection on design criteria for such facilities, taking into account pollutant removal efficiencies and, consistent with the objective of the Act, assuring that water quality will not be adversely affected by deeming such facilities as the equivalent of secondary treatment.

(e) The Administrator, after consultation with appropriate Federal and State agencies and other interested persons, may publish regulations, supplemental to any effluent limitations specified under subsections (b) and (c) of this section for a class or category of point sources, for any specific pollutant which the Administrator is charged with a duty to regulate as a toxic or hazardous pollutant under section 307(a)(1) or 311 of this Act, to control plant site runoff, spillage or leaks, sludge or waste disposal, and drainage from raw material storage which the Administrator determines are associated with or ancillary to the industrial manufacturing or treatment process within such class or category of point sources and may contribute significant amounts of such pollutants, to navigable waters. Any applicable controls established under this subsection shall be included as a requirement for the purposes of section 301, 302, 307, or 403, as the case may be, in any permit issued to a point source pursuant to section 402 of this Act.

(f) The Administrator, after consultation with appropriate Federal and State agencies and other interested persons, shall issue to appropriate Federal agencies, the States, water pollution control agencies, and agencies designated under section 208 of this Act, within one year after the effective date of this subsection (and from time to time thereafter) information including (1) guidelines for identifying and evaluating the nature and extent of nonpoint sources of pollutants, and (2) processes, procedures, and methods to control pollution resulting from—

(A) agricultural and silvicultural activities, including runoff from fields and crop and forest lands;

(B) mining activities, including runoff and siltation from new, currently operating, and abandoned surface and underground mines;

(C) all construction activity, including runoff from the facilities resulting from such construction;

(D) the disposal of pollutants in wells or in subsurface excavations;

(E) salt water intrusion resulting from reductions of fresh water flow from any cause, including extraction of ground water, irrigation, obstruction, and diversion; and

(F) changes in the movement, flow, or circulation of any navigable waters or ground waters, including changes caused by the construction of dams, levees, channels, causeways, or flow diversion facilities.

Such information and revisions thereof shall be published in the Federal Register and otherwise made available to the public.

(g)(1) For the purpose of assisting States in carrying out programs under section 402 of this Act, the Administrator shall publish, within one hundred and twenty days after the date of enactment of this title, and review at least annually thereafter and, if appropriate, revise guidelines for pretreatment of pollutants which he determines are not susceptible to treatment by publicly owned treatment works. Guidelines under this subsection shall be established to control and prevent the discharge into the navigable waters, the contiguous zone, or the ocean (either directly or through publicly owned treatment works) of any pollutant which interferes with, passes through, or otherwise is incompatible with such works.

(2) When publishing guidelines under this subsection, the Administrator shall designate the category or categories of treatment works to which the guidelines shall apply.

(h) The Administrator shall, within one hundred and eighty days from the date of enactment of this title, promulgate guidelines establishing test procedures for the analysis of pollutants that shall include the factors which must be provided in any certification pursuant to section 401 of this Act or permit application pursuant to section 402 of this Act.

(i) The Administrator shall (1) within sixty days after the enactment of this title promulgate guidelines for the purpose of establishing uniform application forms and other minimum requirements for the acquisition of information from owners and operators of point-sources of discharge subject to any State program under section 402 of this Act, and (2) within sixty days from the date of enactment of this title promulgate guidelines establishing the minimum procedural and other elements of any State program under section 402 of this Act which shall include:

(A) monitoring requirements;

(B) reporting requirements (including procedures to make information available to the public);

(C) enforcement provisions; and

(D) funding, personnel qualifications, and manpower requirements (including a requirement that no board or body which approves permit applications or portions thereof shall include, as a member, any person who receives, or has during the previous two years received, a significant portion of his income directly or indirectly from permit holders or applicants for a permit).

(j) LAKE RESTORATION GUIDANCE MANUAL.—The Administrator shall, within 1 year after the date of the enactment of the Water Quality Act of 1987 and biennially thereafter, publish and disseminate a lake restoration guidance manual describing methods, procedures, and processes to guide State and local efforts to improve, restore, and enhance water quality in the Nation's publicly owned lakes.

(k)(1) The Administrator shall enter into agreements with the Secretary of Agriculture, the Secretary of the Army, and the Secretary of the Interior, and the heads of such other departments, agencies, and instrumentalities of the United States as the Administrator determines, to provide for the maximum utilization of other

Federal laws and programs for the purpose of achieving and maintaining water quality through appropriate implementation of plans approved under section 208 of this Act and nonpoint source pollution management programs approved under section 319 of this Act.

(2) The Administrator is authorized to transfer to the Secretary of Agriculture, the Secretary of the Army, and the Secretary of the Interior and the heads of such other departments, agencies, and instrumentalities of the United States as the Administrator determines, any funds appropriated under paragraph (3) of this subsection to supplement funds otherwise appropriated to programs authorized pursuant to any agreement under paragraph (1).

(3) There is authorized to be appropriated to carry out the provisions of this subsection, $100,000,000 per fiscal year for the fiscal years 1979 through 1983 and such sums as may be necessary for fiscal years 1984 through 1990.

(l) INDIVIDUAL CONTROL STRATEGIES FOR TOXIC POLLUTANTS.—

(1) STATE LIST OF NAVIGABLE WATERS AND DEVELOPMENT OF STRATEGIES.—Not later than 2 years after the date of the enactment of this subsection, each State shall submit to the Administrator for review, approval, and implementation under this subsection—

(A) a list of those waters within the State which after the application of effluent limitations required under section 301(b)(2) of this Act cannot reasonably be anticipated to attain or maintain (i) water quality standards for such waters reviewed, revised, or adopted in accordance with section 303(c)(2)(B) of this Act, due to toxic pollutants, or (ii) that water quality which shall assure protection of public health, public water supplies, agricultural and industrial uses, and the protection and propagation of a balanced population of shellfish, fish and wildlife, and allow recreational activities in and on the water;

(B) a list of all navigable waters in such State for which the State does not expect the applicable standard under section 303 of this Act will be achieved after the requirements of sections 301(b), 306, and 307(b) are met, due entirely or substantially to discharges from point sources of any toxic pollutants listed pursuant to section 307(a);

(C) for each segment of the navigable waters included on such lists, a determination of the specific point sources discharging any such toxic pollutant which is believed to be preventing or impairing such water quality and the amount of each toxic pollutant discharged by each such source; and

(D) for each such segment, an individual control strategy which the State determines will produce a reduction in the discharge of toxic pollutants from point sources identified by the State under this paragraph through the establishment of effluent limitations under section 402 of this Act and water quality standards under section 303(c)(2)(B) of this Act, which reduction is sufficient, in combination with existing controls on point and nonpoint sources of pollution, to achieve the applicable water quality standard as soon as possible, but not later than 3 years after the date of the establishment of such strategy.

(2) APPROVAL OR DISAPPROVAL.—Not later than 120 days after the last day of the 2-year period referred to in paragraph (1), the Administrator shall approve or disapprove the control strategies submitted under paragraph (1) by any State.

(3) ADMINISTRATOR'S ACTION.—If a State fails to submit control strategies in accordance with paragraph (1) or the Administrator does not approve the control strategies submitted by such State in accordance with paragraph (1), then, not later than 1 year after the last day of the period referred to in paragraph (2), the Administrator, in cooperation with such State and after notice and opportunity for public comment, shall implement the requirements of paragraph (1) in such State. In the implementation of such requirements, the Administrator shall, at a minimum, consider for listing under this subsection any navigable waters for which any person submits a petition to the Administrator for listing not later than 120 days after such last day.

(m) SCHEDULE FOR REVIEW OF GUIDELINES.—

(1) PUBLICATION.—Within 12 months after the date of the enactment of the Water Quality Act of 1987, and biennially thereafter, the Administrator shall publish in the Federal Register a plan which shall—

(A) establish a schedule for the annual review and revision of promulgated effluent guidelines, in accordance with subsection (b) of this section;

(B) identify categories of sources discharging toxic or nonconventional pollutants for which guidelines under subsection (b)(2) of this section and section 306 have not previously been published; and

(C) establish a schedule for promulgation of effluent guidelines for categories identified in subparagraph (B), under which promulgation of such guidelines shall be no later than 4 years after such date of enactment for categories identified in the first published plan or 3 years after the publication of the plan for categories identified in later published plans.

(2) PUBLIC REVIEW.—The Administrator shall provide for public review and comment on the plan prior to final publication.

(33 U.S.C. 1314)

WATER QUALITY INVENTORY

SEC. 305. (a) The Administrator, in cooperation with the States and with the assistance of appropriate Federal agencies, shall prepare a report to be submitted to the Congress on or before January 1, 1974, which shall—

(1) describe the specific quality, during 1973, with appropriate supplemental descriptions as shall be required to take into account seasonal, tidal, and other variations, of all navigable waters and the waters of the contiguous zone;

(2) include an inventory of all point sources of discharge (based on a qualitative and quantitative analysis of discharges) of pollutants, into all navigable waters and the waters of the contiguous zone; and

(3) identify specifically those navigable waters, the quality of which—

(A) is adequate to provide for the protection and propagation of a balanced population of shellfish, fish, and wildlife and allow recreational activities in and on the water;

(B) can reasonably be expected to attain such level by 1977 or 1983; and

(C) can reasonably be expected to attain such level by any later date.

(b)(1) Each State shall prepare and submit to the Administrator by April 1, 1975, and shall bring up to date by April 1, 1976, and biennially thereafter, a report which shall include—

(A) a description of the water quality of all navigable waters in such State during the preceding year, with appropriate supplemental descriptions as shall be required to take into account seasonal, tidal, and other variations, correlated with the quality of water required by the objective of this Act (as identified by the Administrator pursuant to criteria published under section 304(a) of this Act) and the water quality described in subparagraph (B) of this paragraph;

(B) an analysis of the extent to which all navigable waters of such State provide for the protection and propagation of a balanced population of shellfish, fish, and wildlife, and allow recreational activities in and on the water;

(C) an analysis of the extent to which the elimination of the discharge of pollutants and a level of water quality which provides for the protection and propagation of a balanced population of shellfish, fish, and wildlife and allows recreational activities in and on the water, have been or will be achieved by the requirements of this Act, together with recommendations as to additional action necessary to achieve such objectives and for what waters such additional action is necessary;

(D) an estimate of (i) the environmental impact, (ii) the economic and social costs necessary to achieve the objective of this Act in such State, (iii) the economic and social benefits of such achievement, and (iv) an estimate of the date of such achievement; and

(E) a description of the nature and extent of nonpoint sources of pollutants, and recommendations as to the programs which must be undertaken to control each category of such sources, including an estimate of the costs of implementing such programs.

(2) The Administrator shall transmit such State reports, together with an analysis thereof, to Congress on or before October 1, 1975, and October 1, 1976, and biennially thereafter.

(33 U.S.C. 1315)

NATIONAL STANDARDS OF PERFORMANCE

SEC. 306. (a) For purposes of this section:

(1) The term "standard of performance" means a standard for the control of the discharge of pollutants which reflects the greatest degree of effluent reduction which the Administrator determines to be achievable through application of the best available demonstrated control technology, processes, operating methods, or other alternatives, including, where practicable, a standard permitting no discharge of pollutants.

(2) The term "new source" means any source, the construction of which is commenced after the publication of proposed regulations prescribing a standard of performance under this section which will be applicable to such sources, if such standard is thereafter promulgated in accordance with this section.

(3) The term "source" means any building, structure, facility, or installation from which there is or may be the discharge of pollutants.

(4) The term "owner or operator" means any person who owns, leases, operates, controls, or supervises a source.

(5) The term "construction" means any placement, assembly, or installation of facilities or equipment (including contractual obligations to purchase such facilities or equipment) at the premises where such equipment will be used, including preparation work at such premises.

(b)(1)(A) The Administrator shall, within ninety days after the date of enactment of this title publish (and from time to time thereafter shall revise) a list of categories of sources, which shall, at the minimum, include:

pulp and paper mills;
paperboard, builders paper and board mills;
meat product and rendering processing;
dairy product processing;
grain mills;
canned and preserved fruits and vegetables processing;
canned and preserved seafood processing;
sugar processing;
textile mills;
cement manufacturing;
feedlots;
electroplating;
organic chemicals manufacturing;
inorganic chemicals manufacturing;
plastic and synthetic materials manufacturing;
soap and detergent manufacturing
fertilizer manufacturing;
petroleum refining;
iron and steel manufacturing;
nonferrous metals manufacturing;
phosphate manufacturing;
steam electric powerplants;
ferroalloy manufacturing;
leather tanning and finishing;
glass and asbestos manufacturing;

rubber processing; and
timber products processing.

(B) As soon as practicable, but in no case more than one year, after a category of sources is included in a list under subparagraph (A) of this paragraph, the Administrator shall propose and publish regulations establishing Federal standards of performance for new sources within such category. The Administrator shall afford interested persons an opportunity for written comment on such proposed regulations. After considering such comments, he shall promulgate, within one hundred and twenty days after publication of such proposed regulations, such standards with such adjustments as he deems appropriate. The Administrator shall, from time to time, as technlogy and alternatives change, revise such standards following the procedure required by this subsection for promulgation of such standards. Standards of performance, or revisions thereof, shall become effective upon promulgation. In establishing or revising Federal standards of performance for new sources under this section, the Administrator shall take into consideration the cost of achieving such effluent reduction, and any non-water quality environmental impact and energy requirements.

(2) The Administrator may distinguish among classes, types, and sizes within categories of new sources for the purpose of establishing such standards and shall consider the type of process employed (including whether batch or continuous).

(3) The provisions of this section shall apply to any new source owned or operated by the United States.

(c) Each State may develop and submit to the Administrator a procedure under State law for applying and enforcing standards of performance for new sources located in such State. If the Administrator finds that the procedure and the law of any State require the application and enforcement of standards of performance to at least the same extent as required by this section, such State is authorized to apply and enforce such standards of performance (except with respect to new sources owned or operated by the United States).

(d) Notwithstanding any other provision of this Act, any point source the construction of which is commenced after the date of enactment of the Federal Water Pollution Control Act Amendments of 1972 and which is so constructed as to meet all applicable standards of performance shall not be subject to any more stringent standard of performance during a ten-year period beginning on the date of completion of such construction or during the period of depreciation or amortization of such facility for the purposes of section 167 or 169 (or both) of the Internal Revenue Code of 1954, whichever period ends first.

(e) After the effective date of standards of performance promulgated under this section, it shall be unlawful for any owner or operator of any new source to operate such source in violation of any standard of performance applicable to such source.

(33 U.S.C. 1316)

TOXIC AND PRETREATMENT EFFLUENT STANDARDS

SEC. 307. (a)(1) On and after the date of enactment of the Clean Water Act of 1977, the list of toxic pollutants or combination of pollutants subject to this Act shall consist of those toxic polllutants listed in table 1 of Committee Print Numbered 95–30 of the Committee on Public Works and Transportation of the House of Representatives, and the Administrator shall publish, not later than the thirtieth day after the date of enactment of the Clean Water Act of 1977, that list. From time to time thereafter, the Administrator may revise such list and the Administrator is authorized to add to or remove from such list any pollutant. The Administrator in publishing any revised list, including the addition or removal of any pollutant from such list, shall take into account the toxicity of the pollutant, its persistence, degradability, the usual or potential presence of the affected organisms in any waters, the importance of the affected organisms, and the nature and extent of the effect of the toxic pollutant on such organisms. A determination of the Administrator under this paragraph shall be final except that if, on judicial review, such determination was based on arbitrary and capricious action of the Administrator, the Administrator shall make a redetermination.

(2) Each toxic pollutant listed in accordance with paragraph (1) of this subsection shall be subject to effluent limitations resulting from the application of the best available technology economically achieveable for the applicable category or class of point sources established in accordance with section 301(b)(2)(A) and 304(b)(2) of this Act. The Administrator, in his discretion, may publish in the Federal Register a proposed effluent standard (which may include a prohibition) establishing requirements for a toxic pollutant which, if an effluent limitation is applicable to a class or category of point sources, shall be applicable to such category or class only if such standard imposes more stringent requirements. Such published effluent standard (or prohibition) shall take into account the toxicity of the pollutant, its persistence, degradability, the usual or potential presence of the affected organisms in any waters, the importance of the affected organisms and the nature and extent of the effect of the toxic pollutant on such organisms, and the extent to which effective control is being or may be achieved under other regulatory authority. The Administrator shall allow a period of not less than sixty days following publication of any such proposed effluent standard (or prohibition) for written comment by interested persons on such proposed standard. In addition, if within thirty days of publication of any such proposed effluent standard (or prohibition) any interested person so requests, the Administrator shall hold a public hearing in connection therewith. Such a public hearing shall provide an opportunity for oral and written presentations, such cross-examination as the Administrator determines is appropriate on disputed issues of material fact, and the transcription of a verbatim record which shall be available to the public. After consideration of such comments and any information and material presented at any public hearing held on such proposed standard or prohibition, the Administrator shall promulgate such standards (or prohibition) with such modifications as the Administrator finds are

justified. Such promulgation by the Administrator shall be made within two hundred and seventy days after publication of proposed standard (or prohibition). Such standard (or prohibition) shall be final except that if, on judicial review, such standard was not based on substantial evidence, the Administrator shall promulgate a revised standard. Effluent limitations shall be established in accordance with sections 301(b)(2)(A) and 304(b)(2) for every toxic pollutant referred to in table 1 of Committee Print Numbered 95–30 of the Committee on Public Works and Transportation of the House of Representatives as soon as practicable after the date of enactment of the Clean Water Act of 1977, but no later than July 1, 1980. Such effluent limitations or effluent standards (or prohibitions) shall be established for every other toxic pollutant listed under paragraph (1) of this subsection as soon as practicable after it is so listed.

(3) Each such effluent standard (or prohibition) shall be reviewed and, if appropriate, revised at least every three years.

(4) Any effluent standard promulgated under this section shall be at that level which the Administrator determines provides an ample margin of safety.

(5) When proposing or promulgating any effluent standard (or prohibition) under this section, the Administrator shall designate the category or categories of sources to which the effluent standard (or prohibition) shall apply. Any disposal of dredged material may be included in such a category of sources after consultation with the Secretary of the Army.

(6) Any effluent standard (or prohibition) established pursuant to this section shall take effect on such date or dates as specified in the order promulgating such standard, but in no case, more than one year from the date of such promulgation. If the Administrator determines that compliance within one year from the date of promulgation is technologically infeasible for a category of sources, the Administrator may establish the effective date of the effluent standard (or prohibition) for such category at the earliest date upon which compliance can be feasibly attained by sources within such category, but in no event more than three years after the date of such promulgation.

(7) Prior to publishing any regulations pursuant to this section the Administrator shall, to the maximum extent practicable within the time provided, consult with appropriate advisory committees, States, independent experts, and Federal departments and agencies.

(b)(1) The Administrator shall, within one hundred and eighty days after the date of enactment of this title and from time to time thereafter, publish proposed regulations establishing pretreatment standards for introduction of pollutants into treatment works (as defined in section 212 of this Act) which are publicly owned for those pollutants which are determined not to be susceptible to treatment by such treatment works or which would interfere with the operation of such treatment works. Not later than ninety days after such publication, and after opportunity for public hearing, the Administrator shall promulgate such pretreatment standards. Pretreatment standards under this subsection shall specify a time for compliance not to exceed three years from the date of promulgation and shall be established to prevent the discharge of any pollutant through treatment works (as defined in section 212 of this Act) which are publicly owned, which pollutant interfers with, passes through, or otherwise is incompatible with such works. If, in the case of any toxic pollutant under subsection (a) of this section introduced by a source into a publicly owned treatment works, the treatment by such works removes all or any part of such toxic pollutant and the discharge from such works does not violate that effluent limitation or standard which would be applicable to such toxic pollutant if it were discharged by such source other than through a publicly owned treatment works, and does not prevent sludge use or disposal by such works in accordance with section 405 of this Act, then the pretreatment requirements for the sources actually discharging such toxic pollutant into such publicly owned treatment works may be revised by the owner or operator of such works to reflect the removal of such toxic pollutant by such works.

(2) The Administrator shall, from time to time, as control technology, processes, operating methods, or other alternative change, revise such standards following the procedures established by this subsection for promulgation of such standards.

(3) When proposing or promulgating any pretreatment standard under this section, the Administrator shall designate the category or categories of sources to which such standard shall apply.

(4) Nothing in this subsection shall affect any pretreatment requirement established by any State or local law not in conflict with any pretreatment standard established under this subsection.

(c) In order to ensure that any source introducing pollutants into a publicly owned treatment works, which source would be a new source subject to section 306 if it were to discharge pollutants, will not cause a violation of the effluent limitations established for any such treatment works, the Administrator shall promulgate pretreatment standards for the category of such sources simultaneously with the promulgation of standards of performance under section 306 for the equivalent category of new sources. Such pretreatment standards shall prevent the discharge of any pollutant into such treatment works, which pollutant may interfere with, pass through, or otherwise be incompatible with such works.

(d) After the effective date of any effluent standard or prohibition or pretreatment standard promulgated under this section, it shall be unlawful for any owner or operator of any source to operate any source in violation of any such effluent standard or prohibition or pretreatment standard.

(e) COMPLIANCE DATE EXTENSION FOR INNOVATIVE PRETREATMENT SYSTEMS.—In the case of any existing facility that proposes to comply with the pretreatment standards of subsection (b) of this section by applying an innovative system that meets the requirements of section 301(k) of this Act, the owner or operator of the publicly owned treatment works receiving the treated effluent from such facility may extend the date for compliance with the applicable pretreatment standard established under this section for a period not to exceed 2 years—

(1) if the Administrator determines that the innovative system has the potential for industrywide application, and

(2) if the Administrator (or the State in consultation with the Administrator, in any case in which the State has a pretreatment program approved by the Administrator)—

(A) determines that the proposed extension will not cause the publicly owned treatment works to be in violation of its permit under section 402 or of section 405 or to contribute to such a violation, and

(B) concurs with the proposed extension.

(33 U.S.C. 1317)

INSPECTIONS, MONITORING, AND ENTRY

SEC. 308. (a) Whenever required to carry out the objective of this Act, including but not limited to (1) developing or assisting in the development of any effluent limitation, or other limitation, prohibition, or effluent standard, pretreatment standard, or standard of performance under this Act; (2) determining whether any person is in violation of any such effluent limitation, or other limitation, prohibition or effluent standard, pretreatment standard, or standard of performance; (3) any requirement established under this section; or (4) carrying out sections 305, 311, 402, 404 (relating to State permit programs), 405, and 504 of this Act—

(A) the Administrator shall require the owner or operator of any point source to (i) establish and maintain such records, (ii) make such reports, (iii) install, use, and maintain such monitoring equipment or methods (including where appropriate, biological monitoring methods), (iv) sample such effluents (in accordance with such methods, at such locations, at such intervals, and in such manner as the Administrator shall prescribe), and (v) provide such other information as he may reasonably require; and

(B) the Administrator or his authorized representative (including an authorized contractor acting as a representative of the Administrator), upon presentation of his credentials—

(i) shall have a right of entry to, upon, or through any premises in which an effluent source is located or in which any records required to be maintained under clause (A) of this subsection are located, and

(ii) may at reasonable times have access to and copy any records, inspect any monitoring equipment or method required under clause (A), and sample any effluents which the owner or operator of such source is required to sample under such clause.

(b) Any records, reports, or information obtained under this section (1) shall, in the case of effluent data, be related to any applicable effluent limitations, toxic, pretreatment, or new source performance standards, and (2) shall be available to the public, except that upon a showing satisfactory to the Administrator by any person that records, reports, or information, or particular part thereof (other than effluent data), to which the Administrator has access under this section, if made public would divulge methods or processes entitled to protection as trade secrets of such person, the Administrator shall consider such record, report, or information, or particular portion thereof confidential in accordance with the purposes of section 1905 of title 18 of the United States Code. Any authorized representative of the Administrator (including an authorized contractor acting as a representative of the Administrator) who knowingly or willfully publishes, divulges, discloses, or makes known in any manner or to any extent not authorized by law any information which is required to be considered confidential under this subsection shall be fined not more than $1,000 or imprisoned not more than 1 year, or both. Nothing in this subsection shall prohibit the Administrator or an authorized representative of the Administrator (including any authorized contractor acting as a representative of the Administrator) from disclosing records, reports, or information to other officers, employees, or authorized representatives of the United States concerned with carrying out this Act or when relevant in any proceeding under this Act.

(c) Each State may develop and submit to the Administrator procedures under State law for inspection, monitoring, and entry with respect to point sources located in such State. If the Administrator finds that the procedures and the law of any State relating to inspection, monitoring, and entry are applicable to at least the same extent as those required by this section, such State is authorized to apply and enforce its procedures for inspection, monitoring, and entry with respect to point sources located in such State (except with respect to point sources owned or operated by the United States).

(d) ACCESS BY CONGRESS.—Notwithstanding any limitation contained in this section or any other provision of law, all information reported to or otherwise obtained by the Administrator (or any representative of the Administrator) under this Act shall be made available, upon written request of any duly authorized committee of Congress, to such committee.

(33 U.S.C. 1318)

FEDERAL ENFORCEMENT

SEC. 309. (a)(1) Whenever, on the basis of any information available to him, the Administrator finds that any person is in violation of any condition or limitation which implements section 301, 302, 306, 307, 308, 318, or 405 of this Act in a permit issued by a State under an approved permit program under section 402 or 404 of this Act, he shall proceed under his authority in paragraph (3) of this subsection or he shall notify the person in alleged violation and such State of such finding. If beyond the thirtieth day after the Administrator's notification the State has not commenced appropriate enforcement action, the Administrator shall issue an order requiring such person to comply with such condition or limitation or shall bring a civil action in accordance with subsection (b) of this section.

(2) Whenever, on the basis of information available to him, the Administrator finds that violations of permit conditions or limitations as set forth in paragraph (1) of this subsection are so widespread that such violations appear to result from a failure of the State to enforce such permit conditions or limitations effectively, he shall so notify the State. If the Administrator finds such failure extends beyond the thirtieth day after such notice, he shall give pub-

lic notice of such finding. During the period beginning with such public notice and ending when such State satisfies the Administrator that it will enforce such conditions and limitations (hereafter referred to in this section as the period of "federally assumed enforcement"), except where an extension has been granted under paragraph (5)(B) of this subsection, the Administrator shall enforce any permit condition or limitation with respect to any person—

(A) by issuing an order to comply with such condition or limitation, or

(B) by bringing a civil action under subsection (b) of this section.

(3) Whenever on the basis of any information available to him the Administrator finds that any person is in violation of section 301, 302, 306, 307, 308, 318, or 405 of this Act, or is in violation of any permit condition or limitation implementing any of such sections in a permit issued under section 402 of this Act by him or by a State or in a permit issued under section 404 of this Act by a State, he shall issue an order requiring such person to comply with such section or requirement, or he shall bring a civil action in accordance with subsection (b) of this section.

(4) A copy of any order issued under this subsection shall be sent immediately by the Administrator to the State in which the violation occurs and other affected States. In any case in which an order under this subsection (or notice to a violator under paragraph (1) of this subsection) is issued to a corporation, a copy of such order (or notice) shall be served on any appropriate corporate officers. An order issued under this subsection relating to a violation of section 308 of this Act shall not take effect until the person to whom it is issued has had an opportunity to confer with the Administrator concerning the alleged violation.

(5)(A) Any order issued under this subsection shall be by personal service, shall state with reasonable specificity the nature of the violation, and shall specify a time for compliance not to exceed thirty days in the case of a violation of an interim compliance schedule or operation and maintenance requirement and not to exceed a time the Administrator determines to be reasonable in the case of a violation of a final deadline, taking into account the seriousness of the violation and any good faith efforts to comply with applicable requirements.

(B) The Administrator may, if he determines (i) that any person who is a violator of, or any person who is otherwise not in compliance with, the time requirements under this Act or in any permit issued under this Act, has acted in good faith, and has made a commitment (in the form of contracts or other securities) of necessary resources to achieve compliance by the earliest possible date after July 1, 1977, but not later than April 1, 1979; (ii) that any extension under this provision will not result in the imposition of any additional controls on any other point or nonpoint source; (iii) that an application for a permit under section 402 of this Act was filed for such person prior to December 31, 1974; and (iv) that the facilities necessary for compliance with such requirements are under construction, grant an extension of the date referred to in section 301(b)(1)(A) to a date which will achieve compliance at the earliest time possible but not later than April 1, 1979.

(6) Whenever, on the basis of information available to him, the Administrator finds (A) that any person is in violation of section 301(b)(1) (A) or (C) of this Act, (B) that such person cannot meet the requirements for a time extension under section 301(i)(2) of this Act, and (C) that the most expeditious and appropriate means of compliance with this Act by such person is to discharge into a publicly owned treatment works, then, upon request of such person, the Administrator may issue an order requiring such person to comply with this Act at the earliest date practicable, but not later than July 1, 1983, by discharging into a publicly owned treatment works if such works concur with such order. Such order shall include a schedule of compliance.

(b) The Administrator is authorized to commence a civil action for appropriate relief, including a permanent or temporary injunction, for any violation for which he is authorized to issue a compliance order under subsection (a) of this section. Any action under this subsection may be brought in the district court of the United States for the district in which the defendant is located or resides or is doing business, and such court shall have jurisdiction to restrain such violation and to require compliance. Notice of the commencement of such action shall be given immediately to the appropriate State.

(c) CRIMINAL PENALTIES.—

(1) NEGLIGENT VIOLATIONS.—Any person who—

(A) negligently violates section 301, 302, 306, 307, 308, 311(b)(3), 318, or 405 of this Act, or any permit condition or limitation implementing any of such sections in a permit issued under section 402 of this Act by the Administrator or by a State, or any requirement imposed in a pretreatment program approved under section 402(a)(3) or 402(b)(8) of this Act or in a permit issued under section 404 of this Act by the Secretary of the Army or by a State; or

(B) negligently introduces into a sewer system or into a publicly owned treatment works any pollutant or hazardous substance which such person knew or reasonably should have known could cause personal injury or property damage or, other than in compliance with all applicable Federal, State, or local requirements or permits, which causes such treatment works to violate any effluent limitation or condition in any permit issued to the treatment works under section 402 of this Act by the Administrator or a State;

shall be punished by a fine of not less than $2,500 nor more than $25,000 per day of violation, or by imprisonment for not more than 1 year, or by both. If a conviction of a person is for a violation committed after a first conviction of such person under this paragraph, punishment shall be by a fine of not more than $50,000 per day of violation, or by imprisonment of not more than 2 years, or by both.

(2) KNOWING VIOLATIONS.—Any person who—

(A) knowingly violates section 301, 302, 306, 307, 308, 311(b)(3), 318, or 405 of this Act, or any permit condition or limitation implementing any of such sections in a permit issued under section 402 of this Act by the Administrator or by a State, or any requirement imposed in a pretreatment program approved under section 402(a)(3) or 402(b)(8) of this Act or in a permit issued under section 404 of this Act by the Secretary of the Army or by a State; or

(B) knowingly introduces into a sewer system or into a publicly owned treatment works any pollutant or hazardous substance which such person knew or reasonably should have known could cause personal injury or property damage or, other than in compliance with all applicable Federal, State, or local requirements or permits, which causes such treatment works to violate any effluent limitation or condition in a permit issued to the treatment works under section 402 of this Act by the Administrator or a State;

shall be punished by a fine of not less that $5,000 nor more than $50,000 per day of violation, or by imprisonment for not more than 3 years, or by both. If a conviction of a person is for a violation committed after a first conviction of such person under this paragraph, punishment shall be by a fine of not more than $100,000 per day of violation, or imprisonment of not more than 6 years, or by both.

(3) KNOWING ENDANGERMENT.—

(A) GENERAL RULE.—Any person who knowingly violates section 301, 302, 306, 307, 308, 311(b)(3), 318, or 405 of this Act, or any permit condition or limitation implementing any of such sections in a permit issued under section 402 of this Act by the Administrator or by a State, or in a permit issued under section 404 of this Act by the Secretary of the Army or by a State, and who knows at that time that he thereby places another person in imminent danger of death or serious bodily injury, shall, upon conviction, be subject to a fine of not more than $250,000 or imprisonment of not more than 15 years, or both. A person which is an organization shall, upon conviction of violating this subparagraph, be subject to a fine of not more than $1,000,000. If a conviction of a person is for a violation committed after a first conviction of such person under this paragraph, the maximum punishment shall be doubled with respect to both fine and imprisonment.

(B) ADDITIONAL PROVISIONS.—For the purpose of subparagraph (A) of this paragraph—

(i) in determining whether a defendant who is an individual knew that his conduct placed another person in imminent danger of death or serious bodily injury—

(I) the person is responsible only for actual awareness or actual belief that he possessed; and

(II) knowledge possessed by a person other than the defendant but not by the defendant himself may not be attributed to the defendant;

except that in proving the defendant's possession of actual knowledge, circumstantial evidence may be used, including evidence that the defendant took affirmative steps to shield himself from relevant information;

(ii) it is an affirmative defense to prosecution that the conduct charged was consented to by the person endangered and that the danger and conduct charged were reasonably foreseeable hazards of—

(I) an occupation, a business, or a profession; or

(II) medical treatment or medical or scientific experimentation conducted by professionally approved methods and such other person had been made aware of the risks involved prior to giving consent;

and such defense may be established under this subparagraph by a preponderance of the evidence;

(iii) the term "organization" means a legal entity, other than a government, established or organized for any purpose, and such term includes a corporation, company, association, firm, partnership, joint stock company, foundation, institution, trust, society, union, or any other association of persons; and

(iv) the term "serious bodily injury" means bodily injury which involves a substantial risk of death, unconsciousness, extreme physical pain, protracted and obvious disfigurement, or protracted loss or impairment of the function of a bodily member, organ, or mental faculty.

(4) FALSE STATEMENTS.—Any person who knowingly makes any false material statement, representation, or certification in any application, record, report, plan, or other document filed or required to be maintained under this Act or who knowingly falsifies, tampers with, or renders inaccurate any monitoring device or method required to be maintained under this Act, shall upon conviction, be punished by a fine of not more than $10,000, or by imprisonment for not more than 2 years, or by both. If a conviction of a person is for a violation committed after a first conviction of such person under this paragraph, punishment shall be by a fine of not more than $20,000 per day of violation, or by imprisonment of not more than 4 years, or by both.

(5) TREATMENT OF SINGLE OPERATIONAL UPSET.—For purposes of this subsection, a single operational upset which leads to simultaneous violations of more than one pollutant parameter shall be treated as a single violation.

(6) RESPONSIBLE CORPORATE OFFICER AS "PERSON".—For the purpose of this subsection, the term "person" means, in addition to the definition contained in section 502(5) of this Act, any responsible corporate officer.

(7) HAZARDOUS SUBSTANCE DEFINED.—For the purpose of this subsection, the term "hazardous substance" means (A) any substance designated pursuant to section 311(b)(2)(A) of this Act, (B) any element, compound, mixture, solution, or substance designated pursuant to section 102 of the Comprehen-

sive Environmental Response, Compensation, and Liability Act of 1980, (C) any hazardous waste having the characteristics identified under or listed pursuant to section 3001 of the Solid Waste Disposal Act (but not including any waste the regulation of which under the Solid Waste Disposal Act has been suspended by Act of Congress), (D) any toxic pollutant listed under section 307(a) of this Act, and (E) any imminently hazardous chemical substance or mixture with respect to which the Administrator has taken action pursuant to section 7 of the Toxic Substances Control Act.

(d) Any person who violates section 301, 302, 306, 307, 308, 311(b)(3), 318 or 405 of this Act, or any permit condition or limitation implementing any of such sections in a permit issued under section 402 of this Act by the Administrator, or by a State, or in a permit issued under section 404 of this Act by a State,,[1] or any requirement imposed in a pretreatment program approved under section 402(a)(3) or 402(b)(8) of this Act, and any person who violates any order issued by the Administrator under subsection (a) of this section, shall be subject to a civil penalty not to exceed $25,000 per day for each violation. In determining the amount of a civil penalty the court shall consider the seriousness of the violation or violations, the economic benefit (if any) resulting from the violation, any history of such violations, any good-faith efforts to comply with the applicable requirements, the economic impact of the penalty on the violator, and such other matters as justice may require. For purposes of this subsection, a single operational upset which leads to simultaneous violations of more than one pollutant parameter shall be treated as a single violation.

(e) Whenever a municipality is a party to a civil action brought by the United States under this section, the State in which such municipality is located shall be joined as a party. Such State shall be liable for payment of any judgment, or any expenses incurred as a result of complying with any judgment, entered against the municipality in such action to the extent that the laws of that State prevent the municipality from raising revenues needed to comply with such judgment.

(f) Whenever, on the basis of an information available to him, the Administrator finds that an owner or operator of any source is introducing a pollutant into a treatment works in violation of subsection (d) of section 307, the Administrator may notify the owner or operator of such treatment works and the State of such violation. If the owner or operator of the treatment works does not commence appropriate enforcement action within 30 days of the date of such notification, the Administrator may commence a civil action for appropriate relief, including but not limited to, a permanent or temporary injunction, against the owner or operator of such treatment works. In any such civil action the Administrator shall join the owner or operator of such source as a party to the action. Such action shall be brought in the district court of the United States in the district in which the treatment works is located. Such court shall have jurisdiction to restrain such violation and to require the owner or operator of the treatment works and the owner or oper-

---

[1] So in law. See P.L. 100–4, sec. 313(a)(1), 101 Stat. 45.

ator of the source to take such action as may be necessary to come into compliance with this Act. Notice of commencement of any such action shall be given to the State. Nothing in this subsection shall be construed to limit or prohibit any other authority the Administrator may have under this Act.

(g) ADMINISTRATIVE PENALTIES.—

(1) VIOLATIONS.—Whenever on the basis of any information available—

(A) the Administrator finds that any person has violated section 301, 302, 306, 307, 308, 318, or 405 of this Act, or has violated any permit condition or limitation implementing any of such sections in a permit issued under section 402 of this Act by the Administrator or by a State, or in a permit issued under section 404 by a State, or

(B) the Secretary of the Army (hereinafter in this subsection referred to as the "Secretary") finds that any person has violated any permit condition or limitation in a permit issued under section 404 of this Act by the Secretary,

the Administrator or Secretary, as the case may be, may, after consultation with the State in which the violation occurs, assess a class I civil penalty or a class II civil penalty under this subsection.

(2) CLASSES OF PENALTIES.—

(A) CLASS I.—The amount of a class I civil penalty under paragraph (1) may not exceed $10,000 per violation, except that the maximum amount of any class I civil penalty under this subparagraph shall not exceed $25,000. Before issuing an order assessing a civil penalty under this subparagraph, the Administrator or the Secretary, as the case may be, shall give to the person to be assessed such penalty written notice of the Administrator's or Secretary's proposal to issue such order and the opportunity to request, within 30 days of the date the notice is received by such person, a hearing on the proposed order. Such hearing shall not be subject to section 554 or 556 of title 5, United States Code, but shall provide a reasonable opportunity to be heard and to represent evidence.

(B) CLASS II.—The amount of a class II civil penalty under paragraph (1) may not exceed $10,000 per day for each day during which the violation continues; except that the maximum amount of any class II civil penalty under this subparagraph shall not exceed $125,000. Except as otherwise provided in this subsection, a class II civil penalty shall be assessed and collected in the same manner, and subject to the same provisions, as in the case of civil penalties assessed and collected after notice and opportunity for a hearing on the record in accordance with section 554 of title 5, United States Code. The Administrator and the Secretary may issue rules for discovery procedures for hearings under this subparagraph.

(3) DETERMINING AMOUNT.—In determining the amount of any penalty assessed under this subsection, the Administrator or the Secretary, as the case may be, shall take into account

the nature, circumstances, extent and gravity of the violation, or violations, and, with respect to the violator, ability to pay, any prior history of such violations, the degree of culpability, economic benefit or savings (if any) resulting from the violation, and such other matters as justice may require. For purposes of this subsection, a single operational upset which leads to simultaneous violations of more than one pollutant parameter shall be treated as a single violation.

(4) RIGHTS OF INTERESTED PERSONS.—

(A) PUBLIC NOTICE.—Before issuing an order assessing a civil penalty under this subsection the Administrator or Secretary, as the case may be, shall provide public notice of and reasonable opportunity to comment on the proposed issuance of such order.

(B) PRESENTATION OF EVIDENCE.—Any person who comments on a proposed assessment of a penalty under this subsection shall be given notice of any hearing held under this subsection and of the order assessing such penalty. In any hearing held under this subsection, such person shall have a reasonable opportunity to be heard and to present evidence.

(C) RIGHTS OF INTERESTED PERSONS TO A HEARING.—If no hearing is held under paragraph (2) before issuance of an order assessing a penalty under this subsection, any person who commented on the proposed assessment may petition, within 30 days after the issuance of such order, the Administrator or Secretary, as the case may be, to set aside such order and to provide a hearing on the penalty. If the evidence presented by the petitioner in support of the petition is material and was not considered in the issuance of the order, the Administrator or Secretary shall immediately set aside such order and provide a hearing in accordance with paragraph (2)(A) in the case of a class I civil penalty and paragraph (2)(B) in the case of a class II civil penalty. If the Administrator or Secretary denies a hearing under this subparagraph, the Administrator or Secretary shall provide to the petitioner, and publish in the Federal Register, notice of and the reasons for such denial.

(5) FINALITY OF ORDER.—An order issued under this subsection shall become final 30 days after its issuance unless a petition for judicial review is filed under paragraph (8) or a hearing is requested under paragraph (4)(C). If such a hearing is denied, such order shall become final 30 days after such denial.

(6) EFFECT OF ORDER.—

(A) LIMITATION ON ACTIONS UNDER OTHER SECTIONS.—Action taken by the Administrator or the Secretary, as the case may be, under this subsection shall not affect or limit the Administrator's or Secretary's authority to enforce any provision of this Act; except that any violation—

(i) with respect to which the Administrator or the Secretary has commenced and is diligently prosecuting an action under this subsection,

(ii) with respect to which a State has commenced and is diligently prosecuting an action under a State law comparable to this subsection, or

(iii) for which the Administrator, the Secretary, or the State has issued a final order not subject to further judicial review and the violator has paid a penalty assessed under this subsection, or such comparable State law, as the case may be,

shall not be the subject of a civil penalty action under subsection (d) of this section or section 311(b) or section 505 of this Act.

(B) APPLICABILITY OF LIMITATION WITH RESPECT TO CITIZEN SUITS.—The limitations contained in subparagraph (A) on civil penalty actions under section 505 of this Act shall not apply with respect to any violation for which—

(i) a civil action under section 505(a)(1) of this Act has been filed prior to commencement of an action under this subsection, or

(ii) notice of an alleged violation of section 505(a)(1) of this Act has been given in accordance with section 505(b)(1)(A) prior to commencement of an action under this subsection and an action under section 505(a)(1) with respect to such alleged violation is filed before the 120th day after the date on which such notice is given.

(7) EFFECT OF ACTION ON COMPLIANCE.—No action by the Administrator or the Secretary under this subsection shall affect any person's obligation to comply with any section of this Act or with the terms and conditions of any permit issued pursuant to section 402 or 404 of this Act.

(8) JUDICIAL REVIEW.—Any person against whom a civil penalty is assessed under this subsection or who commented on the proposed assessment of such penalty in accordance with paragraph (4) may obtain review of such assessment—

(A) in the case of assessment of a class I civil penalty, in the United States District Court for the District of Columbia or in the district in which the violation is alleged to have occurred, or

(B) in the case of assessment of a class II civil penalty, in United States Court of Appeals for the District of Columbia Circuit or for any other circuit in which such person resides or transacts business,

by filing a notice of appeal in such court within the 30-day period beginning on the date the civil penalty order is issued and by simultaneously sending a copy of such notice by certified mail to the Administrator or the Secretary, as the case may be, and the Attorney General. The Administrator or the Secretary shall promptly file in such court a certified copy of the record on which the order was issued. Such court shall not set aside or remand such order unless there is not substantial evidence in the record, taken as a whole, to support the finding of a violation or unless the Administrator's or Secretary's assessment of the penalty constitutes an abuse of discretion and shall not impose additional civil penalties for the same violation unless the Administrator's or Secretary's assessment of the penalty constitutes an abuse of discretion.

(9) COLLECTION.—If any person fails to pay an assessment of a civil penalty—

(A) after the order making the assessment has become final, or

(B) after a court in an action brought under paragraph (8) has entered a final judgment in favor of the Administrator or the Secretary, as the case may be,

the Administrator or the Secretary shall request the Attorney General to bring a civil action in an appropriate district court to recover the amount assessed (plus interest at currently prevailing rates from the date of the final order or the date of the final judgment, as the case may be). In such an action, the validity, amount, and appropriateness of such penalty shall not be subject to review. Any person who fails to pay on a timely basis the amount of an assessment of a civil penalty as described in the first sentence of this paragraph shall be required to pay, in addition to such amount and interest, attorneys fees and costs for collection proceedings and a quarterly nonpayment penalty for each quarter during which such failure to pay persists. Such nonpayment penalty shall be in an amount equal to 20 percent of the aggregate amount of such person's penalties and nonpayment penalties which are unpaid as of the beginning of such quarter.

(10) SUBPOENAS.—The Administrator or Secretary, as the case may be, may issue subpoenas for the attendance and testimony of witnesses and the production of relevant papers, books, or documents in connection with hearings under this subsection. In case of contumacy or refusal to obey a subpoena issued pursuant to this paragraph and served upon any person, the district court of the United States for any district in which such person is found, resides, or transacts business, upon application by the United States and after notice to such person, shall have jurisdiction to issue an order requiring such person to appear and give testimony before the administrative law judge or to appear and produce documents before the administrative law judge, or both, and any failure to obey such order of the court may be punished by such court as a contempt thereof.

(11) PROTECTION OF EXISTING PROCEDURES.—Nothing in this subsection shall change the procedures existing on the day before the date of the enactment of the Water Quality Act of 1987 under other subsections of this section for issuance and enforcement of orders by the Administrator.

(33 U.S.C. 1319)

INTERNATIONAL POLLUTION ABATEMENT

SEC. 310. (a) Whenever the Administrator, upon receipts of reports, surveys, or studies from any duly constituted international agency, has reason to believe that pollution is occurring which endangers the health or welfare of persons in a foreign country, and the Secretary of State requests him to abate such pollution, he shall give formal notification thereof to the State water pollution control agency of the State or States in which such discharge or discharges originate and to the appropriate interstate agency, if any. He shall also promptly call such a hearing, if he believes that such pollution is occurring in sufficient quantity to warrant such action, and if such foreign country has given the United States essentially the same rights with respect to the prevention and control of pollution occurring in that country as is given that country by this subsection. The Administrator, through the Secretary of State, shall invite the foreign country which may be adversely affected by the pollution to attend and participate in the hearing, and the representative of such country shall, for the purpose of the hearing and any further proceeding resulting from such hearing, have all the rights of a State water pollution control agency. Nothing in this subsection shall be construed to modify, amend, repeal, or otherwise affect the provisions of the 1909 Boundary Waters Treaty between Canada and the United States or the Water Utilization Treaty of 1944 between Mexico and the United States (59 Stat. 1219), relative to the control and abatement of pollution in waters covered by those treaties.

(b) The calling of a hearing under this section shall not be construed by the courts, the Administrator, or any person as limiting, modifying, or otherwise affecting the functions and responsibilities of the Administrator under this section to establish and enforce water quality requirements under this Act.

(c) The Administrator shall publish in the Federal Register a notice of a public hearing before a hearing board of five or more persons appointed by the Administrator. A majority of the members of the board and the chairman who shall be designated by the Administrator shall not be officers or employees of Federal, State, or local governments. On the basis of the evidence presented at such hearing, the board shall within sixty days after completion of the hearing make findings of fact as to whether or not such pollution is occurring and shall thereupon by decision, incorporating its findings therein, make such recommendations to abate the pollution as may be appropriate and shall transmit such decision and the record of the hearings to the Administrator. All such decisions shall be public. Upon receipt of such decision, the Administrator shall promptly implement the board's decision in accordance with the provisions of this Act.

(d) In connection with any hearing called under this subsection, the board is authorized to require any persons whose alleged activities result in discharges causing or contributing to pollution to file with it in such forms as it may prescribe, a report based on existing data, furnishing such information as may reasonably be required as to the character, kind, and quantity of such discharges and the use of facilities or other means to prevent or reduce such discharges by the person filing such a report. Such report shall be made under oath or otherwise, as the board may prescribe, and shall be filed with the board within such reasonable period as it may prescribe, unless additional time is granted by it. Upon a showing satisfactory to the board by the person filing such report that such report or portion thereof (other than effluent data), to which the Adminsitrator has access under this section, if made public would divulge trade secrets or secret processes of such person, the board shall consider such report or portion thereof confidential for the purposes of section 1905 of title 18 of the United States Code. If any person required to file any report under this paragraph shall fail to do so within the time fixed by the board for filing the same, and such failure shall continue for thirty days after notice of such default, such person shall forfeit to the United States the sum of $1,000 for each and every day of the continuance of such failure, which forfeiture shall be payable into the Treasury of the United States, and shall be recoverable in a civil suit in the name of the United States in the district court of the United States where such person has his principal office or in any district in which he does business. The Administrator may upon application therefor remit or mitigate any forfeiture provided for under this subsection.

(e) Board members, other than officers or employees of Federal, State, or local governments, shall be for each day (including traveltime) during which they are performing board business, entitled to receive compensation at a rate fixed by the Administrator but not in excess of the maximum rate of pay for grade GS–18, as provided in the General Schedule under section 5332 of title 5 of the United States Code, and shall, notwithstanding the limitations of sections 5703 and 5704 of title 5 of the United States Code, be fully reimbursed for travel, subsistence, and related expenses.

(f) When any such recommendation adopted by the Administrator involves the institution of enforcement proceedings against any person to obtain the abatement of pollution subject to such recommendation, the Administrator shall institute such proceedings if he believes that the evidence warrants such proceedings. The district court of the United States shall consider and determine de novo all relevant issues, but shall receive in evidence the record of the proceedings before the conference or hearing board. The court shall have jurisdiction to enter such judgment and orders enforcing such judgment as it deems appropriate or to remand such proceedings to the Administrator for such further action as it may direct.

(33 U.S.C. 1320)

OIL AND HAZARDOUS SUBSTANCE LIABILITY

SEC. 311. (a) For the purpose of this section, the term—

(1) "oil" means oil of any kind or in any form, including, but not limited to, petroleum, fuel oil, sludge, oil refuse, and oil mixed with wastes other than dredged spoil;

(2) "discharge" includes, but is not limited to, any spilling, leaking, pumping, pouring, emitting, emptying or dumping, but excludes (A) discharges in compliance with a permit under section 402 of this Act, (B) discharges resulting from circumstances identified and reviewed and made a part of the public record with respect to a permit issued or modified under section 402 of this Act, and subject to a condition in such permit, ,(C)[1] continuous or anticipated intermittent discharges from a point source, identified in a permit or permit applica-

[1] So in law.

tion under section 402 of this Act, which are caused by events occurring within the scope of relevant operating or treatment systems, and (D) discharges incidental to mechanical removal authorized by the President under subsection (c) of this section;

(3) "vessel" means every description of watercraft or other artificial contrivance used, or capable of being used, as a means of transportation on water other than a public vessel;

(4) "public vessel" means a vessel owned or bareboat-chartered and operated by the United States, or by a State or political subdivision thereof, or by a foreign nation, except when such vessel is engaged in commerce;

(5) "United States" means the States, the District of Columbia, the Commonwealth of Puerto Rico, the Commonwealth of the Northern Mariana Islands, Guam, American Samoa, the Virgin Islands, and the Trust Territory of the Pacific Islands;

(6) "owner or operator" means (A) in the case of a vessel, any person owning, operating, or chartering by demise, such vessel, and (B) in the case of an onshore facility, and an offshore facility, any person owning or operating such onshore facility or offshore facility, and (C) in the case of any abandoned offshore facility, the person who owned or operated such facility immediately prior to such abandonment;

(7) "person" includes an individual, firm, corporation, association, and a partnership;

(8) "remove" or "removal" refers to containment and removal of the oil or hazardous substances from the water and shorelines or the taking of such other actions as may be necessary to prevent, minimize, or mitigate damage to the public health or welfare, including, but not limited to, fish, shellfish, wildlife, and public and private property, shorelines, and beaches;

(9) "contiguous zone" means the entire zone established or to be established by the United States under article 24 of the Convention on the Territorial Sea and the Contiguous Zone;

(10) "onshore facility" means any facility (including, but not limited to, motor vehicles and rolling stock) of any kind located in, on, or under, any land within the United States other than submerged land;

(11) "offshore facility" means any facility of any kind located in, on, or under, any of the navigable waters of the United States, and any facility of any kind which is subject to the jurisdiction of the United States and is located in, on, or under any other waters, other than a vessel or a public vessel;

(12) "act of God" means an act occasioned by an unanticipated grave natural disaster;

(13) "barrel" means 42 United States gallons at 60 degrees Fahrenheit;

(14) "hazardous substance" means any substance designated pursuant to subsection (b)(2) of this section;

(15) "inland oil barge" means a non-self-propelled vessel carrying oil in bulk as cargo and certificated to operate only in the inland waters of the United States, while operating in such waters;

(16) "inland waters of the United States" means those waters of the United States lying inside the baseline from which the territorial sea is measured and those water outside such baseline which are a part of the Gulf Intracoastal Waterway;

(17) "otherwise" subject to the jurisdiction of the United States" means subject to the jurisdiction of the United States by virtue of United States citizenship, United States vessel documentation or numbering, or as provided for by international agreement to which the United States is a party;

(18) "Area Committee" means an Area Committee established under subsection (j);

(19) "Area Contingency Plan" means an Area Contingency Plan prepared under subsection (j);

(20) "Coast Guard District Response Group" means a Coast Guard District Response Group established under subsection (j);

(21) "Federal On-Scene Coordinator" means a Federal On-Scene Coordinator designated in the National Contingency Plan;

(22) "National Contingency Plan" means the National Contingency Plan prepared and published under subsection (d);

(23) "National Response Unit" means the National Response Unit established under subsection (j);

(24) "worst case discharge" means—

(A) in the case of a vessel, a discharge in adverse weather conditions of its entire cargo; and

(B) in the case of an offshore facility or onshore facility, the largest foreseeable discharge in adverse weather conditions; and

(25) "removal costs" means—

(A) the costs of removal of oil or a hazardous substance that are incurred after it is discharged; and

(B) in any case in which there is a substantial threat of a discharge of oil or a hazardous substance, the costs to prevent, minimize, or mitigate that threat.

(b)(1) The Congress hereby declares that it is the policy of the United States that there should be no discharges of oil or hazardous substances into or upon the navigable waters of the United States, adjoining shorelines, or into or upon the waters of the contiguous zone, or in connection with activities under the Outer Continental Shelf Lands Act or the Deepwater Port Act of 1974, or which may affect natural resources belonging to, appertaining to, or under the exclusive management authority of the United States (including resources under the Fishery Conservation and Management Act of 1976).

(2)(A) The Administrator shall develop, promulgate, and revise as may be appropriate, regulations designating as hazardous substances, other than oil as defined in this section, such elements and compounds which, when discharged in any quantity into or upon the navigable waters of the United States or adjoining shorelines or the waters of the contiguous zone or in connection with activities under the Outer Continental Shelf Lands Act or the Deepwater Port Act of 1974, or which may affect natural resources belonging to, appertaining to, or under the exclusive management authority of the United States (including resources under the Fishery Conservation and Management Act of 1976), present an imminent and substantial danger to the public health or welfare, including, but not limited to, fish, shellfish, wildlife, shorelines, and beaches.

(B) The Administrator shall within 18 months after the date of enactment of this paragraph, conduct a study and report to the Congress on methods, mechanisms, and procedures to create incentives to achieve a higher standard of care in all aspects of the management and movement of hazardous substances on the part of owners, operators, or persons in charge of onshore facilities, offshore facilities, or vessels. The Administrator shall include in such study (1) limits of liability, (2) liability for third party damages, (3) penalties and fees, (4) spill prevention plans, (5) current practices in the insurance and banking industries, and (6) whether the penalty enacted in subclause (bb) of clause (iii) of subparagraph (B) of subsection (b)(2) of section 311 of Public Law 92–500 should be enacted.

(3) The discharge of oil or hazardous substances (i) into or upon the navigable waters of the United States, adjoining shorelines, or into or upon the waters of the contiguous zone, or (ii) in connection with activities under the Outer Continental Shelf Lands Act or the Deepwater Port Act of 1974, or which may affect natural resources belonging to, appertaining to, or under the exclusive management authority of the United States (including resources under the Fishery Conservation and Management Act of 1976), in such quantities as may be harmful as determined by the President under paragraph (4) of this subsection, is prohibited, except (A) in the case of such discharges into the waters of the contiguous zone or which may affect natural resources belonging to, appertaining to, or under the exclusive management authority of the United States (including resources under the Fishery Conservation and Management Act of 1976), where permitted under the Protocol of 1978 Relating to the International Convention for the Prevention of Pollution from Ships, 1973, and (B) where permitted in quantities and at times and locations or under such circumstances or conditions as the President may, by regulation, determine not to be harmful. Any regulations issued under this subsection shall be consistent with maritime safety and with marine and navigation laws and regulations and applicable water quality standards.

(4) The President shall by regulation determine for the purposes of this section those quantities of oil and any hazardous substances the discharge of which may be harmful to the public health or welfare or the environment of the United States, including but not limited to fish, shellfish, wildlife, and public and private property, shorelines, and beaches.

(5) Any person in charge of a vessel or of an onshore facility or an offshore facility shall, as soon as he has knowledge of any discharge of oil or a hazardous substance from such vessel or facility in violation of paragraph (3) of this subsection, immediately notify the appropriate agency of the United States Government of such discharge. The Federal agency shall immediately notify the appropriate State agency of any State which is, or may reasonably be expected to be, affected by the discharge of oil or a hazardous substance. Any such person (A) in charge of a vessel from which

oil or a hazardous substance is discharged in violation of pragraph (3)(i) of this subsection, or (B) in charge of a vessel from which oil or a hazardous substance is discharged in violation of paragraph (3)(ii) of this subsection and who is otherwise subject to the jurisdiction of the United States at the time of the discharge, or (C) in charge of an onshore facility or an offshore facility, who fails to notify immediately such agency of such discharge shall, upon conviction, be fined in accordance with title 18, United States Code, or imprisoned for not more than 5 years, or both. Notification received pursuant to this paragraph shall not be used against any such natural person in any criminal case, except a prosecution for perjury or for giving a false statement.

(6) ADMINISTRATIVE PENALTIES.—

(A) VIOLATIONS.—Any owner, operator, or person in charge of any vessel, onshore facility, or offshore facility—

(i) from which oil or a hazardous substance is discharged in violation of paragraph (3), or

(ii) who fails or refuses to comply with any regulation issued under subsection (j) to which that owner, operator, or person in charge is subject,

may be assessed a class I or class II civil penalty by the Secretary of the department in which the Coast Guard is operating or the Administrator.

(B) CLASSES OF PENALTIES.—

(i) CLASS I.—The amount of a class I civil penalty under subparagraph (A) may not exceed $10,000 per violation, except that the maximum amount of any class I civil penalty under this subparagraph shall not exceed $25,000. Before assessing a civil penalty under this clause, the Administrator or Secretary, as the case may be, shall give to the person to be assessed such penalty written notice of the Administrator's or Secretary's proposal to assess the penalty and the opportunity to request, within 30 days of the date the notice is received by such person, a hearing on the proposed penalty. Such hearing shall not be subject to section 554 or 556 of title 5, United States Code, but shall provide a reasonable opportunity to be heard and to present evidence.

(ii) CLASS II.—The amount of a class II civil penalty under subparagraph (A) may not exceed $10,000 per day for each day during which the violation continues; except that the maximum amount of any class II civil penalty under this subparagraph shall not exceed $125,000. Except as otherwise provided in this subsection, a class II civil penalty shall be assessed and collected in the same manner, and subject to the same provisions, as in the case of civil penalties assessed and collected after notice and opportunity for a hearing on the record in accordance with section 554 of title 5, United States Code. The Administrator and Secretary may issue rules for discovery procedures for hearings under this paragraph.

(C) RIGHTS OF INTERESTED PERSONS.—

(i) PUBLIC NOTICE.—Before issuing an order assessing a class II civil penalty under this paragraph the Administrator or Secretary, as the case may be, shall provide public notice of and reasonable opportunity to comment on the proposed issuance of such order.

(ii) PRESENTATION OF EVIDENCE.—Any person who comments on a proposed assessment of a class II civil penalty under this paragraph shall be given notice of any hearing held under this paragraph and of the order assessing such penalty. In any hearing held under this paragraph, such person shall have a reasonable opportunity to be heard and to present evidence.

(iii) RIGHTS OF INTERESTED PERSONS TO A HEARING.—If no hearing is held under subparagraph (B) before issuance of an order assessing a class II civil penalty under this paragraph, any person who commented on the proposed assessment may petition, within 30 days after the issuance of such order, the Administrator or Secretary, as the case may be, to set aside such order and to provide a hearing on the penalty. If the evidence presented by the petitioner in support of the petition is material and was not considered in the issuance of the order, the Administrator or Secretary shall immediately set aside such order and provide a hearing in accordance with subparagraph (B)(ii). If the Administrator or Secretary denies a hearing under this clause, the Administrator or Secretary shall provide to the petitioner, and publish in the Federal Register, notice of and the reasons for such denial.

(D) FINALITY OF ORDER.—An order assessing a class II civil penalty under this paragraph shall become final 30 days after its issuance unless a petition for judicial review is filed under subparagraph (G) or a hearing is requested under subparagraph (C)(iii). If such a hearing is denied, such order shall become final 30 days after such denial.

(E) EFFECT OF ORDER.—Action taken by the Administrator or Secretary, as the case may be, under this paragraph shall not affect or limit the Administrator's or Secretary's authority to enforce any provision of this Act; except that any violation—

(i) with respect to which the Administrator or Secretary has commenced and is diligently prosecuting an action to assess a class II civil penalty under this paragraph, or

(ii) for which the Administrator or Secretary has issued a final order assessing a class II civil penalty not subject to further judicial review and the violator has paid a penalty assessed under this paragraph,

shall not be the subject of a civil penalty action under section 309(d), 309(g), or 505 of this Act or under paragraph (7).

(F) EFFECT OF ACTION ON COMPLIANCE.—No action by the Administrator or Secretary under this paragraph shall affect any person's obligation to comply with any section of this Act.

(G) JUDICIAL REVIEW.—Any person against whom a civil penalty is assessed under this paragraph or who commented on the proposed assessment of such penalty in accordance with subparagraph (C) may obtain review of such assessment—

(i) in the case of assessment of a class I civil penalty, in the United States District Court for the District of Columbia or in the district in which the violation is alleged to have occurred, or

(ii) in the case of assessment of a class II civil penalty, in United States Court of Appeals for the District of Columbia Circuit or for any other circuit in which such person resides or transacts business,

by filing a notice of appeal in such court within the 30-day period beginning on the date the civil penalty order is issued and by simultaneously sending a copy of such notice by certified mail to the Administrator or Secretary, as the case may be, and the Attorney General. The Administrator or Secretary shall promptly file in such court a certified copy of the record on which the order was issued. Such court shall not set aside or remand such order unless there is not substantial evidence in the record, taken as a whole, to support the finding of a violation or unless the Administrator's or Secretary's assessment of the penalty constitutes an abuse of discretion and shall not impose additional civil penalties for the same violation unless the Administrator's or Secretary's assessment of the penalty constitutes an abuse of discretion.

(H) COLLECTION.—If any person fails to pay an assessment of a civil penalty—

(i) after the assessment has become final, or

(ii) after a court in an action brought under subparagraph (G) has entered a final judgment in favor of the Administrator or Secretary, as the case may be,

the Administrator or Secretary shall request the Attorney General to bring a civil action in an appropriate district court to recover the amount assessed (plus interest at currently prevailing rates from the date of the final order or the date of the final judgment, as the case may be). In such an action, the validity, amount, and appropriateness of such penalty shall not be subject to review. Any person who fails to pay on a timely basis the amount of an assessment of a civil penalty as described in the first sentence of this subparagraph shall be required to pay, in addition to such amount and interest, attorneys fees and costs for collection proceedings and a quarterly nonpayment penalty for each quarter during which such failure to pay persists. Such nonpayment penalty shall be in an amount equal to 20 percent of the aggregate amount of such person's penalties and nonpayment penalties which are unpaid as of the beginning of such quarter.

(I) SUBPOENAS.—The Administrator or Secretary, as the case may be, may issue subpoenas for the attendance and testimony of witnesses and the production of relevant papers, books, or documents in connection with hearings under this paragraph. In case of contumacy or refusal to obey a subpoena issued pursuant to this subparagraph and served upon any person, the district court of the United States for any district in which such person is found, resides, or transacts business, upon application by the United States and after notice to such person, shall have jurisdiction to issue an order requiring such person to appear and give testimony before the administrative law judge or to appear and produce documents before the administrative law judge, or both, and any failure to obey such order of the court may be punished by such court as a contempt thereof.

(7) CIVIL PENALTY ACTION.—

(A) DISCHARGE, GENERALLY.—Any person who is the owner, operator, or person in charge of any vessel, onshore facility, or offshore facility from which oil or a hazardous substance is discharged in violation of paragraph (3), shall be subject to a civil penalty in an amount up to $25,000 per day of violation or an amount up to $1,000 per barrel of oil or unit of reportable quantity of hazardous substances discharged.

(B) FAILURE TO REMOVE OR COMPLY.—Any person described in subparagraph (A) who, without sufficient cause—

(i) fails to properly carry out removal of the discharge under an order of the President pursuant to subsection (c); or

(ii) fails to comply with an order pursuant to subsection (e)(1)(B);

shall be subject to a civil penalty in an amount up to $25,000 per day of violation or an amount up to 3 times the costs incurred by the Oil Spill Liability Trust Fund as a result of such failure.

(C) FAILURE TO COMPLY WITH REGULATION.—Any person who fails or refuses to comply with any regulation issued under subsection (j) shall be subject to a civil penalty in an amount up to $25,000 per day of violation.

(D) GROSS NEGLIGENCE.—In any case in which a violation of paragraph (3) was the result of gross negligence or willful misconduct of a person described in subparagraph (A), the person shall be subject to a civil penalty of not less than $100,000, and not more than $3,000 per barrel of oil or unit of reportable quantity of hazardous substance discharged.

(E) JURISDICTION.—An action to impose a civil penalty under this paragraph may be brought in the district court of the United States for the district in which the defendant is located, resides, or is doing business, and such court shall have jurisdiction to assess such penalty.

(F) LIMITATION.—A person is not liable for a civil penalty under this paragraph for a discharge if the person has been assessed a civil penalty under paragraph (6) for the discharge.

(8) DETERMINATION OF AMOUNT.—In determining the amount of a civil penalty under paragraphs (6) and (7), the Administrator, Secretary, or the court, as the case may be, shall consider the seriousness of the violation or violations, the economic benefit to the violator, if any, resulting from the violation, the degree of culpability involved, any other penalty for the same incident, any history of prior violations, the nature, extent, and degree of success of any efforts of the violator to minimize or mitigate the effects of the discharge, the economic impact of the penalty on the violator, and any other matters as justice may require.

(9) MITIGATION OF DAMAGE.—In addition to establishing a penalty for the discharge of oil or a hazardous substance, the Administrator or the Secretary of the department in which the Coast Guard is operating may act to mitigate the damage to the public health or welfare caused by such discharge. The cost of such mitigation shall be deemed a cost incurred under subsection (c) of this section for the removal of such substance by the United States Government.

(10) RECOVERY OF REMOVAL COSTS.—Any costs of removal incurred in connection with a discharge excluded by subsection (a)(2)(C) of this section shall be recoverable from the owner or operator of the source of the discharge in an action brought under section 309(b) of this Act.

(11) LIMITATION.—Civil penalties shall not be assessed under both this section and section 309 for the same discharge.

(12)[1] WITHHOLDING CLEARANCE.—If any owner, operator, or person in charge of a vessel is liable for a civil penalty under this subsection, or if reasonable cause exists to believe that the owner, operator, or person in charge may be subject to a civil penalty under this subsection, the Secretary of the Treasury, upon the request of the Secretary of the department in which the Coast Guard is operating or the Administrator, shall with respect to such vessel refuse or revoke—

(A) the clearance required by section 4197 of the Revised Statutes of the United States (46 U.S.C. App. 91);

(B) a permit to proceed under section 4367 of the Revised Statutes of the United States (46 U.S.C. App. 313); and

(C) a permit to depart required under section 443 of the Tariff Act of 1930 (19 U.S.C. 1443);

as applicable. Clearance or a permit refused or revoked under this paragraph may be granted upon the filing of a bond or other surety satisfactory to the Secretary of the department in which the Coast Guard is operating or the Administrator.

(c) FEDERAL REMOVAL AUTHORITY.—

[1] Indentation so in law.

(1) GENERAL REMOVAL REQUIREMENT.—(A) The President shall, in accordance with the National Contingency Plan and any appropriate Area Contingency Plan, ensure effective and immediate removal of a discharge, and mitigation or prevention of a substantial threat of a discharge, of oil or a hazardous substance—

(i) into or on the navigable waters;

(ii) on the adjoining shorelines to the navigable waters;

(iii) into or on the waters of the exclusive economic zone; or

(iv) that may affect natural resources belonging to, appertaining to, or under the exclusive management authority of the United States.

(B) In carrying out this paragraph, the President may—

(i) remove or arrange for the removal of a discharge, and mitigate or prevent a substantial threat of a discharge, at any time;

(ii) direct or monitor all Federal, State, and private actions to remove a discharge; and

(iii) remove and, if necessary, destroy a vessel discharging, or threatening to discharge, by whatever means are available.

(2) DISCHARGE POSING SUBSTANTIAL THREAT TO PUBLIC HEALTH OR WELFARE.—(A) If a discharge, or a substantial threat of a discharge, of oil or a hazardous substance from a vessel, offshore facility, or onshore facility is of such a size or character as to be a substantial threat to the public health or welfare of the United States (including but not limited to fish, shellfish, wildlife, other natural resources, and the public and private beaches and shorelines of the United States), the President shall direct all Federal, State, and private actions to remove the discharge or to mitigate or prevent the threat of the discharge.

(B) In carrying out this paragraph, the President may, without regard to any other provision of law governing contracting procedures or employment of personnel by the Federal Government—

(i) remove or arrange for the removal of the discharge, or mitigate or prevent the substantial threat of the discharge; and

(ii) remove and, if necessary, destroy a vessel discharging, or threatening to discharge, by whatever means are available.

(3) ACTIONS IN ACCORDANCE WITH NATIONAL CONTINGENCY PLAN.—(A) Each Federal agency, State, owner or operator, or other person participating in efforts under this subsection shall act in accordance with the National Contingency Plan or as directed by the President.

(B) An owner or operator participating in efforts under this subsection shall act in accordance with the National Contingency Plan and the applicable response plan required under subsection (j), or as directed by the President, except that the owner or operator may deviate from the applicable response plan if the President or the Federal On-Scene Coordinator determines that deviation from the response plan would provide for a more expeditious or effective response to the spill or mitigation of its environmental effects.

(4) EXEMPTION FROM LIABILITY.—(A) A person is not liable for removal costs or damages which result from actions taken or omitted to be taken in the course of rendering care, assistance, or advice consistent with the National Contingency Plan or as otherwise directed by the President relating to a discharge or a substantial threat of a discharge of oil or a hazardous substance.

(B) Subparagraph (A) does not apply—

(i) to a responsible party;

(ii) to a response under the Comprehensive Environmental Response, Compensation, and Liability Act of 1980 (42 U.S.C. 9601 et seq.);

(iii) with respect to personal injury or wrongful death; or

(iv) if the person is grossly negligent or engages in willful misconduct.

(C) A responsible party is liable for any removal costs and damages that another person is relieved of under subparagraph (A).

(5) OBLIGATION AND LIABILITY OF OWNER OR OPERATOR NOT AFFECTED.—Nothing in this subsection affects—

(A) the obligation of an owner or operator to respond immediately to a discharge, or the threat of a discharge, of oil; or

(B) the liability of a responsible party under the Oil Pollution Act of 1990.

(6) RESPONSIBLE PARTY DEFINED.—For purposes of this subsection, the term "responsible party" has the meaning given that term under section 1001 of the Oil Pollution Act of 1990.

(d) NATIONAL CONTINGENCY PLAN.—

(1) PREPARATION BY PRESIDENT.—The President shall prepare and publish a National Contingency Plan for removal of oil and hazardous substances pursuant to this section.

(2) CONTENTS.—The National Contingency Plan shall provide for efficient, coordinated, and effective action to minimize damage from oil and hazardous substance discharges, including containment, dispersal, and removal of oil and hazardous substances, and shall include, but not be limited to, the following:

(A) Assignment of duties and responsibilities among Federal departments and agencies in coordination with State and local agencies and port authorities including, but not limited to, water pollution control and conservation and trusteeship of natural resources (including conservation of fish and wildlife).

(B) Identification, procurement, maintenance, and storage of equipment and supplies.

(C) Establishment or designation of Coast Guard strike teams, consisting of—

(i) personnel who shall be trained, prepared, and available to provide necessary services to carry out the National Contingency Plan;

(ii) adequate oil and hazardous substance pollution control equipment and material; and

(iii) a detailed oil and hazardous substance pollution and prevention plan, including measures to protect fisheries and wildlife.

(D) A system of surveillance and notice designed to safeguard against as well as ensure earliest possible notice of discharges of oil and hazardous substances and imminent threats of such discharges to the appropriate State and Federal agencies.

(E) Establishment of a national center to provide coordination and direction for operations in carrying out the Plan.

(F) Procedures and techniques to be employed in identifying, containing, dispersing, and removing oil and hazardous substances.

(G) A schedule, prepared in cooperation with the States, identifying—

(i) dispersants, other chemicals, and other spill mitigating devices and substances, if any, that may be used in carrying out the Plan,

(ii) the waters in which such dispersants, other chemicals, and other spill mitigating devices and substances may be used, and

(iii) the quantities of such dispersant, other chemicals, or other spill mitigating device or substance which can be used safely in such waters,

which schedule shall provide in the case of any dispersant, chemical, spill mitigating device or substance, or waters not specifically identified in such schedule that the President, or his delegate, may, on a case-by-case basis, identify the dispersants, other chemicals, and other spill mitigating devices and substances which may be used, the waters in which they may be used, and the quantities which can be used safely in such waters.

(H) A system whereby the State or States affected by a discharge of oil or hazardous substance may act where necessary to remove such discharge and such State or States may be reimbursed in accordance with the Oil Pollution Act of 1990, in the case of any discharge of oil from a vessel or facility, for the reasonable costs incurred for that removal, from the Oil Spill Liability Trust Fund.

(I) Establishment of criteria and procedures to ensure immediate and effective Federal identification of, and response to, a discharge, or the threat of a discharge, that results in a substantial threat to the public health or welfare of the United States, as required under subsection (c)(2).

(J) Establishment of procedures and standards for removing a worst case discharge of oil, and for mitigating or preventing a substantial threat of such a discharge.

(K) Designation of the Federal official who shall be the Federal On-Scene Coordinator for each area for which an Area Contingency Plan is required to be prepared under subsection (j).

(L) Establishment of procedures for the coordination of activities of—

(i) Coast Guard strike teams established under subparagraph (C);

(ii) Federal On-Scene Coordinators designated under subparagraph (K);

(iii) District Response Groups established under subsection (j); and

(iv) Area Committees established under subsection (j).

(M) A fish and wildlife response plan, developed in consultation with the United States Fish and Wildlife Service, the National Oceanic and Atmospheric Administration, and other interested parties (including State fish and wildlife conservation officials), for the immediate and effective protection, rescue, and rehabilitation of, and the minimization of risk of damage to, fish and wildlife resources and their habitat that are harmed or that may be jeopardized by a discharge.

(3) REVISIONS AND AMENDMENTS.—The President may, from time to time, as the President deems advisable, revise or otherwise amend the National Contingency Plan.

(4) ACTIONS IN ACCORDANCE WITH NATIONAL CONTINGENCY PLAN.—After publication of the National Contingency Plan, the removal of oil and hazardous substances and actions to minimize damage from oil and hazardous substance discharges shall, to the greatest extent possible, be in accordance with the National Contingency Plan.

(e) CIVIL ENFORCEMENT.—

(1) ORDERS PROTECTING PUBLIC HEALTH.—In addition to any action taken by a State or local government, when the President determines that there may be an imminent and substantial threat to the public health or welfare of the United States, including fish, shellfish, and wildlife, public and private property, shorelines, beaches, habitat, and other living and nonliving natural resources under the jurisdiction or control of the United States, because of an actual or threatened discharge of oil or a hazardous substance from a vessel or facility in violation of subsection (b), the President may—

(A) require the Attorney General to secure any relief from any person, including the owner or operator of the vessel or facility, as may be necessary to abate such endangerment; or

(B) after notice to the affected State, take any other action under this section, including issuing administrative orders, that may be necessary to protect the public health and welfare.

(2) JURISDICTION OF DISTRICT COURTS.—The district courts of the United States shall have jurisdiction to grant any relief

under this subsection that the public interest and the equities of the case may require.

(f)(1) Except where an owner or operator can prove that a discharge was caused solely by (A) an act of God, (B) an act of war, (C) negligence on the part of the United States Government, or (D) an act or omission of a third party without regard to whether any such act or omission was or was not negligent, or any combination of the foregoing clauses, such owner or operator of any vessel from which oil or a hazardous substance is discharged in violation of subsection (b)(3) of this section shall, notwithstanding any other provision of law, be liable to the United States Government for the actual costs incurred under subsection (c) for the removal of such oil or substance by the United States Government in an amount not to exceed, in the case of an inland oil barge $125 per gross ton of such barge, or $125,000, whichever is greater, and in the case of any other vessel, $150 per gross ton of such vessel (or, for a vessel carrying oil or hazardous substances as cargo, $250,000), whichever is greater, except that where the United States can show that such discharge was the result of willful negligence or willful misconduct within the privity and knowledge of the owner, such owner or operator shall be liable to the United States Government for the full amount of such costs. Such costs shall constitute a maritime lien on such vessel which may be recovered in an action in rem in the district court of the United States for any district within which any vessel may be found. The United States may also bring an action against the owner or operator of such vessel in any court of competent jurisdiction to recover such costs.

(2) Except where an owner or operator of an onshore facility can prove that a discharge was caused solely by (A) an act of God, (B) an act of war, (C) negligence on the part of the United States Government, or (D) an act or omission of a third party without regard to whether any such act or omission was or was not negligent, or any combination of the foregoing clauses, such owner or operator of any such facility from which oil or a hazardous substance is discharged in violation of subsection (b)(3) of this section shall be liable to the United States Government for the actual costs incurred under subsection (c) for the removal of such oil or substance by the United States Government in an amount not to exceed $50,000,000, except that where the United States can show that such discharge was the result of willful negligence or willful misconduct within the privity and knowledge of the owner, such owner or operator shall be liable to the United States Government for the full amount of such costs. The United States may bring an action against the owner or operator of such facility in any court of competent jurisdiction to recover such costs. The Administrator is authorized, by regulation, after consultation with the Secretary of Commerce and the Small Business Administration, to establish reasonable and equitable classifications, of those onshore facilities having a total fixed storage capacity of 1,000 barrels or less which he determines because of size, type, and location do not present a substantial risk of the discharge of oil or hazardous substance in violation of subsection (b)(3) of this section, and apply with respect to such classifications differing limits of liability which may be less than the amount contained in this paragraph.

(3) Except where an owner or operator of an onshore facility can prove that a discharge was caused solely by (A) an act of God, (B) an act of war, (C) negligence on the part of the United States Government, or (D) an act or omission of a third party without regard to whether any such act or omission was or was not negligent, or any combination of the foregoing clauses, such owner or operator of any such facility from which oil or a hazardous substance is discharged in violation of subsection (b)(3) of this section shall, notwithstanding any other provision of law, be liable to the United States Government for the actual costs incurred under subsection (c) for the removal of such oil or substance by the United States Government in an amount not to exceed $50,000,000, except that where the United States can show that such discharge was the result of willful negligence or willful misconduct within the privity and knowledge of the owner, such owner or operator shall be liable to the United States Government for the full amount of such costs. The United States may bring an action against the owner or operator of such facility in any court of competent jurisdiction to recover such costs.

(4) The costs of removal of oil or a hazardous substance for which the owner or operator of a vessel or onshore or offshore facility is liable under subsection (f) of this section shall include any costs or expenses incurred by the Federal Government or any State government in the restoration or replacement of natural resources damaged or destroyed as a result of a discharge of oil or a hazardous substance in violation of subsection (b) of this section.

(5) The President, or the authorized representative of any State, shall act on behalf of the public as trustee of the natural resources to recover for the costs of replacing or restoring such resources. Sums recovered shall be used to restore, rehabilitate, or acquire the equivalent of such natural resources by the appropriate agencies of the Federal Government, or the State government.

(g) Where the owner or operator of a vessel (other than an inland oil barge) carrying oil or hazardous substances as cargo or an onshore or offshore facility which handles or stores oil or hazardous substances in bulk, from which oil or a hazardous substance is discharged in violation of subsection (b) of this section, alleges that such discharge was caused solely by an act or omission of a third party, such owner or operator shall pay to the United States Government the actual costs incurred under subsection (c) for removal of such oil or substance and shall be entitled by subrogation to all rights of the United States Government to recover such costs from such third party under this subsection. In any case where an owner or operator of a vessel, of an onshore facility, or of an offshore facility, from which oil or a hazardous substance is discharged in violation of subsection (b)(3) of this section, proves that such discharge of oil or hazardous substance was caused solely by an act or omission of a third party, or was caused solely by such an act or omission in combination with an act of God, an act of war, or negligence on the part of the United States Government, such third party shall, not withstanding any other provision of law, be liable to the United States Government for the actual costs incurred under subsection (c) for removal of such oil or substance by the United States Government, except where such third party can prove that such discharge was caused solely by (A) an act of God, (B) an act of war, (C) negligence on the part of the United States Government, or (D) an act or omission of another party without regard to whether such an act or omission was or was not negligent, or any combination of the foregoing clauses. If such third party was the owner or operator of a vessel which caused the discharge of oil or a hazardous substance in violation of subsection (b)(3) of this section, the liability of such third party under this subsection shall not exceed, in the case of an inland oil barge $125 per gross ton of such barge, $125,000, whichever is greater, and in the case of any other vessel, $150 per gross ton of such vessel (or, for a vessel carrying oil or hazardous substances as cargo, $250,000), whichever is greater. In any other case the liability of such third party shall not exceed the limitation which would have been applicable to the owner or operator of the vessel or the onshore or offshore facility from which the discharge actually occurred if such owner or operator were liable. If the United States can show that the discharge of oil or a hazardous substance in violation of subsection (b)(3) of this section was the result of willful negligence or willful misconduct within the privity and knowledge of such third party, such third party shall be liable to the United States Government for the full amount of such removal costs. The United States may bring an action against the third party in any court of competent jurisdiction to recover such removal costs.

(h) The liabilities established by this section shall in no way affect any rights which (1) the owner or operator of a vessel or of an onshore facility or an offshore facility may have against any third party whose acts may in any way have caused or contributed to such discharge, or (2) The [1] United States Government may have against any third party whose actions may in any way have caused or contributed to the discharge of oil or hazardous substance.

(i) In any case where an owner or operator of a vessel or an onshore facility or an offshore facility from which oil or a hazardous substance is discharged in violation of subsection (b)(3) of this section acts to remove such oil or substance in accordance with regulations promulgated pursuant to this section, such owner or operator shall be entitled to recover the reasonable costs incurred in such removal upon establishing, in a suit which may be brought against the United States Government in the United States Claims Court, that such discharge was caused solely by (A) an act of God, (B) an act of war, (C) negligence on the part of the United States Government, or (D) an act or omission of a third party without regard to whether such act or omission was or was not negligent, or of any combination of the foregoing clauses.

(j) National Response System.—

(1) In general.—Consistent with the National Contingency Plan required by subsection (c)(2) of this section, as soon as practicable after the effective date of this section, and from time to time thereafter, the President shall issue regulations consistent with maritime safety and with marine and navigation laws (A) establishing methods and procedures for removal of discharged oil and hazardous substances, (B) establishing

[1] So in law. Should not be capitalized.

criteria for the development and implementation of local and regional oil and hazardous substance removal contingency plans, (C) establishing procedures, methods, and equipment and other requirements for equipment to prevent discharges of oil and hazardous substances from vessels and from onshore facilities and offshore facilities, and to contain such discharges, and (D) governing the inspection of vessels carrying cargoes of oil and hazardous substances and the inspection of such cargoes in order to reduce the likelihood of discharges of oil from vessels in violation of this section.

(2) NATIONAL RESPONSE UNIT.—The Secretary of the department in which the Coast Guard is operating shall establish a National Response Unit at Elizabeth City, North Carolina. The Secretary, acting through the National Response Unit—

(A) shall compile and maintain a comprehensive computer list of spill removal resources, personnel, and equipment that is available worldwide and within the areas designated by the President pursuant to paragraph (4), and of information regarding previous spills, including data from universities, research institutions, State governments, and other nations, as appropriate, which shall be disseminated as appropriate to response groups and area committees, and which shall be available to Federal and State agencies and the public;

(B) shall provide technical assistance, equipment, and other resources requested by a Federal On-Scene Coordinator;

(C) shall coordinate use of private and public personnel and equipment to remove a worst case discharge, and to mitigate or prevent a substantial threat of such a discharge, from a vessel, offshore facility, or onshore facility operating in or near an area designated by the President pursuant to paragraph (4);

(D) may provide technical assistance in the preparation of Area Contingency Plans required under paragraph (4);

(E) shall administer Coast Guard strike teams established under the National Contingency Plan;

(F) shall maintain on file all Area Contingency Plans approved by the President under this subsection; and

(G) shall review each of those plans that affects its responsibilities under this subsection.

(3) COAST GUARD DISTRICT RESPONSE GROUPS.—(A) The Secretary of the department in which the Coast Guard is operating shall establish in each Coast Guard district a Coast Guard District Response Group.

(B) Each Coast Guard District Response Group shall consist of—

(i) the Coast Guard personnel and equipment, including firefighting equipment, of each port within the district;

(ii) additional prepositioned equipment; and

(iii) a district response advisory staff.

(C) Coast Guard district response groups—

(i) shall provide technical assistance, equipment, and other resources when required by a Federal On-Scene Coordinator;

(ii) shall maintain all Coast Guard response equipment within its district;

(iii) may provide technical assistance in the preparation of Area Contingency Plans required under paragraph (4); and

(iv) shall review each of those plans that affect its area of geographic responsibility.

(4) AREA COMMITTEES AND AREA CONTINGENCY PLANS.—(A) There is established for each area designated by the President an Area Committee comprised of members appointed by the President from qualified personnel of Federal, State, and local agencies.

(B) Each Area Committee, under the direction of the Federal On-Scene Coordinator for its area, shall—

(i) prepare for its area the Area Contingency Plan required under subparagraph (C);

(ii) work with State and local officials to enhance the contingency planning of those officials and to assure preplanning of joint response efforts, including appropriate procedures for mechanical recovery, dispersal, shoreline cleanup, protection of sensitive environmental areas, and protection, rescue, and rehabilitation of fisheries and wildlife; and

(iii) work with State and local officials to expedite decisions for the use of dispersants and other mitigating substances and devices.

(C) Each Area Committee shall prepare and submit to the President for approval an Area Contingency Plan for its area. The Area Contingency Plan shall—

(i) when implemented in conjunction with the National Contingency Plan, be adequate to remove a worst case discharge, and to mitigate or prevent a substantial threat of such a discharge, from a vessel, offshore facility, or onshore facility operating in or near the area;

(ii) describe the area covered by the plan, including the areas of special economic or environmental importance that might be damaged by a discharge;

(iii) describe in detail the responsibilities of an owner or operator and of Federal, State, and local agencies in removing a discharge, and in mitigating or preventing a substantial threat of a discharge;

(iv) list the equipment (including firefighting equipment), dispersants or other mitigating substances and devices, and personnel available to an owner or operator and Federal, State, and local agencies, to ensure an effective and immediate removal of a discharge, and to ensure mitigation or prevention of a substantial threat of a discharge;

(v) compile a list of local scientists, both inside and outside Federal Government service, with expertise in the environmental effects of spills of the types of oil typically transported in the area, who may be contacted to provide information or, where appropriate, participate in meetings of the scientific support team convened in response to a spill, and describe the procedures to be followed for obtaining an expedited decision regarding the use of dispersants;

(vi) describe in detail how the plan is integrated into other Area Contingency Plans and vessel, offshore facility, and onshore facility response plans approved under this subsection, and into operating procedures of the National Response Unit;

(vii) include any other information the President requires; and

(viii) be updated periodically by the Area Committee.

(D) The President shall—

(i) review and approve Area Contingency Plans under this paragraph; and

(ii) periodically review Area Contingency Plans so approved.

(5) TANK VESSEL AND FACILITY RESPONSE PLANS.—(A) The President shall issue regulations which require an owner or operator of a tank vessel or facility described in subparagraph (B) to prepare and submit to the President a plan for responding, to the maximum extent practicable, to a worst case discharge, and to a substantial threat of such a discharge, of oil or a hazardous substance.

(B) The tank vessels and facilities referred to in subparagraph (A) are the following:

(i) A tank vessel, as defined under section 2101 of title 46, United States Code.

(ii) An offshore facility.

(iii) An onshore facility that, because of its location, could reasonably be expected to cause substantial harm to the environment by discharging into or on the navigable waters, adjoining shorelines, or the exclusive economic zone.

(C) A response plan required under this paragraph shall—

(i) be consistent with the requirements of the National Contingency Plan and Area Contingency Plans;

(ii) identify the qualified individual having full authority to implement removal actions, and require immediate communications between that individual and the appropriate Federal official and the persons providing personnel and equipment pursuant to clause (iii);

(iii) identify, and ensure by contract or other means approved by the President the availability of, private personnel and equipment necessary to remove to the maximum extent practicable a worst case discharge (including a discharge resulting from fire or explosion), and to mitigate or prevent a substantial threat of such a discharge;

(iv) describe the training, equipment testing, periodic unannounced drills, and response actions of persons on the vessel or at the facility, to be carried out under the plan to ensure the safety of the vessel or facility and to mitigate or prevent the discharge, or the substantial threat of a discharge;

(v) be updated periodically; and

(vi) be resubmitted for approval of each significant change.

(D) With respect to any response plan submitted under this paragraph for an onshore facility that, because of its location, could reasonably be expected to cause significant and substantial harm to the environment by discharging into or on the navigable waters or adjoining shorelines or the exclusive economic zone, and with respect to each response plan submitted under this paragraph for a tank vessel or offshore facility, the President shall—

(i) promptly review such response plan;

(ii) require amendments to any plan that does not meet the requirements of this paragraph;

(iii) approve any plan that meets the requirements of this paragraph; and

(iv) review each plan periodically thereafter.

(E) [1] A tank vessel, offshore facility, or onshore facility required to prepare a response plan under this subsection may not handle, store, or transport oil unless—

(i) in the case of a tank vessel, offshore facility, or onshore facility for which a response plan is reviewed by the President under subparagraph (D), the plan has been approved by the President; and

(ii) the vessel or facility is operating in compliance with the plan.

(F) Notwithstanding subparagraph (E), the President may authorize a tank vessel, offshore facility, or onshore facility to operate without a response plan approved under this paragraph, until not later than 2 years after the date of the submission to the President of a plan for the tank vessel or facility, if the owner or operator certifies that the owner or operator has ensured by contract or other means approved by the President the availability of private personnel and equipment necessary to respond, to the maximum extent practicable, to a worst case discharge or a substantial threat of such a discharge.

(G) The owner or operator of a tank vessel, offshore facility, or onshore facility may not claim as a defense to liability under title I of the Oil Pollution Act of 1990 that the owner or operator was acting in accordance with an approved response plan.

(H) The Secretary shall maintain, in the Vessel Identification System established under chapter 125 of title 46, United States Code, the dates of approval and review of a response plan under this paragraph for each tank vessel that is a vessel of the United States.

(6) EQUIPMENT REQUIREMENTS AND INSPECTION.—Not later than 2 years after the date of enactment of this section, the President shall require—

[1] Subparagraph (E) of section 311(j)(5) shall take effect 36 months (August 18, 1993) after the date of the enactment of Public Law 101-380. See P.L. 101-380, sec. 4202(b)(4)(C), 104 Stat. 532.

(A) periodic inspection of containment booms, skimmers, vessels, and other major equipment used to remove discharges; and

(B) vessels operating on navigable waters and carrying oil or a hazardous substance in bulk as cargo to carry appropriate removal equipment that employs the best technology economically feasible and that is compatible with the safe operation of the vessel.

(7) AREA DRILLS.—The President shall periodically conduct drills of removal capability, without prior notice, in areas for which Area Contingency Plans are required under this subsection and under relevant tank vessel and facility response plans. The drills may include participation by Federal, State, and local agencies, the owners and operators of vessels and facilities in the area, and private industry. The President may publish annual reports on these drills, including assessments of the effectiveness of the plans and a list of amendments made to improve plans.

(8) UNITED STATES GOVERNMENT NOT LIABLE.—The United States Government is not liable for any damages arising from its actions or omissions relating to any response plan required by this section.

[Subsection (k) was repealed by sec. 2002(b)(2) of P.L. 101-380.]

(l) The President is authorized to delegate the administration of this section to the heads of those Federal departments, agencies, and instrumentalities which he determines to be appropriate. Each such department, agency, and instrumentality, in order to avoid duplication of effort, shall, whenever appropriate, utilize the personnel, services, and facilities of other Federal departments, agencies, and instrumentalities.

(m) ADMINISTRATIVE PROVISIONS.—

(1) FOR VESSELS.—Anyone authorized by the President to enforce the provisions of this section with respect to any vessel may, except as to public vessels—

(A) board and inspect any vessel upon the navigable waters of the United States or the waters of the contiguous zone,

(B) with or without a warrant, arrest any person who in the presence or view of the authorized person violates the provisions of this section or any regulation issued thereunder, and

(C) execute any warrant or other process issued by an officer or court of competent jurisdiction.

(2) FOR FACILITIES.—

(A) RECORDKEEPING.—Whenever required to carry out the purposes of this section, the Administrator or the Secretary of the Department in which the Coast Guard is operating shall require the owner or operator of a facility to which this section applies to establish and maintain such records, make such reports, install, use, and maintain such monitoring equipment and methods, and provide such other information as the Administrator or Secretary, as the case may be, may require to carry out the objectives of this section.

(B) ENTRY AND INSPECTION.—Whenever required to carry out the purposes of this section, the Administrator or the Secretary of the Department in which the Coast Guard is operating or an authorized representative of the Administrator or Secretary, upon presentation of appropriate credentials, may—

(i) enter and inspect any facility to which this section applies, including any facility at which any records are required to be maintained under subparagraph (A); and

(ii) at reasonable times, have access to and copy any records, take samples, and inspect any monitoring equipment or methods required under subparagraph (A).

(C) ARRESTS AND EXECUTION OF WARRANTS.—Anyone authorized by the Administrator or the Secretary of the department in which the Coast Guard is operating to enforce the provisions of this section with respect to any facility may—

(i) with or without a warrant, arrest any person who violates the provisions of this section or any regulation issued thereunder in the presence or view of the person so authorized; and

(ii) execute any warrant or process issued by an officer or court of competent jurisdiction.

(D) PUBLIC ACCESS.—Any records, reports, or information obtained under this paragraph shall be subject to the same public access and disclosure requirements which are applicable to records, reports, and information obtained pursuant to section 308.

(n) The several district courts of the United States are invested with jurisdiction for any actions, other than actions pursuant to subsection (i)(1), arising under this section. In the case of Guam and the Trust Territory of the Pacific Islands, such actions may be brought in the district court of Guam, and in the case of the Virgin Islands such actions may be brought in the district court of the Virgin Islands. In the case of American Samoa and the Trust Territory of the Pacific Islands, such actions may be brought in the District Court of the United States for the District of Hawaii and such court shall have jurisdiction of such actions. In the case of the Canal Zone, such actions may be brought in the United States District Court for the District of the Canal Zone.

(o)(1) Nothing in this section shall affect or modify in any way the obligations of any owner or operator of any vessel, or of any owner or operator of any onshore facility or offshore facility to any person or agency under any provision of law for damages to any publicly owned or privately owned property resulting from a discharge of any oil or hazardous substance or from the removal of any such oil or hazardous substance.

(2) Nothing in this section shall be construed as preempting any State or political subdivision thereof from imposing any requirement or liability with respect to the discharge of oil or haz-

ardous substance into any waters within such State, or with respect to any removal activites related to such discharge.

(3) Nothing in this section shall be construed as affecting or modifying any other existing authority of any Federal department, agency, or instrumentality, relative to onshore or offshore facilities under this Act or any other provision of law, or to affect any State or local law not in conflict with this section.

[Subsection (p) was repealed by sec. 2002(b)(4) of Public Law 101-380, 104 Stat. 507.]

(q) The President is authorized to establish, with repect to any class or category of onshore or offshore facilities, a maximum limit of liability under subsections (f)(2) and (3) of this section of less than $50,000,0000, but not less than, $8,000,000.

(r) Nothing in this section shall be construed to impose, or authorize the imposition of any limitation on liability under the Outer Continental Shelf Lands Act or the Deepwater Port Act of 1974.

(s) The Oil Spill Liability Trust Fund established under section 9509 of the Internal Revenue Code of 1986 (26 U.S.C. 9509) shall be available to carry out subsections (b), (c), (d), (j), and (l) as those subsections apply to discharges, and substantial threats of discharges, of oil. Any amounts received by the United States under this section shall be deposited in the Oil Spill Liability Trust Fund.

(33 U.S.C. 1321)

MARINE SANITATION DEVICES

SEC. 312. (a) For the purpose of this section, the term—

(1) "new vessel" includes every description of watercraft or other artificial contrivance used, or capable of being used, as a means of transportation on the navigable waters, the construction of which is initiated after promulgation of standards and regulations under this section;

(2) "existing vessel" includes every description of watercraft or other artificial contrivance used, or capable of being used, as a means of transportation on the navigable waters, the construction of which is initiated before promulgation of standards and regulations under this section;

(3) "public vessel" means a vessel owned or bareboat chartered and operated by the United States, by a State or political subdivision thereof, or by a foreign nation, except when such vessel is engaged in commerce;

(4) "United States" includes the States, the District of Columbia, the Commonwealth of Puerto Rico, the Virgin Islands, Guam, American Samoa, the Canal Zone, and the Trust Territory of the Pacific Islands;

(5) "marine sanitation device" includes any equipment for installation on board a vessel which is designed to receive, retain, treat, or discharge sewage, and any process to treat such sewage;

(6) "sewage" means human body wastes and the wastes from toilets and other receptacles intended to receive or retain body wastes except that, with respect to commercial vessels on the Great Lakes, such term shall include graywater;

(7) “manufacture” means any person engaged in the manufacturing, assembling, or importation of marine sanitation devices or of vessels subject to standards and regulations promulgated under this section;

(8) “person” means an individual, partnership, firm, corporation, association, or agency of the United States, but does not include an individual on board a public vessel;

(9) “discharge” includes, but is not limited to, any spilling, leaking, pumping, pouring, emitting, emptying or dumping;

(10) “commercial vessels” means those vessels used in the business of transporting property for compensation or hire, or in transporting property in the business of the owner, lessee, or operator of the vessel;

(11) “graywater” means galley, bath, and shower water;

(12) “discharge incidental to the normal operation of a vessel”—

(A) means a discharge, including—

(i) graywater, bilge water, cooling water, weather deck runoff, ballast water, oil water separator effluent, and any other pollutant discharge from the operation of a marine propulsion system, shipboard maneuvering system, crew habitability system, or installed major equipment, such as an aircraft carrier elevator or a catapult, or from a protective, preservative, or absorptive application to the hull of the vessel; and

(ii) a discharge in connection with the testing, maintenance, and repair of a system described in clause (i) whenever the vessel is waterborne; and

(B) does not include—

(i) a discharge of rubbish, trash, garbage, or other such material discharged overboard;

(ii) an air emission resulting from the operation of a vessel propulsion system, motor driven equipment, or incinerator; or

(iii) a discharge that is not covered by part 122.3 of title 40, Code of Federal Regulations (as in effect on the date of the enactment of subsection (n));

(13) “marine pollution control device” means any equipment or management practice, for installation or use on board a vessel of the Armed Forces, that is—

(A) designed to receive, retain, treat, control, or discharge a discharge incidental to the normal operation of a vessel; and

(B) determined by the Administrator and the Secretary of Defense to be the most effective equipment or management practice to reduce the environmental impacts of the discharge consistent with the considerations set forth in subsection (n)(2)(B); and

(14) “vessel of the Armed Forces” means—

(A) any vessel owned or operated by the Department of Defense, other than a time or voyage chartered vessel; and

(B) any vessel owned or operated by the Department of Transportation that is designated by the Secretary of the department in which the Coast Guard is operating as a vessel equivalent to a vessel described in subparagraph (A).

(b)(1) As soon as possible, after the enactment of this section and subject to the provisions of section 104(j) of this Act, the Administrator, after consultation with the Secretary of the department in which the Coast Guard is operating, after giving appropriate consideration to the economic costs involved, and within the limits of available technology, shall promulgate Federal standards of performance for marine sanitation devices (hereinafter in this section referred to as "standards") which shall be designed to prevent the discharge of untreated or inadequately treated sewage into or upon the navigable waters from new vessels and existing vessels, except vessels not equipped with installed toilet facilities. Such standards and standards established under subsection (c)(1)(B) of this section shall be consistent with maritime safety and the marine and navigation laws and regulations and shall be coordinated with the regulations issued under this subsection by the Secretary of the department in which the Coast Guard is operating. The Secretary of the department in which the Coast Guard is operating shall promulgate regulations, which are consistent with standards promulgated under this subsection and subsection (c) of this section and with maritime safety and the marine and navigation laws and regulations governing the design, construction, installation, and operation of any marine sanitation device on board such vessels.

(2) Any existing vessel equipped with a marine sanitation device on the date of promulgation of initial standards and regulations under this section, which device is in compliance with such initial standards and regulations, shall be deemed in compliance with this section until such time as the device is replaced or is found not to be in compliance with such initial standards and regulations.

(c)(1)(A) Initial standards and regulations under this section shall become effective for new vessels two years after promulgation; and for existing vessels five years after promulgation. Revisions of standards and regulations shall be effective upon promulgation, unless another effective date is specified, except that no revision shall take effect before the effective date of the standard or regulation being revised.

(B) The Administrator shall, with respect to commercial vessels on the Great Lakes, establish standards which require at a minimum the equivalent of secondary treatment as defined under section 304(d) of this Act. Such standards and regulations shall take effect for existing vessels after such time as the Administrator determines to be reasonable for the upgrading of marine sanitation devices to attain such standard.

(2) The Secretary of the department in which the Coast Guard is operating with regard to his regulatory authority established by this section, after consultation with the Administrator, may distinguish among classes, types, and sizes of vessels as well as between new and existing vessels, and may waive applicability of standards and regulations as necessary or appropriate for such classes, types, and sizes of vessels (including existing vessels equipped with ma-

rine sanitation devices on the date of promulgation of the initial standards required by this section), and, upon application, for individual vessels.

(d) The provisions of this section and the standards and regulations promulgated hereunder apply to vessels owned and operated by the United States unless the Secretary of Defense finds that compliance would not be in the interest of national security. With respect to vessels owned and operated by the Department of Defense, regulations under the last sentence of subsection (b)(1) of this section and certifications under subsection (g)(2) of this section shall be promulgated and issued by the Secretary of Defense.

(e) Before the standards and regulations under this section are promulgated, the Administrator and the Secretary of the department in which the Coast Guard is operating shall consult with the Secretary of State; the Secretary of Health, Education, and Welfare; the Secretary of Defense; the Secretary of the Treasury; the Secretary of Commerce; other interested Federal agencies; and the States and industries interested; and otherwise comply with the requirements of section 553 of title 5 of the United States Code.

(f)(1)(A) Except as provided in subparagraph (B), after the effective date of the initial standards and regulations promulgated under this section, no State or political subdivision thereof shall adopt or enforce any statute or regulation of such State or political subdivision with respect to the design, manufacture, or installation or use of any marine sanitation device on any vessel subject to the provisions of this section.

(B) A State may adopt and enforce a statute or regulation with respect to the design, manufacture, or installation or use of any marine sanitation device on a houseboat, if such statute or regulation is more stringent than the standards and regulations promulgated under this section. For purposes of this paragraph, the term "houseboat" means a vessel which, for a period of time determined by the State in which the vessel is located, is used primarily as a residence and is not used primarily as a means of transportation.

(2) If, after promulgation of the initial standards and regulations and prior to their effective date, a vessel is equipped with a marine sanitation device in compliance with such standards and regulations and the installation and operation of such device is in accordance with such standards and regulations, such standards and regulations shall, for the purposes of paragraph (1) of this subsection, become effective with respect to such vessel on the date of such compliance.

(3) After the effective date of the initial standards and regulations promulgated under this section, if any State determines that the protection and enhancement of the quality of some or all of the waters within such State require greater environmental protection, such State may completely prohibit the discharge from all vessels of any sewage, whether treated or not, into such waters, except that no such prohibition shall apply until the Administrator determines that adequate facilities for the safe and sanitary removal and treatment of sewage from all vessels are reasonably available for such water to which such prohibition would apply. Upon application of the State, the Administrator shall make such determination within 90 days of the date of such application.

(4)(A) If the Administrator determines upon application by a State that the protection and enhancement of the quality of specified waters within such State requires such a prohibition, he shall by regulation completely prohibit the discharge from a vessel of any sewage (whether treated or not) into such waters.

(B) Upon application by a State, the Administrator shall, by regulation, establish a drinking water intake zone in any waters within such State and prohibit the discharge of sewage from vessels within that zone.

(g)(1) No manufacturer of a marine sanitation device shall sell, offer for sale, or introduce or deliver for introduction in interstate commerce, or import into the United States for sale or resale any marine sanitation device manufactured after the effective date of the standards and regulations promulgated under this section unless such device is in all material respects substantially the same as a test device certified under this subsection.

(2) Upon application of the manufacturer, the Secretary of the department in which the Coast Guard is operating shall so certify a marine sanitation device if he determines, in accordance with the provisions of this paragraph, that it meets the appropriate standards and regulations promulgated under this section. The Secretary of the department in which the Coast Guard is operating shall test or require such testing of the device in accordance with procedures set forth by the Administrator as to standards of performance and for such other purposes as may be appropriate. If the Secretary of the department in which the Coast Guard is operating determines that the device is satisfactory from the standpoint of safety and any other requirements of maritime law or regulation, and after consideration of the design, installation, operation, material, or other appropriate factors, he shall certify the device. Any device manufactured by such manufacturer which is in all material respects substantially the same as the certified test device shall be deemed to be in conformity with the appropriate standards and regulations established under this section.

(3) Every manufacturer shall establish and maintain such records, make such reports, and provide such information as the Administrator or the Secretary of the department in which the Coast Guard is operating may reasonably require to enable him to determine whether such manufacturer has acted or is acting in compliance with this section and regulations issued thereunder and shall, upon request of an officer or employee duly designated by the Administrator or the Secretary of the department in which the Coast Guard is operating, permit such officer or employee at reasonable times to have access to and copy such records. All information reported to or otherwise obtained by the Administrator or the Secretary of the department in which the Coast Guard is operating or their representatives pursuant to this subsection which contains or relates to a trade secret or other matter referred in section 1905 of title 18 of the United States Code shall be considered confidential for the purpose of that section, except that such information may be disclosed to other officers or employees concerned with carrying out this section. This paragraph shall not apply in the case of the construction of a vessel by an individual for his own use.

(h) After the effective date of standards and regulations promulgated under this section, it shall be unlawful—

(1) for the manufacturer of any vessel subject to such standards and regulations to manufacture for sale, to sell or offer for sale, or to distribute for sale or resale any such vessel unless it is equipped with a marine sanitation device which is in all material respects substantially the same as the appropriate test device certified pursuant to this section;

(2) for any person, prior to the sale or delivery of a vessel subject to such standards and regulations to the ultimate purchaser, wrongfully to remove or render inoperative any certified marine sanitation device or element of design of such device installed in such vessel;

(3) for any person to fail or refuse to permit access to or copying of records or to fail to make reports or provide information required under this section; and

(4) for a vessel subject to such standards and regulations to operate on the navigable waters of the United States, if such vessel is not equipped with an operable marine sanitation device certified pursuant to this section.

(i) The district courts of the United States shall have jurisdictions to restrain violations of subsection (g)(1) of this section and subsections (h)(1) through (3) of this section. Actions to restrain such violations shall be brought by, and in, the name of the United States. In case of contumacy or refusal to obey a subpena served upon any person under this subsection, the district court of the United States for any district in which such person is found or resides or transacts business, upon application by the United States and after notice to such person, shall have jurisdiction to issue an order requiring such person to appear and give testimony or to appear and produce documents, and any failure to obey such order of the court may be punished by such court as a contempt thereof.

(j) Any person who violates subsection (g)(1), clause (1) or (2) of subsection (h), or subsection (n)(8) shall be liable to a civil penalty of not more than $5,000 for each violation. Any person who violates clause (4) of subsection (h) of this section or any regulation issued pursuant to this section shall be liable to a civil penalty of not more than $2,000 for each violation. Each violation shall be a separate offense. The Secretary of the department in which the Coast Guard is operating may assess and compromise any such penalty. No penalty shall be assessed until the person charged shall have been given notice and an opportunity for a hearing on such charge. In determining the amount of the penalty, or the amount agreed upon in compromise, the gravity of the violation, and the demonstrated good faith of the person charged in attempting to achieve rapid compliance, after notification of a violation, shall be considered by said Secretary.

(k) The provisions of this section shall be enforced by the Secretary of the department in which the Coast Guard is operating and he may utilize by agreement, with or without reimbursement, law enforcement officers or other personnel and facilities of the Administrator, other Federal agencies, or the States to carry out the provisions of this section. The provisions of this section may also be enforced by a State.

(l) Anyone authorized by the Secretary of the department in which the Coast Guard is operating to enforce the provisions of this section may, except as to public vessels, (1) board and inspect any vessel upon the navigable waters of the United States and (2) execute any warrant or other process issued by an officer or court of competent jurisdiction.

(m) In the case of Guam and the Trust Territory of the Pacific Islands, actions arising under this section may be brought in the district court of Guam, and in the case of the Virgin Islands such actions may be brought in the district court of the Virgin Islands. In the case of American Samoa and the Trust Territory of the Pacific Islands, such actions may be brought in the District Court of the United States for the District of Hawaii and such court shall have jurisdiction of such actions. In the case of the Canal Zone, such actions may be brought in the District Court for the District of the Canal Zone.

(n) UNIFORM NATIONAL DISCHARGE STANDARDS FOR VESSELS OF THE ARMED FORCES.—

(1) APPLICABILITY.—This subsection shall apply to vessels of the Armed Forces and discharges, other than sewage, incidental to the normal operation of a vessel of the Armed Forces, unless the Secretary of Defense finds that compliance with this subsection would not be in the national security interests of the United States.

(2) DETERMINATION OF DISCHARGES REQUIRED TO BE CONTROLLED BY MARINE POLLUTION CONTROL DEVICES.—

(A) IN GENERAL.—The Administrator and the Secretary of Defense, after consultation with the Secretary of the department in which the Coast Guard is operating, the Secretary of Commerce, and interested States, shall jointly determine the discharges incidental to the normal operation of a vessel of the Armed Forces for which it is reasonable and practicable to require use of a marine pollution control device to mitigate adverse impacts on the marine environment. Notwithstanding subsection (a)(1) of section 553 of title 5, United States Code, the Administrator and the Secretary of Defense shall promulgate the determinations in accordance with such section. The Secretary of Defense shall require the use of a marine pollution control device on board a vessel of the Armed Forces in any case in which it is determined that the use of such a device is reasonable and practicable.

(B) CONSIDERATIONS.—In making a determination under subparagraph (A), the Administrator and the Secretary of Defense shall take into consideration—

(i) the nature of the discharge;

(ii) the environmental effects of the discharge;

(iii) the practicability of using the marine pollution control device;

(iv) the effect that installation or use of the marine pollution control device would have on the operation or operational capability of the vessel;

(v) applicable United States law;

(vi) applicable international standards; and

(vii) the economic costs of the installation and use of the marine pollution control device.

(3) PERFORMANCE STANDARDS FOR MARINE POLLUTION CONTROL DEVICES.—

(A) IN GENERAL.—For each discharge for which a marine pollution control device is determined to be required under paragraph (2), the Administrator and the Secretary of Defense, in consultation with the Secretary of the department in which the Coast Guard is operating, the Secretary of State, the Secretary of Commerce, other interested Federal agencies, and interested States, shall jointly promulgate Federal standards of performance for each marine pollution control device required with respect to the discharge. Notwithstanding subsection (a)(1) of section 553 of title 5, United States Code, the Administrator and the Secretary of Defense shall promulgate the standards in accordance with such section.

(B) CONSIDERATIONS.—In promulgating standards under this paragraph, the Administrator and the Secretary of Defense shall take into consideration the matters set forth in paragraph (2)(B).

(C) CLASSES, TYPES, AND SIZES OF VESSELS.—The standards promulgated under this paragraph may—

(i) distinguish among classes, types, and sizes of vessels;

(ii) distinguish between new and existing vessels; and

(iii) provide for a waiver of the applicability of the standards as necessary or appropriate to a particular class, type, age, or size of vessel.

(4) REGULATIONS FOR USE OF MARINE POLLUTION CONTROL DEVICES.—The Secretary of Defense, after consultation with the Administrator and the Secretary of the department in which the Coast Guard is operating, shall promulgate such regulations governing the design, construction, installation, and use of marine pollution control devices on board vessels of the Armed Forces as are necessary to achieve the standards promulgated under paragraph (3).

(5) DEADLINES; EFFECTIVE DATE.—

(A) DETERMINATIONS.—The Administrator and the Secretary of Defense shall—

(i) make the initial determinations under paragraph (2) not later than 2 years after the date of the enactment of this subsection; and

(ii) every 5 years—

(I) review the determinations; and

(II) if necessary, revise the determinations based on significant new information.

(B) STANDARDS.—The Administrator and the Secretary of Defense shall—

(i) promulgate standards of performance for a marine pollution control device under paragraph (3) not later than 2 years after the date of a determination under paragraph (2) that the marine pollution control device is required; and

(ii) every 5 years—

(I) review the standards; and

(II) if necessary, revise the standards, consistent with paragraph (3)(B) and based on significant new information.

(C) REGULATIONS.—The Secretary of Defense shall promulgate regulations with respect to a marine pollution control device under paragraph (4) as soon as practicable after the Administrator and the Secretary of Defense promulgate standards with respect to the device under paragraph (3), but not later than 1 year after the Administrator and the Secretary of Defense promulgate the standards. The regulations promulgated by the Secretary of Defense under paragraph (4) shall become effective upon promulgation unless another effective date is specified in the regulations.

(D) PETITION FOR REVIEW.—The Governor of any State may submit a petition requesting that the Secretary of Defense and the Administrator review a determination under paragraph (2) or a standard under paragraph (3), if there is significant new information, not considered previously, that could reasonably result in a change to the particular determination or standard after consideration of the matters set forth in paragraph (2)(B). The petition shall be accompanied by the scientific and technical information on which the petition is based. The Administrator and the Secretary of Defense shall grant or deny the petition not later than 2 years after the date of receipt of the petition.

(6) EFFECT ON OTHER LAWS.—

(A) PROHIBITION ON REGULATION BY STATES OR POLITICAL SUBDIVISIONS OF STATES.—Beginning on the effective date of—

(i) a determination under paragraph (2) that it is not reasonable and practicable to require use of a marine pollution control device regarding a particular discharge incidental to the normal operation of a vessel of the Armed Forces; or

(ii) regulations promulgated by the Secretary of Defense under paragraph (4);

except as provided in paragraph (7), neither a State nor a political subdivision of a State may adopt or enforce any statute or regulation of the State or political subdivision with respect to the discharge or the design, construction, installation, or use of any marine pollution control device required to control discharges from a vessel of the Armed Forces.

(B) FEDERAL LAWS.—This subsection shall not affect the application of section 311 to discharges incidental to the normal operation of a vessel.

(7) ESTABLISHMENT OF STATE NO-DISCHARGE ZONES.—

(A) STATE PROHIBITION.—

(i) IN GENERAL.—After the effective date of—

(I) a determination under paragraph (2) that it is not reasonable and practicable to require use of a marine pollution control device regarding a particular discharge incidental to the normal operation of a vessel of the Armed Forces; or

(II) regulations promulgated by the Secretary of Defense under paragraph (4);

if a State determines that the protection and enhancement of the quality of some or all of the waters within the State require greater environmental protection, the State may prohibit 1 or more discharges incidental to the normal operation of a vessel, whether treated or not treated, into the waters. No prohibition shall apply until the Administrator makes the determinations described in subclauses (II) and (III) of subparagraph (B)(i).

(ii) DOCUMENTATION.—To the extent that a prohibition under this paragraph would apply to vessels of the Armed Forces and not to other types of vessels, the State shall document the technical or environmental basis for the distinction.

(B) PROHIBITION BY THE ADMINISTRATOR.—

(i) IN GENERAL.—Upon application of a State, the Administrator shall by regulation prohibit the discharge from a vessel of 1 or more discharges incidental to the normal operation of a vessel, whether treated or not treated, into the waters covered by the application if the Administrator determines that—

(I) the protection and enhancement of the quality of the specified waters within the State require a prohibition of the discharge into the waters;

(II) adequate facilities for the safe and sanitary removal of the discharge incidental to the normal operation of a vessel are reasonably available for the waters to which the prohibition would apply; and

(III) the prohibition will not have the effect of discriminating against a vessel of the Armed Forces by reason of the ownership or operation by the Federal Government, or the military function, of the vessel.

(ii) APPROVAL OR DISAPPROVAL.—The Administrator shall approve or disapprove an application submitted under clause (i) not later than 90 days after the date on which the application is submitted to the Administrator. Notwithstanding clause (i)(II), the Administrator shall not disapprove an application for the sole reason that there are not adequate facilities to remove any discharge incidental to the normal operation of a vessel from vessels of the Armed Forces.

(C) APPLICABILITY TO FOREIGN FLAGGED VESSELS.—A prohibition under this paragraph—

(i) shall not impose any design, construction, manning, or equipment standard on a foreign flagged vessel engaged in innocent passage unless the prohibition implements a generally accepted international rule or standard; and

(ii) that relates to the prevention, reduction, and control of pollution shall not apply to a foreign flagged vessel engaged in transit passage unless the prohibition implements an applicable international regulation regarding the discharge of oil, oily waste, or any other noxious substance into the waters.

(8) PROHIBITION RELATING TO VESSELS OF THE ARMED FORCES.—After the effective date of the regulations promulgated by the Secretary of Defense under paragraph (4), it shall be unlawful for any vessel of the Armed Forces subject to the regulations to—

(A) operate in the navigable waters of the United States or the waters of the contiguous zone, if the vessel is not equipped with any required marine pollution control device meeting standards established under this subsection; or

(B) discharge overboard any discharge incidental to the normal operation of a vessel in waters with respect to which a prohibition on the discharge has been established under paragraph (7).

(9) ENFORCEMENT.—This subsection shall be enforceable, as provided in subsections (j) and (k), against any agency of the United States responsible for vessels of the Armed Forces notwithstanding any immunity asserted by the agency.

(33 U.S.C. 1322)

FEDERAL FACILITIES POLLUTION CONTROL

SEC. 313. (a) Each department, agency, or instrumentality of the executive, legislative, and judicial branches of the Federal Government (1) having jurisdiction over any property or facility, or (2) engaged in any activity resulting, or which may result, in the discharge or runoff of pollutants, and each officer, agent, or employee thereof in the performance of his official duties, shall be subject to, and comply with, all Federal, State, interstate, and local requirements, administrative authority, and process and sanctions respecting the control and abatement of water pollution in the same manner, and to the same extent as any nongovernmental entity including the payment of reasonable service charges. The preceding sentence shall apply (A) to any requirement whether substantive or procedural (including any recordkeeping or reporting requirement, any requirement respecting permits and any other requirement, whatsoever), (B) to the exercise of any Federal, State, or local administrative authority, and (C) to any process and sanction, whether enforced in Federal, State, or local courts or in any other manner. This subsection shall apply notwithstanding any immunity of such agencies, officers, agents, or employees under any law or rule of law. Nothing in this section shall be construed to prevent any department, agency, or instrumentality of the Federal Government, or any officer, agent, or employee thereof in the performance of his official duties, from removing to the appropriate Federal district court any proceeding to which the department, agency, or instrumentality or officer, agent, or employee thereof is subject pursuant to this section, and any such proceeding may be removed in accordance with 28 U.S.C. 1441 et seq. No officer, agent, or employee of the United States shall be personally liable for any civil penalty arising from the performance of his official duties, for which he is not otherwise liable, and the United States shall be liable only for those civil penalties arising under Federal law or imposed by a State or local court to enforce an order or the process of such court. The President may exempt any effluent source of any department, agency, or instrumentality in the executive branch from compliance with any such a requirement if he determines it to be in the paramount interest of the United States to do so; except that no exemption may be granted from the requirements of section 306 or 307 of this Act. No such exemptions shall be granted due to lack of appropriation unless the President shall have specifically requested such appropriation as a part of the budgetary process and the Congress shall have failed to make available such requested appropriation. Any exemption shall be for a period not in excess of one year, but additional exemptions may be granted for periods of not to exceed one year upon the President's making a new determination. The President shall report each January to the Congress all exemptions from the requirements of this section granted during the preceding calendar year, together with his reason for granting such exemption. In addition to any such exemption of a particular effluent source, the President may, if he determines it to be in the paramount interest of the United States to do so, issue regulations exempting from compliance with the requirements of this section any weaponry, equipment, aircraft, vessels, vehicles, or other classes or categories of property, and access to such property, which are owned or operated by the Armed Forces of the United States (including the Coast Guard) or by the National Guard of any State and which are uniquely military in nature. The President shall reconsider the need for such regulations at three-year intervals.

(b)(1) The Administrator shall coordinate with the head of each department, agency, or instrumentality of the Federal Government having jurisdiction over any property or facility utilizing federally owned wastewater facilities to develop a program of cooperation for utilizing wastewater control systems utilizing those innovative treatment processes and techniques for which guidelines have been promulgated under section 304(d)(3). Such program shall include an inventory of property and facilities which could utilize such processes and techniques.

(2) Construction shall not be initiated for facilities for treatment of wastewater at any Federal property or facility after September 30, 1979, if alternative methods for wastewater treatment at such property or facility utilizing innovative treatment processes and techniques, including but not limited to methods utilizing recycle and reuse techniques and land treatment are not utilized, unless the life cycle cost of the alternative treatment works exceeds the life cycle cost of the most cost effective alternative by more than 15 per centum. The Administrator may waive the application

of this paragraph in any case where the Administrator determines it to be in the public interest, or that compliance with this paragraph would interfere with the orderly compliance with the conditions of a permit issued pursuant to section 402 of this Act.

(33 U.S.C. 1323)

CLEAN LAKES

SEC. 314. (a) ESTABLISHMENT AND SCOPE OF PROGRAM.—

(1) STATE PROGRAM REQUIREMENTS.—Each State on a biennial basis shall prepare and submit to the Administrator for his approval—

(A) an identification and classification according to eutrophic condition of all publicly owned lakes in such State;

(B) a description of procedures, processes, and methods (including land use requirements), to control sources of pollution of such lakes;

(C) a description of methods and procedures, in conjunction with appropriate Federal agencies, to restore the quality of such lakes;

(D) methods and procedures to mitigate the harmful effects of high acidity, including innovative methods of neutralizing and restoring buffering capacity of lakes and methods of removing from lakes toxic metals and other toxic substances mobilized by high acidity;

(E) a list and description of those publicily owned lakes in such State for which uses are known to be impaired, including those lakes which are known not to meet applicable water quality standards or which require implementation of control programs to maintain compliance with applicable standards and those lakes in which water quality has deteriorated as a result of high acidity that may reasonably be due to acid deposition; and

(F) an assessment of the status and trends of water quality in lakes in such State, including but not limited to, the nature and extent of pollution loading from point and nonpoint sources and the extent to which the use of lakes is impaired as a result of such pollution, particularly with respect to toxic pollution.

(2) SUBMISSION AS PART OF 305(b)(1) REPORT.—The information required under paragraph (1) shall be included in the report required under section 305(b)(1) of this Act, beginning with the report required under such section by April 1, 1988.

(3) ELIGIBILITY REQUIREMENT.—Beginning after April 1, 1988, a State must have submitted the information required under paragraph (1) in order to receive grant assistance under this section.

(b) The Administrator shall provide financial assistance to States in order to carry out methods and procedures approved by him under subsection (a) of this section. The Administrator shall provide financial assistance to States to prepare the identification and classification surveys required in subsection (a)(1) of this section.

(c)(1) The amount granted to any State for any fiscal year under subsection (b) of this section shall not exceed 70 per centum of the funds expended by such State in such year for carrying out approved methods and procedures under subsection (a) of this section.

(2) There is authorized to be appropriated $50,000,000 for the fiscal year ending June 30, 1973; $100,000,000 for the fiscal year 1974; $150,000,000 for the fiscal year 1975, $50,000,000 for fiscal year 1977, $60,000,000 for fiscal year 1978, $60,000,000 for fiscal year 1979, $60,000,000 for fiscal year 1980, $30,000,000 for fiscal year 1981, $30,000,000 for fiscal year 1982, such sums as may be necessary for fiscal years 1983 through 1985, and $30,000,000 per fiscal year for each of the fiscal years 1986 through 1990 for grants to States under subsection (b) of this section which such sums shall remain available until expended. The Administrator shall provide for an equitable distribution of such sums to the States with approved methods and procedures under subsection (a) of this section.

(d) DEMONSTRATION PROGRAM.—

(1) GENERAL REQUIREMENTS.—The Administrator is authorized and directed to establish and conduct at locations throughout the Nation a lake water quality demonstration program. The program shall, at a minimum—

(A) develop cost effective technologies for the control of pollutants to preserve or enhance lake water quality while optimizing multiple lakes uses;

(B) control nonpoint sources of pollution which are contributing to the degradation of water quality in lakes;

(C) evaluate the feasibility of implementing regional consolidated pollution control strategies;

(D) demonstrate environmentally preferred techniques for the removal and disposal of contaminated lake sediments;

(E) develop improved methods for the removal of silt, stumps, aquatic growth, and other obstructions which impair the quality of lakes;

(F) construct and evaluate silt traps and other devices or equipment to prevent or abate the deposit of sediment in lakes; and

(G) demonstrate the costs and benefits of utilizing dredged material from lakes in the reclamation of despoiled land.

(2) GEOGRAPHICAL REQUIREMENTS.—Demonstration projects authorized by this subsection shall be undertaken to reflect a variety of geographical and environmental conditions. As a priority, the Administrator shall undertake demonstration projects at Lake Champlain, New York and Vermont; Lake Houston, Texas; Beaver Lake, Arkansas; Greenwood Lake and Belcher Creek, New Jersey; Deal Lake, New Jersey; Alcyon Lake, New Jersey; Gorton's Pond, Rhode Island; Lake Washington, Rhode Island; Lake Bomoseen, Vermont; Sauk Lake, Minnesota; and Lake Worth, Texas.

(3) REPORTS.—By January 1, 1997, and January 1 of every odd-numbered year thereafter, the Administrator shall report to the Committee on Transportation and Infrastructure of the

House of Representatives and the Committee on Environment and Public Works of the Senate on work undertaken pursuant to this subsection. Upon completion of the program authorized by this subsection, the Administrator shall submit to such committees a final report on the results of such program, along with recommendations for further measures to improve the water quality of the Nation's lakes.

(4) AUTHORIZATION OF APPROPRIATIONS.—

(A) IN GENERAL.—There is authorized to be appropriated to carry out this subsection not to exceed $40,000,000 for fiscal years beginning after September 30, 1986, to remain available until expended.

(B) SPECIAL AUTHORIZATIONS.—

(i) AMOUNT.—There is authorized to be appropriated to carry out subsection (b) with respect to subsection (a)(1)(D) not to exceed $15,000,000 for fiscal years beginning after September 30, 1986, to remain available until expended.

(ii) DISTRIBUTION OF FUNDS.—The Administrator shall provide for an equitable distribution of sums appropriated pursuant to this subparagraph among States carrying out approved methods and procedures. Such distribution shall be based on the relative needs of each such State for the mitigation of the harmful effects on lakes and other surface waters of high acidity that may reasonably be due to acid deposition or acid mine drainage.

(iii) GRANTS AS ADDITIONAL ASSISTANCE.—The amount of any grant to a State under this subparagraph shall be in addition to, and not in lieu of, any other Federal financial assistance.

(33 U.S.C. 1324)

NATIONAL STUDY COMMISSION

SEC. 315. (a) There is established a National Study Commission, which shall make a full and complete investigation and study of all of the technological aspects of achieving, and all aspects of the total economic, social, and environmental effects of achieving or not achieving, the effluent limitations and goals set forth for 1983 in section 301(b)(2) of this Act.

(b) Such Commission shall be composed of fifteen members, including five members of the Senate, who are members of the Public Works committee, appointed by the President of the Senate, five members of the House, who are members of the Public Works committee, appointed by the Speaker of the House, and five members of the public appointed by the President. The Chairman of such Commission shall be elected from among its members.

(c) In the conduct of such study, the Commission is authorized to contract with the National Academy of Sciences and the National Academy of Engineering (acting through the National Research Council), the National Institute of Ecology, Brookings Institution, and other nongovernmental entities, for the investigation of matters within their competence.

(d) The heads of the departments, agencies and instrumentalities of the executive branch of the Federal Government shall cooperate with the Commission in carrying out the requirements of this section, and shall furnish to the Commission such information as the Commission deems necessary to carry out this section.

(e) A report shall be submitted to the Congress of the results of such investigation and study, together with recommendations, not later than three years after the date of enactment of this title.

(f) The members of the Commission who are not officers or employees of the United States, while attending conferences or meetings of the Commission or while otherwise serving at the request of the Chairman shall be entitled to receive compensation at a rate not in excess of the maximum rate of pay for grade GS–18, as provided in the General Schedule under section 5332 of title V of the United States Code, including traveltime and while away from their homes or regular places of business they may be allowed travel expenses, including per diem in lieu of subsistence as authorized by law (5 U.S.C. 73b–2) for persons in the Government service employed intermittently.

(g) In addition to authority to appoint personnel subject to the provisions of title 5, United States Code, governing appointments in the competitive service, and to pay such personnel in accordance with the provisions of chapter 51 and subchapter III of chapter 53 of such title relating to classification and General Schedule pay rates, the Commission shall have authority to enter into contracts with private or public organizations who shall furnish the Commission with such administrative and technical personnel as may be necessary to carry out the purpose of this section. Personnel furnished by such organizations under this subsection are not, and shall not be considered to be, Federal employees for any purposes, but in the performance of their duties shall be guided by the standards which apply to employees of the legislative branches under rules 41 and 43 of the Senate and House of Representatives, respectively.

(h) There is authorized to be appropriated, for use in carrying out this section, not to exceed $17,250,000.

(33 U.S.C. 1325)

THERMAL DISCHARGES

SEC. 316. (a) With respect to any point source otherwise subject to the provisions of section 301 or section 306 of this Act, whenever the owner or operator of any such source, after opportunity for public hearing, can demonstrate to the satisfaction of the Administrator (or, if appropriate, the State) that any effluent limitation proposed for the control of the thermal component of any discharge from such source will require effluent limitations more stringent than necessary to assure the projection and propagation of a balanced, indigenous population of shellfish, fish, and wildlife in and on the body of water into which the discharge is to be made, the Administrator (or, if appropriate, the State) may impose an effluent limitation under such sections for such plant, with respect to the thermal component of such discharge (taking into account the interaction of such thermal component with other pollutants), that will assure the projection and propagation of a balanced, indigenous population of shellfish, fish, and wildlife in and on that body of water.

(b) Any standard established pursuant to section 301 or section 306 of this Act and applicable to a point source shall require that the location, design, construction, and capacity of cooling water intake structures reflect the best technology available for minimizing adverse environmental impact.

(c) Notwithstanding any other provision of this Act, any point source of a discharge having a thermal component, the modification of which point source is commenced after the date of enactment of the Federal Water Pollution Control Act Amendments of 1972 and which, as modified, meets effluent limitations established under section 301 or, if more stringent, effluent limitations established under section 303 and which effluent limitations will assure protection and propagation of a balanced, indigenous population of shellfish, fish, and wildlife in or on the water into which the discharge is made, shall not be subject to any more stringent effluent limitation with respect to the thermal component of its discharge during a ten year period beginning on the date of completion of such modification or during the period of depreciation or amortization of such facility for the purpose of section 167 or 169 (or both) of the Internal Revenue Code of 1954, whichever period ends first.

(33 U.S.C. 1326)

FINANCING STUDY

SEC. 317. (a) The Administrator shall continue to investigate and study the feasibility of alternate methods of financing the cost of preventing, controlling and abating pollution as directed in the Water Quality Improvement Act of 1970 (Public Law 91–224), including, but not limited to, the feasibility of establishing a pollution abatement trust fund. The results of such investigation and study shall be reported to the Congress not later than two years after enactment of this title, together with recommendations of the Administrator for financing the programs for preventing, controlling and abating pollution for the fiscal years beginning after fiscal year 1976, including any necessary legislation.

(b) There is authorized to be appropriated for use in carrying out this section, not to exceed $1,000,000.

(33 U.S.C. 1327)

AQUACULTURE

SEC. 318. (a) The Administrator is authorized, after public hearings, to permit the discharge of a specific pollutant or pollutants under controlled conditions associated with an approved aquaculture project under Federal or State supervision pursuant to section 402 of this Act.

(b) The Administrator shall by regulation establish any procedures and guidelines which the Administrator deems necessary to carry out this section. Such regulations shall require the application to such discharge of each criterion, factor, procedure, and requirement applicable to a permit issued under section 402 of this title, as the Administrator determines necessary to carry out the objective of this Act.

(c) Each State desiring to administer its own permit program within its jurisdiction for discharge of a specific pollutant or pollutants under controlled conditions associated with an approved aquaculture project may do so if upon submission of such program the Administrator determines such program is adequate to carry out the objective of this Act.

(33 U.S.C. 1328)

SEC. 319. NONPOINT SOURCE MANAGEMENT PROGRAMS.

(a) STATE ASSESSMENT REPORTS.—

(1) CONTENTS.—The Governor of each State shall, after notice and opportunity for public comment, prepare and submit to the Administrator for approval, a report which—

(A) identifies those navigable waters within the State which, without additional action to control nonpoint sources of pollution, cannot reasonably be expected to attain or maintain applicable water quality standards or the goals and requirements of this Act;

(B) identifies those categories and subcategories of nonpoint sources or, where appropriate, particular nonpoint sources which add significant pollution to each portion of the navigable waters identified under subparagraph (A) in amounts which contribute to such portion not meeting such water quality standards or such goals and requirements;

(C) describes the process, including intergovernmental coordination and public participation, for identifying best management practices and measures to control each category and subcategory of nonpoint sources and, where appropriate, particular nonpoint sources identified under subparagraph (B) and to reduce, to the maximum extent practicable, the level of pollution resulting from such category, subcategory, or source; and

(D) identifies and describes State and local programs for controlling pollution added from nonpoint sources to, and improving the quality of, each such portion of the navigable waters, including but not limited to those programs which are receiving Federal assistance under subsections (h) and (i).

(2) INFORMATION USED IN PREPARATION.—In developing the report required by this section, the State (A) may rely upon information developed pursuant to sections 208, 303(e), 304(f), 305(b), and 314, and other information as appropriate, and (B) may utilize appropriate elements of the waste treatment management plans developed pursuant to sections 208(b) and 303, to the extent such elements are consistent with and fulfill the requirements of this section.

(b) STATE MANAGEMENT PROGRAMS.—

(1) IN GENERAL.—The Governor of each State, for that State or in combination with adjacent States, shall, after notice and opportunity for public comment, prepare and submit to the Administrator for approval a management program which such

State proposes to implement in the first four fiscal years beginning after the date of submission of such management program for controlling pollution added from nonpoint sources to the navigable waters within the State and improving the quality of such waters.

(2) SPECIFIC CONTENTS.—Each management program proposed for implementation under this subsection shall include each of the following:

(A) An identification of the best management practices and measures which will be undertaken to reduce pollutant loadings resulting from each category, subcategory, or particular nonpoint source designated under paragraph (1)(B), taking into account the impact of the practice on ground water quality.

(B) An identification of programs (including, as appropriate, nonregulatory or regulatory programs for enforcement, technical assistance, financial assistance, education, training, technology transfer, and demonstration projects) to achieve implementation of the best management practices by the categories, subcategories, and particular nonpoint sources designated under subparagraph (A).

(C) A schedule containing annual milestones for (i) utilization of the program implementation methods identified in subparagraph (B), and (ii) implementation of the best management practices identified in subparagraph (A) by the categories, subcategories, or particular nonpoint sources designated under paragraph (1)(B). Such schedule shall provide for utilization of the best management practices at the earliest practicable date.

(D) A certification of the attorney general of the State or States (or the chief attorney of any State water pollution control agency which has independent legal counsel) that the laws of the State or States, as the case may be, provide adequate authority to implement such management program or, if there is not such adequate authority, a list of such additional authorities as will be necessary to implement such management program. A schedule and commitment by the State or States to seek such additional authorities as expeditiously as practicable.

(E) Sources of Federal and other assistance and funding (other than assistance provided under subsections (h) and (i)) which will be available in each of such fiscal years for supporting implementation of such practices and measures and the purposes for which such assistance will be used in each of such fiscal years.

(F) An identification of Federal financial assistance programs and Federal development projects for which the State will review individual assistance applications or development projects for their effect on water quality pursuant to the procedures set forth in Executive Order 12372 as in effect on September 17, 1983, to determine whether such assistance applications or development projects would be consistent with the program prepared under this subsection; for the purposes of this subparagraph, identification shall not be limited to the assistance programs or development projects subject to Executive Order 12372 but may include any programs listed in the most recent Catalog of Federal Domestic Assistance which may have an effect on the purposes and objectives of the State's nonpoint source pollution management program.

(3) UTILIZATION OF LOCAL AND PRIVATE EXPERTS.—In developing and implementing a management program under this subsection, a State shall, to the maximum extent practicable, involve local public and private agencies and organizations which have expertise in control of nonpoint sources of pollution.

(4) DEVELOPMENT ON WATERSHED BASIS.—A State shall, to the maximum extent practicable, develop and implement a management program under this subsection on a watershed-by-watershed basis within such State.

(c) ADMINISTRATIVE PROVISIONS.—

(1) COOPERATION REQUIREMENT.—Any report required by subsection (a) and any management program and report required by subsection (b) shall be developed in cooperation with local, substate regional, and interstate entities which are actively planning for the implementation of nonpoint source pollution controls and have either been certified by the Administrator in accordance with section 208, have worked jointly with the State on water quality management planning under section 205(j), or have been designated by the State legislative body or Governor as water quality management planning agencies for their geographic areas.

(2) TIME PERIOD FOR SUBMISSION OF REPORTS AND MANAGEMENT PROGRAMS.—Each report and management program shall be submitted to the Administrator during the 18-month period beginning on the date of the enactment of this section.

(d) APPROVAL OR DISAPPROVAL OF REPORTS AND MANAGEMENT PROGRAMS.—

(1) DEADLINE.—Subject to paragraph (2), not later than 180 days after the date of submission to the Administrator of any report or management program under this section (other than subsections (h), (i), and (k)), the Administrator shall either approve or disapprove such report or management program, as the case may be. The Administrator may approve a portion of a management program under this subsection. If the Administrator does not disapprove a report, management program, or portion of a management program in such 180-day period, such report, management program, or portion shall be deemed approved for purposes of this section.

(2) PROCEDURE FOR DISAPPROVAL.—If, after notice and opportunity for public comment and consultation with appropriate Federal and State agencies and other interested persons, the Administrator determines that—

(A) the proposed management program or any portion thereof does not meet the requirements of subsection (b)(2) of this section or is not likely to satisfy, in whole or in part, the goals and requirements of this Act;

(B) adequate authority does not exist, or adequate resources are not available, to implement such program or portion;

(C) the schedule for implementing such program or portion is not sufficiently expeditious; or

(D) the practices and measures proposed in such program or portion are not adequate to reduce the level of pollution in navigable waters in the State resulting from nonpoint sources and to improve the quality of navigable waters in the State;

the Administrator shall within 6 months of the receipt of the proposed program notify the State of any revisions or modifications necessary to obtain approval. The State shall thereupon have an additional 3 months to submit its revised management program and the Administrator shall approve or disapprove such revised program within three months of receipt.

(3) FAILURE OF STATE TO SUBMIT REPORT.—If a Governor of a State does not submit the report required by subsection (a) within the period specified by subsection (c)(2), the Administrator shall, within 30 months after the date of the enactment of this section, prepare a report for such State which makes the identifications required by paragraphs (1)(A) and (1)(B) of subsection (a). Upon completion of the requirement of the preceding sentence and after notice and opportunity for comment, the Administrator shall report to Congress on his actions pursuant to this section.

(e) LOCAL MANAGEMENT PROGRAMS; TECHNICAL ASSISTANCE.—If a State fails to submit a management program under subsection (b) or the Administrator does not approve such a management program, a local public agency or organization which has expertise in, and authority to, control water pollution resulting from nonpoint sources in any area of such State which the Administrator determines is of sufficient geographic size may, with approval of such State, request the Administrator to provide, and the Administrator shall provide, technical assistance to such agency or organization in developing for such area a management program which is described in subsection (b) and can be approved pursuant to subsection (d). After development of such management program, such agency or organization shall submit such management program to the Administrator for approval. If the Administrator approves such management program, such agency or organization shall be eligible to receive financial assistance under subsection (h) for implementation of such management program as if such agency or organization were a State for which a report submitted under subsection (a) and a management program submitted under subsection (b) were approved under this section. Such financial assistance shall be subject to the same terms and conditions as assistance provided to a State under subsection (h).

(f) TECHNICAL ASSISTANCE FOR STATE.—Upon request of a State, the Administrator may provide technical assistance to such State in developing a management program approved under subsection (b) for those portions of the navigable waters requested by such State.

(g) INTERSTATE MANAGEMENT CONFERENCE.—

(1) CONVENING OF CONFERENCE; NOTIFICATION; PURPOSE.—If any portion of the navigable waters in any State which is implementing a management program approved under this section is not meeting applicable water quality standards or the goals and requirements of this Act as a result, in whole or in part, of pollution from nonpoint sources in another State, such State may petition the Administrator to convene, and the Administrator shall convene, a management conference of all States which contribute significant pollution resulting from nonpoint sources to such portion. If, on the basis of information available, the Administrator determines that a State is not meeting applicable water quality standards or the goals and requirements of this Act as a result, in whole or in part, of significant pollution from nonpoint sources in another State, the Administrator shall notify such States. The Administrator may convene a management conference under this paragraph not later than 180 days after giving such notification, whether or not the State which is not meeting such standards requests such conference. The purpose of such conference shall be to develop an agreement among such States to reduce the level of pollution in such portion resulting from nonpoint sources and to improve the water quality of such portion. Nothing in such agreement shall supersede or abrogate rights to quantities of water which have been established by interstate water compacts, Supreme Court decrees, or State water laws. This subsection shall not apply to any pollution which is subject to the Colorado River Basin Salinity Control Act. The requirement that the Administrator convene a management conference shall not be subject to the provisions of section 505 of this Act.

(2) STATE MANAGEMENT PROGRAM REQUIREMENT.—To the extent that the States reach agreement through such conference, the management programs of the States which are parties to such agreements and which contribute significant pollution to the navigable waters or portions thereof not meeting applicable water quality standards or goals and requirements of this Act will be revised to reflect such agreement. Such management programs shall be consistent with Federal and State law.

(h) GRANT PROGRAM.—

(1) GRANTS FOR IMPLEMENTATION OF MANAGEMENT PROGRAMS.—Upon application of a State for which a report submitted under subsection (a) and a management program submitted under subsection (b) is approved under this section, the Administrator shall make grants, subject to such terms and conditions as the Administrator considers appropriate, under this subsection to such State for the purpose of assisting the State in implementing such management program. Funds reserved pursuant to section 205(j)(5) of this Act may be used to develop and implement such management program.

(2) APPLICATIONS.—An application for a grant under this subsection in any fiscal year shall be in such form and shall contain such other information as the Administrator may require, including an identification and description of the best management practices and measures which the State proposes

to assist, encourage, or require in such year with the Federal assistance to be provided under the grant.

(3) FEDERAL SHARE.—The Federal share of the cost of each management program implemented with Federal assistance under this subsection in any fiscal year shall not exceed 60 percent of the cost incurred by the State in implementing such management program and shall be made on condition that the non-Federal share is provided from non-Federal sources.

(4) LIMITATION ON GRANT AMOUNTS.—Notwithstanding any other provision of this subsection, not more than 15 percent of the amount appropriated to carry out this subsection may be used to make grants to any one State, including any grants to any local public agency or organization with authority to control pollution from nonpoint sources in any area of such State.

(5) PRIORITY FOR EFFECTIVE MECHANISMS.—For each fiscal year beginning after September 30, 1987, the Administrator may give priority in making grants under this subsection, and shall give consideration in determining the Federal share of any such grant, to States which have implemented or are proposing to implement management programs which will—

(A) control particularly difficult or serious nonpoint source pollution problems, including, but not limited to, problems resulting from mining activities;

(B) implement innovative methods or practices for controlling nonpoint sources of pollution, including regulatory programs where the Administrator deems appropriate;

(C) control interstate nonpoint source pollution problems; or

(D) carry out ground water quality protection activities which the Administrator determines are part of a comprehensive nonpoint source pollution control program, including research, planning, ground water assessments, demonstration programs, enforcement, technical assistance, education, and training to protect ground water quality from nonpoint sources of pollution.

(6) AVAILABILITY FOR OBLIGATION.—The funds granted to each State pursuant to this subsection in a fiscal year shall remain available for obligation by such State for the fiscal year for which appropriated. The amount of any such funds not obligated by the end of such fiscal year shall be available to the Administrator for granting to other States under this subsection in the next fiscal year.

(7) LIMITATION ON USE OF FUNDS.—States may use funds from grants made pursuant to this section for financial assistance to persons only to the extent that such assistance is related to the costs of demonstration projects.

(8) SATISFACTORY PROGRESS.—No grant may be made under this subsection in any fiscal year to a State which in the preceding fiscal year received a grant under this subsection unless the Administrator determines that such State made satisfactory progress in such preceding fiscal year in meeting the schedule specified by such State under subsection (b)(2).

(9) MAINTENANCE OF EFFORT.—No grant may be made to a State under this subsection in any fiscal year unless such

State enters into such agreements with the Administrator as the Administrator may require to ensure that such State will maintain its aggregate expenditures from all other sources for programs for controlling pollution added to the navigable waters in such State from nonpoint sources and improving the quality of such waters at or above the average level of such expenditures in its two fiscal years preceding the date of enactment of this subsection.

(10) REQUEST FOR INFORMATION.—The Administrator may request such information, data, and reports as he considers necessary to make the determination of continuing eligibility for grants under this section.

(11) REPORTING AND OTHER REQUIREMENTS.—Each State shall report to the Administrator on an annual basis concerning (A) its progress in meeting the schedule of milestones submitted pursuant to subsection (b)(2)(C) of this section, and (B) to the extent that appropriate information is available, reductions in nonpoint source pollutant loading and improvements in water quality for those navigable waters or watersheds within the State which were identified pursuant to subsection (a)(1)(A) of this section resulting from implementation of the management program.

(12) LIMITATION ON ADMINISTRATIVE COSTS.—For purposes of this subsection, administrative costs in the form of salaries, overhead, or indirect costs for services provided and charged against activities and programs carried out with a grant under this subsection shall not exceed in any fiscal year 10 percent of the amount of the grant in such year, except that costs of implementing enforcement and regulatory activities, education, training, technical assistance, demonstration projects, and technology transfer programs shall not be subject to this limitation.

(i) GRANTS FOR PROTECTING GROUNDWATER QUALITY.—

(1) ELIGIBLE APPLICANTS AND ACTIVITIES.—Upon application of a State for which a report submitted under subsection (a) and a plan submitted under subsection (b) is approved under this section, the Administrator shall make grants under this subsection to such State for the purpose of assisting such State in carrying out groundwater quality protection activities which the Administrator determines will advance the State toward implementation of a comprehensive nonpoint source pollution control program. Such activities shall include, but not be limited to, research, planning, groundwater assessment, demonstration programs, enforcement, technical assistance, education and training to protect the quality of groundwater and to prevent contamination of groundwater from nonpoint sources of pollution.

(2) APPLICATIONS.—An application for a grant under this subsection shall be in such form and shall contain such information as the Administrator may require.

(3) FEDERAL SHARE; MAXIMUM AMOUNT.—The Federal share of the cost of assisting a State in carrying out groundwater protection activities in any fiscal year under this subsection shall be 50 percent of the costs incurred by the State

in carrying out such activities, except that the maximum amount of Federal assistance which any State may receive under this subsection in any fiscal year shall not exceed $150,000.

(j) AUTHORIZATION OF APPROPRIATIONS.—There is authorized to be appropriated to carry out subsections (h) and (i) not to exceed $70,000,000 for fiscal year 1988, $100,000,000 per fiscal year for each of fiscal years 1989 and 1990, and $130,000,000 for fiscal year 1991; except that for each of such fiscal years not to exceed $7,500,000 may be made available to carry out subsection (i). Sums appropriated pursuant to this subsection shall remain available until expended.

(k) CONSISTENCY OF OTHER PROGRAMS AND PROJECTS WITH MANAGEMENT PROGRAMS.—The Administrator shall transmit to the Office of Management and Budget and the appropriate Federal departments and agencies a list of those assistance programs and development projects identified by each State under subsection (b)(2)(F) for which individual assistance applications and projects will be reviewed pursuant to the procedures set forth in Executive Order 12372 as in effect on September 17, 1983. Beginning not later than sixty days after receiving notification by the Administrator, each Federal department and agency shall modify existing regulations to allow States to review individual development projects and assistance applications under the identified Federal assistance programs and shall accommodate, according to the requirements and definitions of Executive Order 12372, as in effect on September 17, 1983, the concerns of the State regarding the consistency of such applications or projects with the State nonpoint source pollution management program.

(l) COLLECTION OF INFORMATION.—The Administrator shall collect and make available, through publications and other appropriate means, information pertaining to management practices and implementation methods, including, but not limited to, (1) information concerning the costs and relative efficiencies of best management practices for reducing nonpoint source pollution; and (2) available data concerning the relationship between water quality and implementation of various management practices to control nonpoint sources of pollution.

(m) SET ASIDE FOR ADMINISTRATIVE PERSONNEL.—Not less than 5 percent of the funds appropriated pursuant to subsection (j) for any fiscal year shall be available to the Administrator to maintain personnel levels at the Environmental Protection Agency at levels which are adequate to carry out this section in such year.

(33 U.S.C. 1329)

SEC. 320. NATIONAL ESTUARY PROGRAM.

(a) MANAGEMENT CONFERENCE.—

(1) NOMINATION OF ESTUARIES.—The Governor of any State may nominate to the Administrator an estuary lying in whole or in part within the State as an estuary of national significance and request a management conference to develop a comprehensive management plan for the estuary. The nomination shall document the need for the conference, the likelihood of

success, and information relating to the factors in paragraph (2).

(2) CONVENING OF CONFERENCE.—

(A) IN GENERAL.—In any case where the Administrator determines, on his own initiative or upon nomination of a State under paragraph (1), that the attainment or maintenance of that water quality in an estuary which assures protection of public water supplies and the protection and propagation of a balanced, indigenous population of shellfish, fish, and wildlife and allows recreational activities, in and on the water, requires the control of point and nonpoint sources of pollution to supplement existing controls of pollution in more than one State, the Administrator shall select such estuary and convene a management conference.

(B) PRIORITY CONSIDERATION.—The Administrator shall give priority consideration under this section to Long Island Sound, New York and Connecticut; Narragansett Bay, Rhode Island; Buzzards Bay, Massachusetts; Massachusetts Bay, Massachusetts (including Cape Cod Bay and Boston Harbor);[1] Puget Sound, Washington; New York-New Jersey Harbor, New York and New Jersey; Delaware Bay, Delaware and New Jersey; Delaware Inland Bays, Delaware; Albermarle Sound, North Carolina; Sarasota Bay, Florida; San Francisco Bay, California; Santa Monica Bay, California; Galveston Bay, Texas;[2] Barataria-Terrebonne Bay estuary complex, Louisiana; Indian River Lagoon, Florida; and Peconic Bay, New York.

(3) BOUNDARY DISPUTE EXCEPTION.—In any case in which a boundary between two States passes through an estuary and such boundary is disputed and is the subject of an action in any court, the Administrator shall not convene a management conference with respect to such estuary before a final adjudication has been made of such dispute.

(b) PURPOSES OF CONFERENCE.—The purposes of any management conference convened with respect to an estuary under this subsection shall be to—

(1) assess trends in water quality, natural resources, and uses of the estuary;

(2) collect, characterize, and assess data on toxics, nutrients, and natural resources within the estuarine zone to identify the causes of environmental problems;

(3) develop the relationship between the inplace loads and point and nonpoint loadings of pollutants to the estuarine zone and the potential uses of the zone, water quality, and natural resources;

(4) develop a comprehensive conservation and management plan that recommends priority corrective actions and compliance schedules addressing point and nonpoint sources of pollution to restore and maintain the chemical, physical, and bio-

[1] Both P.L. 100–653 and P.L. 100–658 inserted the same Massachusetts Bay phrase after Buzzards Bay; so that the phrase appears twice.

[2] P.L. 100–688, section 2001(3) inserted the Louisiana, Florida, New York bays after "Galveston, Texas;" which technically could not be executed.

logical integrity of the estuary, including restoration and maintenance of water quality, a balanced indigenous population of shellfish, fish and wildlife, and recreational activities in the estuary, and assure that the designated uses of the estuary are protected;

(5) develop plans for the coordinated implementation of the plan by the States as well as Federal and local agencies participating in the conference;

(6) monitor the effectiveness of actions taken pursuant to the plan; and

(7) review all Federal financial assistance programs and Federal development projects in accordance with the requirements of Executive Order 12372, as in effect on September 17, 1983, to determine whether such assistance program or project would be consistent with and further the purposes and objectives of the plan prepared under this section.

For purposes of paragraph (7), such programs and projects shall not be limited to the assistance programs and development projects subject to Executive Order 12372, but may include any programs listed in the most recent Catalog of Federal Domestic Assistance which may have an effect on the purposes and objectives of the plan developed under this section.

(c) MEMBERS OF CONFERENCE.—The members of a management conference convened under this section shall include, at a minimum, the Administrator and representatives of—

(1) each State and foreign nation located in whole or in part in the estuarine zone of the estuary for which the conference is convened;

(2) international, interstate, or regional agencies or entities having jurisdiction over all or a significant part of the estuary;

(3) each interested Federal agency, as determined appropriate by the Administrator;

(4) local governments having jurisdiction over any land or water within the estuarine zone, as determined appropriate by the Administrator; and

(5) affected industries, public and private educational institutions, and the general public, as determined appropriate by the Administrator.

(d) UTILIZATION OF EXISTING DATA.—In developing a conservation and management plan under this section, the management conference shall survey and utilize existing reports, data, and studies relating to the estuary that have been developed by or made available to Federal, interstate, State, and local agencies.

(e) PERIOD OF CONFERENCE.—A management conference convened under this section shall be convened for a period not to exceed 5 years. Such conference may be extended by the Administrator, and if terminated after the initial period, may be reconvened by the Administrator at any time thereafter, as may be necessary to meet the requirements of this section.

(f) APPROVAL AND IMPLEMENTATION OF PLANS.—

(1) APPROVAL.—Not later than 120 days after the completion of a conservation and management plan and after providing for public review and comment, the Administrator shall

approve such plan if the plan meets the requirements of this section and the affected Governor or Governors concur.

(2) IMPLEMENTATION.—Upon approval of a conservation and management plan under this section, such plan shall be implemented. Funds authorized to be appropriated under titles II and VI and section 319 of this Act may be used in accordance with the applicable requirements of this Act to assist States with the implementation of such plan.

(g) GRANTS.—

(1) RECIPIENTS.—The Administrator is authorized to make grants to State, interstate, and regional water pollution control agencies and entities, State coastal zone management agencies, interstate agencies, other public or nonprofit private agencies, institutions, organizations, and individuals.

(2) PURPOSES.—Grants under this subsection shall be made to pay for assisting research, surveys, studies, and modeling and other technical work necessary for the development of a conservation and management plan under this section.

(3) FEDERAL SHARE.—The amount of grants to any person (including a State, interstate, or regional agency or entity) under this subsection for a fiscal year shall not exceed 75 percent of the costs of such research, survey, studies, and work and shall be made on condition that the non-Federal share of such costs are provided from non-Federal sources.

(h) GRANT REPORTING.—Any person (including a State, interstate, or regional agency or entity) that receives a grant under subsection (g) shall report to the Administrator not later than 18 months after receipt of such grants and biennially there after on the progress being made under this section.

(i) AUTHORIZATION OF APPROPRIATIONS.—There are authorized to be appropriated to the Administrator not to exceed $12,000,000 per fiscal year for each of fiscal years 1987, 1988, 1989, 1990, and 1991 for—

(1) expenses related to the administration of management conferences under this section, not to exceed 10 percent of the amount appropriated under this subsection;

(2) making grants under subsection (g); and

(3) monitoring the implementation of a conservation and management plan by the management conference or by the Administrator, in any case in which the conference has been terminated.

The Administrator shall provide up to $5,000,000 per fiscal year of the sums authorized to be appropriated under this subsection to the Administrator of the National Oceanic and Atmospheric Administration to carry out subsection (j).

(j) RESEARCH.—

(1) PROGRAMS.—In order to determine the need to convene a management conference under this section or at the request of such a management conference, the Administrator shall coordinate and implement, through the National Marine Pollution Program Office and the National Marine Fisheries Service of the National Oceanic and Atmospheric Administration, as appropriate, for one or more estuarine zones—

(A) a long-term program of trend assessment monitoring measuring variations in pollutant concentrations, marine ecology, and other physical or biological environmental paramenters which may affect estuarine zones, to provide the Administrator the capacity to determine the potential and actual effects of alternative management strategies and measures;

(B) a program of ecosystem assessment assisting in the development of (i) baseline studies which determine the state of estuarine zones and the effects of natural and anthropogenic changes, and (ii) predictive models capable of translating information on specific discharges or general pollutant loadings within estuarine zones into a set of probable effects on such zones;

(C) a comprehensive water quality sampling program for the continuous monitoring of nutrients, chlorine, acid precipitation dissolved oxygen, and potentially toxic pollutants (including organic chemicals and metals) in estuarine zones, after consultation with interested State, local, interstate, or international agencies and review and analysis of all environmental sampling data presently collected from estuarine zones; and

(D) a program of research to identify the movements of nutrients, sediments and pollutants through estuarine zones and the impact of nutrients, sediments, and pollutants on water quality, the ecosystem, and designated or potential uses of the estuarine zones.

(2) REPORTS.—The Administrator, in cooperation with the Administrator of the National Oceanic and Atmospheric Administration, shall submit to the Congress no less often than biennially a comprehensive report on the activities authorized under this subsection including—

(A) a listing of priority monitoring and research needs;

(B) an assessment of the state and health of the Nation's estuarine zones, to the extent evaluated under this subsection;

(C) a discussion of pollution problems and trends in pollutant concentrations with a direct or indirect effect on water quality, the ecosystem, and designated or potential uses of each estuarine zone, to the extent evaluated under this subsection; and

(D) an evaluation of pollution abatement activities and management measures so far implemented to determine the degree of improvement toward the objectives expressed in subsection (b)(4) of this section.

(k) DEFINITIONS.—For purposes of this section, the terms "estuary" and "estuarine zone" have the meanings such terms have in section 104(n)(3) of this Act, except that the term "estuarine zone" shall also include associated aquatic ecosystems and those portions of tributaries draining into the estuary up to the historic height of migration of anadromous fish or the historic head of tidal influence, whichever is higher.

(33 U.S.C. 1330)

TITLE IV—PERMITS AND LICENSES

CERTIFICATION

SEC. 401. (a)(1) Any applicant for a Federal license or permit to conduct any activity including, but not limited to, the construction or operation of facilities, which may result in any discharge into the navigable waters, shall provide the licensing or permitting agency a certification from the State in which the discharge originates or will originate, or, if appropriate, from the interstate water pollution control agency having jurisdiction over the navigable waters at the point where the discharge originates or will originate, that any such discharge will comply with the applicable provisions of sections 301, 302, 303, 306, and 307 of this Act. In the case of any such activity for which there is not an applicable effluent limitation or other limitation under sections 301(b) and 302, and there is not an applicable standard under sections 306 and 307, the State shall so certify, except that any such certification shall not be deemed to satisfy section 511(c) of this Act. Such State or interstate agency shall establish procedures for public notice in the case of all applications for certification by it and, to the extent it deems appropriate, procedures for public hearings in connection with specific applications. In any case where a State or interstate agency has no authority to give such a certification, such certification shall be from the Administrator. If the State, interstate agency, or Administrator, as the case may be, fails or refuses to act on a request for certification, within a reasonable period of time (which shall not exceed one year) after receipt of such request, the certification requirements of this subsection shall be waived with respect to such Federal application. No license or permit shall be granted until the certification required by this section has been obtained or has been waived as provided in the preceding sentence. No license or permit shall be granted if certification has been denied by the State, interstate agency, or the Administrator, as the case may be.

(2) Upon receipt of such application and certification the licensing or permitting agency shall immediately notify the Administrator of such application and certification. Whenever such a discharge may affect, as determined by the Administrator, the quality of the waters of any other State, the Administrator within thirty days of the date of notice of application for such Federal license or permit shall so notify such other State, the licensing or permitting agency, and the applicant. If, within sixty days after receipt of such notification, such other State determines that such discharge will affect the quality of its waters so as to violate any water quality requirement in such State, and within such sixty-day period notifies the Administrator and the licensing or permitting agency in writing of its objection to the issuance of such license or permit and requests a public hearing on such objection, the licensing or permitting agency shall hold such a hearing. The Administrator shall at such hearing submit his evaluation and recommendations with respect to any such objection to the licensing or permitting agency. Such agency, based upon the recommendations of such State, the Administrator, and upon any additional evidence, if any, presented to the agency at the hearing, shall condition such license or permit in such manner as may be necessary to insure compliance with ap-

plicable water quality requirements. If the imposition of conditions cannot insure such compliance such agency shall not issue such license or permit.

(3) The certification obtained pursuant to paragraph (1) of this subsection with respect to the construction of any facility shall fulfill the requirements of this subsection with respect to certification in connection with any other Federal license or permit required for the operation of such facility unless, after notice to the certifying State, agency, or Administrator, as the case may be, which shall be given by the Federal agency to whom application is made for such operating license or permit, the State, or if appropriate, the interstate agency or the Administrator, notifies such agency within sixty days after receipt of such notice that there is no longer reasonable assurance that there will be compliance with the applicable provisions of sections 301, 302, 303, 306, and 307 of this Act because of changes since the construction license or permit certification was issued in (A) the construction or operation of the facility, (B) the characteristics of the waters into which such discharge is made, (C) the water quality criteria applicable to such waters or (D) applicable effluent limitations or other requirements. This paragraph shall be inapplicable in any case where the applicant for such operating license or permit has failed to provide the certifying State, or, if appropriate, the interstate agency or the Administrator, with notice of any proposed changes in the construction or operation of the facility with respect to which a construction license or permit has been granted, which changes may result in violation of section 301, 302, 303, 306, or 307 of this Act.

(4) Prior to the initial operation of any federally licensed or permitted facility or activity which may result in any discharge into the navigable waters and with respect to which a certification has been obtained pursuant to paragraph (1) of this subsection, which facility or activity is not subject to a Federal operating license or permit, the licensee or permittee shall provide an opportunity for such certifying State, or, if appropriate, the interstate agency or the Administrator to review the manner in which the facility or activity shall be operated or conducted for the purposes of assuring that applicable effluent limitations or other limitations or other applicable water quality requirements will not be violated. Upon notification by the certifying State, or if appropriate, the interstate agency or the Administrator that the operation of any such federally licensed or permitted facility or activity will violate applicable effluent limitations or other limitations or other water quality requirements such Federal agency may, after public hearing, suspend such license or permit. If such license or permit is suspended, it shall remain suspended until notification is received from the certifying State, agency, or Administrator, as the case may be, that there is reasonable assurance that such facility or activity will not violate the applicable provisions of section 301, 302, 303, 306, or 307 of this Act.

(5) Any Federal license or permit with respect to which a certification has been obtained under paragraph (1) of this subsection may be suspended or revoked by the Federal agency issuing such license or permit upon the entering of a judgment under this Act that such facility or activity has been operated in violation of the

applicable provisions of section 301, 302, 303, 306, or 307 of this Act.

(6) Except with respect to a permit issued under section 402 of this Act, in any case where actual construction of a facility has been lawfully commenced prior to April 3, 1970, no certification shall be required under this subsection for a license or permit issued after April 3, 1970, to operate such facility, except that any such license or permit issued without certification shall terminate April 3, 1973, unless prior to such termination date the person having such license or permit submits to the Federal agency which issued such license or permit a certification and otherwise meets the requirements of this section.

(b) Nothing in this section shall be construed to limit the authority of any department or agency pursuant to any other provision of law to require compliance with any applicable water quality requirements. The Administrator shall, upon the request of any Federal department or agency, or State or interstate agency, or applicant, provide, for the purpose of this section, any relevant information on applicable effluent limitations, or other limitations, standards, regulations, or requirements, or water quality criteria, and shall, when requested by any such department or agency or State or interstate agency, or applicant, comment on any methods to comply with such limitations, standards, regulations, requirements, or criteria.

(c) In order to implement the provisions of this section, the Secretary of the Army, acting throught the Chief of Engineers, is authorized, if he deems it to be in the public interest, to permit the use of spoil disposal areas under his jurisdiction by Federal licenses or permittees, and to make an appropriate charge for such use. Moneys received from such licensees or permittees shall be deposited in the Treasury as miscellaneous receipts.

(d) Any certification provided under this section shall set forth any effluent limitations and other limitations, and monitoring requirements necessary to assure that any applicant for a Federal license or permit will comply with any applicable effluent limitations and other limitations, under section 301 or 302 of this Act, standard of performance under section 306 of this Act, or prohibition, effluent standard, or pretreatment standard under section 307 of this Act, and with any other appropriate requirement of State law set forth in such certification, and shall become a condition on any Federal license or permit subject to the provisions of this section.

(33 U.S.C. 1341)

NATIONAL POLLUTANT DISCHARGE ELIMINATION SYSTEM

SEC. 402. (a)(1) Except as provided in sections 318 and 404 of this Act, the Administrator may, after opportunity for public hearing, issue a permit for the discharge of any pollutant, or combination of pollutants, notwithstanding section 301(a), upon condition that such discharge will meet either (A) all applicable requirements under sections 301, 302, 306, 307, 308, and 403 of this Act, or (B) prior to the taking of necessary implementing actions relating to all such requirements, such conditions as the Administrator determines are necessary to carry out the provisions of this Act.

(2) The Administrator shall prescribe conditions for such permits to assure compliance with the requirements of paragraph (1) of this subsection, including conditions on data and information collection, reporting, and such other requirements as he deems appropriate.

(3) The permit program of the Administrator under paragraph (1) of this subsection, and permits issued thereunder, shall be subject to the same terms, conditions, and requirements as apply to a State permit program and permits issued thereunder under subsection (b) of this section.

(4) All permits for discharges into the navigable waters issued pursuant to section 13 of the Act of March 3, 1899, shall be deemed to be permits issued under this title, and permits issued under this title shall be deemed to be permits issued under section 13 of the Act of March 3, 1899, and shall continue in force and effect for their term unless revoked, modified, or suspended in accordance with the provisions of this Act.

(5) No permit for a discharge into the navigable waters shall be issued under section 13 of the Act of March 3, 1899, after the date of enactment of this title. Each application for a permit under section 13 of the Act of March 3, 1899, pending on the date of enactment of this Act shall be deemed to be an application for a permit under this section. The Administrator shall authorize a State, which he determines has the capability of administering a permit program which will carry out the objective of this Act, to issue permits for discharges into the navigable waters within the jurisdiction of such State. The Administrator may exercise the authority granted him by the preceding sentence only during the period which begins on the date of enactment of this Act and ends either on the ninetieth day after the date of the first promulgation of guidelines required by section 304(h)(2) of this Act, or the date of approval by the Administrator of a permit program for such State under subsection (b) of this section, whichever date first occurs, and no such authorization to a State shall extend beyond the last day of such period. Each such permit shall be subject to such conditions as the Administrator determines are necessary to carry out the provisions of this Act. No such permit shall issue if the Administrator objects to such issuance.

(b) At any time after the promulgation of the guidelines required by subsection (h)(2) of section 304 of this Act, the Governor of each State desiring to administer its own permit program for discharges into navigable waters within its jurisdiction may submit to the Administrator a full and complete discription of the program it proposes to establish and administer under State law or under an interstate compact. In addition, such State shall submit a statement from the attorney general (or the attorney for those State water pollution control agencies which have independent legal counsel), or from the chief legal officer in the case of an interstate agency, that the laws of such State, or the interstate compact, as the case may be, provide adequate authority to carry out the described program. The Administrator shall approve each such submitted program unless he determines that adequate authority does not exist:

(1) To issue permits which—

(A) apply, and insure compliance with, any applicable requirements of sections 301, 302, 306, 307, and 403;

(B) are for fixed terms not exceeding five years; and

(C) can be terminated or modified for cause including, but not limited to, the following:

(i) violation of any condition of the permit;

(ii) obtaining a permit by misrepresentation, or failure to disclose fully all relevant facts;

(iii) change in any condition that requires either a temporary or permanent reduction or elimination of the permitted discharge;

(D) control the disposal of pollutants into wells;

(2)(A) To issue permits which apply, and insure compliance with, all applicable requirements of section 308 of this Act, or

(B) To inspect, monitor, enter, and require reports to at least the same extent as required in section 308 of this Act;

(3) To insure that the public, and any other State the waters of which may be affected, receive notice of each application for a permit and to provide an opportunity for public hearing before a ruling on each such application;

(4) To insure that the Administrator receives notice of each application (including a copy thereof) for a permit;

(5) To insure that any State (other than the permitting State), whose waters may be affected by the issuance of a permit may submit written recommendations to the permitting State (and the Administrator) with respect to any permit application and, if any part of such written recommendations are not accepted by the permitting State, that the permitting State will notify such affected State (and the Administrator) in writing of its failure to so accept such recommendations together with its reasons for so doing;

(6) To insure that no permit will be issued if, in the judgment of the Secretary of the Army acting through the Chief of Engineers, after consultation with the Secretary of the department in which the Coast Guard is operating, anchorage and navigation of any of the navigable waters would be substantially impaired thereby;

(7) To abate violations of the permit or the permit program, including civil and criminal penalties and other ways and means of enforcement;

(8) To insure that any permit for a discharge from a publicly owned treatment works includes conditions to require the identification in terms of character and volume of pollutants of any significant source introducing pollutants subject to pretreatment standards under section 307(b) of this Act into such works and a program to assure compliance with such pretreatment standards by each such source, in addition to adequate notice to the permitting agency of (A) new introductions into such works of pollutants from any source which would be a new source as defined in section 306 if such source were discharging pollutants, (B) new introductions of pollutants into such works from a source which would be subject to section 301 if it were discharging such pollutants, or (C) a substantial change in volume or character of pollutants being introduced into such works by a source introducing pollutants into such works at the time of issuance of the permit. Such notice shall include information on the quality and quantity of effluent to be in-

troduced into such treatment works and any anticipated impact of such change in the quantity or quality of effluent to be discharged from such publicly owned treatment works; and

(9) To insure that any industrial user of any publicly owned treatment works will comply with sections 204(b), 307, and 308.

(c)(1) Not later than ninety days after the date on which a State has submitted a program (or revision thereof) pursuant to subsection (b) of this section, the Administrator shall suspend the issuance of permits under subsection (a) of this section as to those discharges subject to such program unless he determines that the State permit program does not meet the requirements of subsection (b) of this section or does not conform to the guidelines issued under section 304(i)(2) of this Act. If the Administrator so determines, he shall notify the State or any revisons or modifications necessary to conform to such requirements or guidelines.

(2) Any State permit program under this section shall at all times be in accordance with this section and guidelines promulgated pursuant to section 304(h)(2) of this Act.

(3) Whenever the Administrator determines after public hearing that a State is not administering a program approved under this section in accordance with requirements of this section, he shall so notify the State and, if appropriate corrective action is not taken within a reasonable time, not to exceed ninety days, the Administrator shall withdraw approval of such program. The Administrator shall not withdraw approval of any such program unless he shall first have notified the State, and made public, in writing, the reasons for such withdrawal.

(4) LIMITATIONS ON PARTIAL PERMIT PROGRAM RETURNS AND WITHDRAWALS.—A State may return to the Administrator administration, and the Administrator may withdraw under paragraph (3) of this subsection approval, of—

(A) a State partial permit program approved under subsection (n)(3) only if the entire permit program being administered by the State department or agency at the time is returned or withdrawn; and

(B) a State partial permit program approved under subsection (n)(4) only if an entire phased component of the permit program being administered by the State at the time is returned or withdrawn.

(d)(1) Each State shall transmit to the Administrator a copy of each permit application received by such State and provide notice to the Administrator of every action related to the consideration of such permit application, including each permit proposed to be issued by such State.

(2) No permit shall issue (A) if the Administrator within ninety days of the date of his notification under subsection (b)(5) of this section objects in writing to the issuance of such permit, or (B) if the Administrator within ninety days of the date of transmittal of the proposed permit by the State objects in writing to the issuance of such permit as being outside the guidelines and requirements of this Act. Whenever the Administrator objects to the issuance of a permit under this paragraph such written objection shall contain a statement of the reasons for such objection and the effluent limita-

tions and conditions which such permit would include if it were issued by the Administrator.

(3) The Administrator may, as to any permit application, waive paragraph (2) of this subsection.

(4) In any case where, after the date of enactment of this paragraph, the Administrator, pursuant to paragraph (2) of this subsection, objects to the issuance of a permit, on request of the State, a public hearing shall be held by the Administrator on such objection. If the State does not resubmit such permit revised to meet such objection within 30 days after completion of the hearing, or, if no hearing is requested within 90 days after the date of such objection, the Administrator may issue the permit pursuant to subsection (a) of this section for such source in accordance with the guidelines and requirements of this Act.

(e) In accordance with guidelines promulgated pursuant to subsection (h)(2) of section 304 of this Act, the Administrator is authorized to waive the requirements of subsection (d) of this section at the time he approves a program pursuant to subsection (b) of this section for any category (including any class, type, or size within such category) of point sources within the State submitting such program.

(f) The Administrator shall promulgate regulations establishing categories of point sources which he determines shall not be subject to the requirements of subsection (d) of this section in any State with a program approved pursuant to subsection (b) of this section. The Administrator may distinguish among classes, types, and sizes within any category of point sources.

(g) Any permit issued under this section for the discharge of pollutants into the navigable waters from a vessel or other floating craft shall be subject to any applicable regulations promulgated by the Secretary of the Department in which the Coast Guard is operating, establishing specifications for safe transportation, handling, carriage, storage, and stowage of pollutants.

(h) In the event any condition of a permit for discharges from a treatment works (as defined in section 212 of this Act) which is publicly owned is violated, a State with a program approved under subsection (b) of this section or the Administrator, where no State program is approved or where the Administrator determines pursuant to section 309(a) of this Act that a State with an approved program has not commenced appropriate enforcement action with respect to such permit, may proceed in a court of competent jurisdiction to restrict or prohibit the introduction of any pollutant into such treatment works by a source not utilizing such treatment works prior to the finding that such condition was violated.

(i) Nothing in this section shall be construed to limit the authority of the Administrator to take action pursuant to section 309 of this Act.

(j) A copy of each permit application and each permit issued under this section shall be available to the public. Such permit application or permit, or portion thereof, shall further be available on request for the purpose of reproduction.

(k) Compliance with a permit issued pursuant to this section shall be deemed compliance, for purposes of sections 309 and 505, with sections 301, 302, 306, 307, and 403, except any standard imposed under section 307 for a toxic pollutant injurious to human health. Until December 31, 1974, in any case where a permit for discharge has been applied for pursuant to this section, but final administrative disposition of such application has not been made, such discharge shall not be a violation of (1) section 301, 306, or 402 of this Act, or (2) section 13 of the Act of March 3, 1899, unless the Administrator or other plaintiff proves that final administrative disposition of such application has not been made because of the failure of the applicant to furnish information reasonably required or requested in order to process the application. For the 180-day period beginning on the date of enactment of the Federal Water Pollution Control Act Amendments of 1972, in the case of any point source discharging any pollutant or combination of pollutants immediately prior to such date of enactment which source is not subject to section 13 of the Act of March 3, 1899, the discharge by such source shall not be a violation of this Act if such a source applies for a permit for discharge pursuant to this section within such 180-day period.

(l) LIMITATION ON PERMIT REQUIREMENT.—

(1) AGRICULTURAL RETURN FLOWS.—The Administrator shall not require a permit under this section for discharges composed entirely of return flows from irrigated agriculture, nor shall the Administrator directly or indirectly, require any State to require such a permit.

(2) STORMWATER RUNOFF FROM OIL, GAS, AND MINING OPERATIONS.—The Administrator shall not require a permit under this section, nor shall the Administrator directly or indirectly require any State to require a permit, for discharges of stormwater runoff from mining operations or oil and gas exploration, production, processing, or treatment operations or transmission facilities, composed entirely of flows which are from conveyances or systems of conveyances (including but not limited to pipes, conduits, ditches, and channels) used for collecting and conveying precipitation runoff and which are not contaminated by contact with, or do not come into contact with, any overburden, raw material, intermediate products, finished product, byproduct, or waste products located on the site of such operations.

(m) ADDITIONAL PRETREATMENT OF CONVENTIONAL POLLUTANTS NOT REQUIRED.—To the extent a treatment works (as defined in section 212 of this Act) which is publicly owned is not meeting the requirements of a permit issued under this section for such treatment works as a result of inadequate design or operation of such treatment works, the Administrator, in issuing a permit under this section, shall not require pretreatment by a person introducing conventional pollutants identified pursuant to a section 304(a)(4) of this Act into such treatment works other than pretreatment required to assure compliance with pretreatment standards under subsection (b)(8) of this section and section 307(b)(1) of this Act. Nothing in this subsection shall affect the Administrator's authority under sections 307 and 309 of this Act, affect State and local authority under sections 307(b)(4) and 510 of this Act, relieve such treatment works of its obligations to meet requirements established under this Act, or otherwise preclude such works from pursuing whatever feasible options are available to meet its responsibility to comply with its permit under this section.

(n) PARTIAL PERMIT PROGRAM.—

(1) STATE SUBMISSION.—The Governor of a State may submit under subsection (b) of this section a permit program for a portion of the discharges into the navigable waters in such State.

(2) MINIMUM COVERAGE.—A partial permit program under this subsection shall cover, at a minimum, administration of a major category of the discharges into the navigable waters of the State or a major component of the permit program required by subsection (b).

(3) APPROVAL OF MAJOR CATEGORY PARTIAL PERMIT PROGRAMS.—The Administrator may approve a partial permit program covering administration of a major category of discharges under this subsection if—

(A) such program represents a complete permit program and covers all of the discharges under the jurisdiction of a department or agency of the State; and

(B) the Administrator determines that the partial program represents a significant and identifiable part of the State program required by subsection (b).

(4) APPROVAL OF MAJOR COMPONENT PARTIAL PERMIT PROGRAMS.—The Administrator may approve under this subsection a partial and phased permit program covering administration of a major component (including discharge categories) of a State permit program required by subsection (b) if—

(A) the Administrator determines that the partial program represents a significant and identifiable part of the State program required by subsection (b); and

(B) the State submits, and the Administrator approves, a plan for the State to assume administration by phases of the remainder of the State program required by subsection (b) by a specified date not more than 5 years after submission of the partial program under this subsection and agrees to make all reasonable efforts to assume such administration by such date.

(o) ANTI-BACKSLIDING.—

(1) GENERAL PROHIBITION.—In the case of effluent limitations established on the basis of subsection (a)(1)(B) of this section, a permit may not be renewed, reissued, or modified on the basis of effluent guidelines promulgated under section 304(b) subsequent to the original issuance of such permit, to contain effluent limitations which are less stringent than the comparable effluent limitations in the previous permit. In the case of effluent limitations established on the basis of section 301(b)(1)(C) or section 303(d) or (e), a permit may not be renewed, reissued, or modified to contain effluent limitations which are less stringent than the comparable effluent limitations in the previous permit except in compliance with section 303(d)(4).

(2) EXCEPTIONS.—A permit with respect to which paragraph (1) applies may be renewed, reissued, or modified to con-

tain a less stringent effluent limitation applicable to a pollutant if—

(A) material and substantial alterations or additions to the permitted facility occurred after permit issuance which justify the application of a less stringent effluent limitation;

(B)(i) information is available which was not available at the time of permit issuance (other than revised regulations, guidance, or test methods) and which would have justified the application of a less stringent effluent limitation at the time of permit issuance; or

(ii) the Administrator determines that technical mistakes or mistaken interpretations of law were made in issuing the permit under subsection (a)(1)(B);

(C) a less stringent effluent limitation is necessary because of events over which the permittee has no control and for which there is no reasonably available remedy;

(D) the permittee has received a permit modification under section 301(c), 301(g), 301(h), 301(i), 301(k), 301(n), or 316(a); or

(E) the permittee has installed the treatment facilities required to meet the effluent limitations in the previous permit and has properly operated and maintained the facilities but has nevertheless been unable to achieve the previous effluent limitations, in which case the limitations in the reviewed, reissued, or modified permit may reflect the level of pollutant control actually achieved (but shall not be less stringent than required by effluent guidelines in effect at the time of permit renewal, reissuance, or modification).

Subparagraph (B) shall not apply to any revised waste load allocations or any alternative grounds for translating water quality standards into effluent limitations, except where the cumulative effect of such revised allocations results in a decrease in the amount of pollutants discharged into the concerned waters, and such revised allocations are not the result of a discharger eliminating or substantially reducing its discharge of pollutants due to complying with the requirements of this Act or for reasons otherwise unrelated to water quality.

(3) LIMITATIONS.—In no event may a permit with respect to which paragraph (1) applies be renewed, reissued, or modified to contain an effluent limitation which is less stringent than required by effluent guidelines in effect at the time the permit is renewed, reissued, or modified. In no event may such a permit to discharge into waters be renewed, reissued, or modified to contain a less stringent effluent limitation if the implementation of such limitation would result in a violation of a water quality standard under section 303 applicable to such waters.

(p) MUNICIPAL AND INDUSTRIAL STORMWATER DISCHARGES.—

(1) GENERAL RULE.—Prior to October 1, 1994, the Administrator or the State (in the case of a permit program approved under section 402 of this Act) shall not require a permit under this section for discharges composed entirely of stormwater.

(2) EXCEPTIONS.—Paragraph (1) shall not apply with respect to the following stormwater discharges:

(A) A discharge with respect to which a permit has been issued under this section before the date of the enactment of this subsection.

(B) A discharge associated with industrial activity.

(C) A discharge from a municipal separate storm sewer system serving a population of 250,000 or more.

(D) A discharge from a municipal separate storm sewer system serving a population of 100,000 or more but less than 250,000.

(E) A discharge for which the Administrator or the State, as the case may be, determines that the stormwater discharge contributes to a violation of a water quality standard or is a significant contributor of pollutants to waters of the United States.

(3) PERMIT REQUIREMENTS.—

(A) INDUSTRIAL DISCHARGES.—Permits for discharges associated with industrial activity shall meet all applicable provisions of this section and section 301.

(B) MUNICIPAL DISCHARGE.—Permits for discharges from municipal storm sewers—

(i) may be issued on a system- or jurisdiction-wide basis;

(ii) shall include a requirement to effectively prohibit non-stormwater discharges into the storm sewers; and

(iii) shall require controls to reduce the discharge of pollutants to the maximum extent practicable, including management practices, control techniques and system, design and engineering methods, and such other provisions as the Administrator or the State determines appropriate for the control of such pollutants.

(4) PERMIT APPLICATION REQUIREMENTS.—

(A) INDUSTRIAL AND LARGE MUNICIPAL DISCHARGES.—Not later than 2 years after the date of the enactment of this subsection, the Administrator shall establish regulations setting forth the permit application requirements for stormwater discharges described in paragraphs (2)(B) and (2)(C). Applications for permits for such discharges shall be filed no later than 3 years after such date of enactment. Not later than 4 year after such date of enactment the Administrator or the State, as the case may be, shall issue or deny each such permit. Any such permit shall provide for compliance as expeditiously as practicable, but in no event later than 3 years after the date of issuance of such permit.

(B) OTHER MUNICIPAL DISCHARGES.—Not later than 4 years after the date of the enactment of this subsection, the Administrator shall establish regulations setting forth the permit application requirements for stormwater discharges described in paragraph (2)(D). Applications for permits for such discharges shall be filed no later than 5 years after such date of enactment. Not later than 6 years after such date of enactment, the Administrator or the State, as the case may be, shall issue or deny each such permit. Any such permit shall provide for compliance as expeditiously as practicable, but in no event later than 3 years after the date of issuance of such permit.

(5) STUDIES.—The Administrator, in consultation with the States, shall conduct a study for the purposes of—

(A) identifying those stormwater discharges or classes of stormwater discharges for which permits are not required pursuant to paragraphs (1) and (2) of this subsection;

(B) determining, to the maximum extent practicable, the nature and extent of pollutants in such discharges; and

(C) establishing procedures and methods to control stormwater discharges to the extent necessary to mitigate impacts on water quality.

Not later than October 1, 1988, the Administrator shall submit to Congress a report on the results of the study described in subparagraphs (A) and (B). Not later than October 1, 1989, the Administrator shall submit to Congress a report on the results of the study described in subparagraph (C).

(6) REGULATIONS.—Not later than October 1, 1993, the Administrator, in consultation with State and local officials, shall issue regulations (based on the results of the studies conducted under paragraph (5)) which designate stormwater discharges, other than those discharges described in paragraph (2), to be regulated to protect water quality and shall establish a comprehensive program to regulate such designated sources. The program shall, at a minimum, (A) establish priorities, (B) establish requirements for State stormwater management programs, and (C) establish expeditious deadlines. The program may include performance standards, guidelines, guidance, and management practices and treatment requirements, as appropriate.

(q) COMBINED SEWER OVERFLOWS.—

(1) REQUIREMENT FOR PERMITS, ORDERS, AND DECREES.—Each permit, order, or decree issued pursuant to this Act after the date of enactment of this subsection for a discharge from a municipal combined storm and sanitary sewer shall conform to the Combined Sewer Overflow Control Policy signed by the Administrator on April 11, 1994 (in this subsection referred to as the "CSO control policy").

(2) WATER QUALITY AND DESIGNATED USE REVIEW GUIDANCE.—Not later than July 31, 2001, and after providing notice and opportunity for public comment, the Administrator shall issue guidance to facilitate the conduct of water quality and designated use reviews for municipal combined sewer overflow receiving waters.

(3) REPORT.—Not later than September 1, 2001, the Administrator shall transmit to Congress a report on the progress made by the Environmental Protection Agency, States, and municipalities in implementing and enforcing the CSO control policy.

(33 U.S.C. 1342)

OCEAN DISCHARGE CRITERIA

SEC. 403. (a) No permit under section 402 of this Act for a discharge into the territorial sea, the waters of the contiguous zone, or the oceans shall be issued, after promulgation of guidelines established under subsection (c) of this section, except in compliance with such guidelines. Prior to the promulgation of such guidelines, a permit may be issued under such section 402 if the Administrator determines it to be in the public interest.

(b) The requirements of subsection (d) of section 402 of this Act may not be waived in the case of permits for discharges into the territorial sea.

(c)(1) The Administrator shall, within one hundred and eighty days after enactment of this Act (and from time to time thereafter), promulgate guidelines for determining the degradation of the waters of the territorial seas, the contiguous zone, and the oceans, which shall include:

(A) the effect of disposal of pollutants on human health or welfare, including but not limited to plankton, fish, shellfish, wildlife, shorelines, and beaches;

(B) the effect of disposal of pollutants on marine life including the transfer, concentration, and dispersal of pollutants or their byproducts through biological, physical, and chemical processes; changes in marine ecosystem diversity, productivity, and stability; and species and community population changes;

(C) the effect of disposal, of pollutants on esthetic, recreation, and economic values;

(D) the persistence and permanence of the effects of disposal of pollutants;

(E) the effect of the disposal at varying rates, of particular volumes and concentrations of pollutants;

(F) other possible locations and methods of disposal or recycling of pollutants including land-based alternatives; and

(G) the effect on alternate uses of the oceans, such as mineral exploitation and scientific study.

(2) In any event where insufficient information exists on any proposed discharge to make a reasonable judgment on any of the guidelines established pursuant to this subsection no permit shall be issued under section 402 of this Act.

(33 U.S.C. 1343)

PERMITS FOR DREDGED OR FILL MATERIAL

SEC. 404. (a) The Secretary may issue permits, after notice and opportunity for public hearings for the discharge of dredged or fill material into the navigable waters at specified disposal sites. Not later than the fifteenth day after the date an applicant submits all the information required to complete an application for a permit under this subsection, the Secretary shall publish the notice required by this subsection.

(b) Subject to subsection (c) of this section, each such disposal site shall be specified for each such permit by the Secretary (1) through the application of guidelines developed by the Administrator, in conjunction with the Secretary which guidelines shall be based upon criteria comparable to the criteria applicable to the territorial seas, the contiguous zone, and the ocean under section 403(c), and (2) in any case where such guidelines under clause (1) alone would prohibit the specification of a site, through the application additionally of the economic impact of the site on navigation and anchorage.

(c) The Administrator is authorized to prohibit the specification (including the withdrawal of specification) of any defined area as a disposal site, and he is authorized to deny or restrict the use of any defined area for specification (including the withdrawal of specification) as a disposal site, whenever he determines, after notice and opportunity for public hearings, that the discharge of such materials into such area will have an unacceptable adverse effect on municipal water supplies, shellfish beds and fishery areas (including spawning and breeding areas), wildlife, or recreational areas. Before making such determination, the Administrator shall consult with the Secretary. The Administrator shall set forth in writing and make public his findings and his reasons for making any determination under this subsection.

(d) The term "Secretary" as used in this section means the Secretary of the Army, acting through the Chief of Engineers.

(e)(1) In carrying out his functions relating to the discharge of dredged or fill material under this section, the Secretary may, after notice and opportunity for public hearing, issue general permits on a State, regional, or nationwide basis for any category of activities involving discharges of dredged or fill material if the Secretary determines that the activities in such category are similar in nature, will cause only minimal adverse environmental effects when performed separately, and will have only minimal cumulative adverse effect on the environment. Any general permit issued under this subsection shall (A) be based on the guidelines described in subsection (b)(1) of this section, and (B) set forth the requirements and standards which shall apply to any activity authorized by such general permit.

(2) No general permit issued under this subsection shall be for a period of more than five years after the date of its issuance and such general permit may be revoked or modified by the Secretary if, after opportunity for public hearing, the Secretary determines that the activities authorized by such general permit have an adverse impact on the environment or such activities are more appropriately authorized by individual permits.

(f)(1) Except as provided in paragraph (2) of this subsection, the discharge of dredge or fill material—

(A) from normal farming, silviculture, and ranching activities such as plowing, seeding, cultivating, minor drainage, harvesting for the production of food, fiber, and forest products, or upland soil and water conservation practices;

(B) for the purpose of maintenance, including emergency reconstruction of recently damaged parts, of currently serviceable structures such as dikes, dams, levees, groins, riprap, breakwaters, causeways, and bridge abutments or approaches, and transportation structures;

(C) for the purpose of construction or maintenance of farm or stock ponds or irrigation ditches, or the maintenance of drainage ditches;

(D) for the purpose of construction of temporary sedimentation basins on a construction site which does not include placement of fill material into the navigable waters;

(E) for the purpose of construction or maintenance of farm roads or forest roads, or temporary roads for moving mining equipment, where such roads are constructed and maintained, in accordance with best management practices, to assure that flow and circulation patterns and chemical and biological characteristics of the navigable waters are not impaired, that the reach of the navigable waters is not reduced, and that any adverse effect on the aquatic environment will be otherwise minimized;

(F) resulting from any activity with respect to which a State has an approved program under section 208(b)(4) which meets the requirements of subparagraphs (B) and (C) of such section,

is not prohibited by or otherwise subject to regulation under this section or section 301(a) or 402 of this Act (except for effluent standards or prohibitions under section 307).

(2) Any discharge of dredged or fill material into the navigable waters incidental to any activity having as its purpose bringing an area of the navigable waters into a use to which it was not previously subject, where the flow or circulation of navigable waters may be impaired or the reach of such waters be reduced, shall be required to have a permit under this section.

(g)(1) The Governor of any State desiring to administer its own individual and general permit program for the discharge of dredged or fill material into the navigable waters (other than those waters which are presently used, or are susceptible to use in their natural condition or by reasonable improvement as a means to transport interstate or foreign commerce shoreward to their ordinary high water mark, including all waters which are subject to the ebb and flow of the tide shoreward to their mean high water mark, or mean higher high water mark on the west coast, including wetlands adjacent thereto), within its jurisdiction may submit to the Administrator a full and complete description of the program it proposes to establish and administer under State law or under an interstate compact. In addition, such State shall submit a statement from the attorney general (or the attorney for those State agencies which have independent legal counsel), or from the chief legal officer in the case of an interstate agency, that the laws of such State, or the interstate compact, as the case may be, provide adequate authority to carry out the described program.

(2) Not later than the tenth day after the date of the receipt of the program, and statement submitted by any State under paragraph (1) of this subsection, the Administrator shall provide copies of such program and statement to the Secretary and the Secretary of the Interior, acting through the Director of the United States Fish and Wildlife Service.

(3) Not later than the ninetieth day after the date of the receipt by the Administrator of the program and statement submitted by any State, under paragraph (1) of this subsection, the Secretary and the Secretary of the Interior, acting through the Director of the United States Fish and Wildlife Service, shall submit any com-

ments with respect to such program and statement to the Administrator in writing.

(h)(1) Not later than the one-hundred-twentieth day after the date of the receipt by the Administrator of a program and statement submitted by any State under paragraph (1) of this subsection, the Administrator shall determine, taking into account any comments submitted by the Secretary and the Secretary of the Interior, acting through the Director of the United States Fish and Wildlife Service, pursuant to subsection (g) of this section, whether such State has the following authority with respect to the issuance of permits pursuant to such program:

(A) To issue permits which—

(i) apply, and assure compliance with, any applicable requirements of this section, including, but not limited to, the guidelines established under subsection (b)(1) of this section, and sections 307 and 403 of this Act;

(ii) are for fixed terms not exceeding five years; and

(iii) can be terminated or modified for cause including, but not limited to, the following:

(I) violation of any condition of the permit;

(II) obtaining a permit by misrepresentation, or failure to disclose fully all relevant facts;

(III) change in any condition that requires either a temporary or permanent reduction or elimination of the permitted discharge.

(B) To issue permits which apply, and assure compliance with, all applicable requirements of section 308 of this Act, or to inspect, monitor, enter, and requrie reports to at least the same extent as required in section 308 of this Act.

(C) To assure that the public, and any other State the waters of which may be affected, receive notice of each application for a permit and to provide an opportunity for public hearing before a ruling on each such application.

(D) To assure that the Administrator receives notice of each application (including a copy thereof) for a permit.

(E) To assure that any State (other than the permitting State), whose waters may be affected by the issuance of a permit may submit written recommendation to the permitting State (and the Administrator) with respect to any permit application and, if any part of such written recommendations are not accepted by the permitting State, that the permitting State will notify such affected State (and the Administrator) in writing of its failure to so accept such recommendations together with its reasons for so doing.

(F) To assure that no permit will be issued if, in the judgment of the Secretary, after consultation with the Secretary of the department in which the Coast Guard is operating, anchorage and navigation of any of the navigable waters would be substantially impaired thereby.

(G) To abate violations of the permit or the permit program, including civil and criminal penalties and other ways and means of enforcement.

(H) To assure continued coordination with Federal and Federal-State water-related planning and review processes.

(2) If, with respect to a State program submitted under subsection (g)(1) of this section, the Administrator determines that such State—

(A) has the authority set forth in paragraph (1) of this subsection, the Administrator shall approve the program and so notify (i) such State, and (ii) the Secretary, who upon subsequent notification from such State that it is administering such program, shall suspend the issuance of permits under subsection (a) and (e) of this section for activities with respect to which a permit may be issued pursuant to such State program; or

(B) does not have the authority set forth in paragraph (1) of this subsection, the Administrator shall so notify such State, which notification shall also describe the revisions or modifications necessary so that such State may resubmit such program for a determination by the Administrator under this subsection.

(3) If the Administrator fails to make a determination with respect to any program submitted by a State under subsection (g)(1) of this section within one-hundred-twenty days after the date of the receipt of such program, such program shall be deemed approved pursuant to paragraph (2)(A) of this subsection and the Administrator shall so notify such State and the Secretary who, upon subsequent notification from such State that it is administering such program, shall suspend the issuance of permits under subsection (a) and (e) of this section for activities with respect to which a permit may be issued by such State.

(4) After the Secretary receives notification from the Administrator under paragraph (2) or (3) of this subsection that a State permit program has been approved, the Secretary shall transfer any applications for permits pending before the Secretary for activities with respect to which a permit may be issued pursuant to such State program to such State for appropriate action.

(5) Upon notification from a State with a permit program approved under this subsection that such State intends to administer and enforce the terms and conditions of a general permit issed by the Secretary under subsection (e) of this section with respect to activities in such State to which such general permit applies, the Secretary shall suspend the administration and enforcement of such general permit with respect to such activities.

(i) Whenever the Administrator determines after public hearing that a State is not administering a program approved under section (h)(2)(A) of this section, in accordance with this section, including, but not limited to, the guidelines established under subsection (b)(1) of this section, the Administrator shall so notify the State, and, if appropriate corrective action is not taken within a reasonable time, not to exceed ninety days after the date of the receipt of such notification, the Administrator shall (1) withdraw approval of such program until the Administrator determines such corrective action has been taken, and (2) notify the Secretary that the Secretary shall resume the program for the issuance of permits under subsections (a) and (e) of this section for activities with respect to which the State was issuing permits and that such authority of the Secretary shall continue in effect until such time as the

Administrator makes the determination described in clause (1) of this subsection and such State again has an approved program.

(j) Each State which is administering a permit program pursuant to this section shall transmit to the Administrator (1) a copy of each permit application received by such State and provide notice to the Administrator of every action related to the consideration of such permit application, including each permit proposed to be issued by such State, and (2) a copy of each proposed general permit which such State intends to issue. Not later than the tenth day after the date of the receipt of such permit application or such proposed general permit, the Administrator shall provide copies of such permit application or such proposed general permit to the Secretary and the Secretary of the Interior, acting through the Director of the United States Fish and Wildlife Service. If the Administrator intends to provide written comments to such State with respect to such permit application or such proposed general permit, he shall so notify such State not later than the thirtieth day after the date of the receipt of such application or such proposed general permit and provide such written comments to such State, after consideration of any comments made in writing with respect to such application or such proposed general permit by the Secretary and the Secretary of the Interior, acting through the Director of the United States Fish and Wildlife Service, not later than the ninetieth day after the date of such receipt. If such State is so notified by the Administrator, it shall not issue the proposed permit until after the receipt of such comments from the Administrator, or after such ninetieth day, whichever first occurs. Such State shall not issue such proposed permit after such ninetieth day if it has received such written comments in which the Administrator objects (A) to the issuance of such proposed permit and such proposed permit is one that has been submitted to the Administrator pursuant to subsection (h)(1)(E), or (B) to the issuance of such proposed permit as being outside the requirements of this section, including, but not limited to, the guidelines developed under subsection (b)(1) of this section unless it modifies such proposed permit in accordance with such comments. Whenever the Administrator objects to the issuance of a permit under the preceding sentence such written objection shall contain a statement of the reasons for such objection and the conditions which such permit would include if it were issued by the Administrator. In any case where the Administrator objects to the issuance of a permit, on request of the State, a public hearing shall be held by the Administrator on such objection. If the State does not resubmit such permit revised to meet such objection within 30 days after completion of the hearing or, if no hearing is requested within 90 days after the date of such objection, the Secretary may issue the permit pursuant to subsection (a) or (e) of this section, as the case may be, for such source in accordance with the guidelines and requirements of this Act.

(k) In accordance with guidelines promulgated pursuant to subsection (i)(2) of section 304 of this Act, the Administrator is authorized to waive the requirements of subsection (j) of this section at the time of the approval of a program pursuant to subsection (h)(2)(A) of this section for any category (including any class, type, or size within such category) of discharge within the State submitting such program.

(l) The Administrator shall promulgate regulations establishing categories of discharges which he determines shall not be subject to the requirements of subsection (j) of this section in any State with a program approved pursuant to subsection (h)(2)(A) of this section. The Administrator may distinguish among classes, types, and sizes within any category of discharges.

(m) Not later than the ninetieth day after the date on which the Secretary notifies the Secretary of the Interior, acting through the Director of the United States Fish and Wildlife Service that (1) an application for a permit under subsection (a) of this section has been received by the Secretary, or (2) the Secretary proposes to issue a general permit under subsection (e) of this section, the Secretary of the Interior, acting through the Director of the United States Fish and Wildlife Service, shall submit any comments with respect to such application or such proposed general permit in writing to the Secretary.

(n) Nothing in this section shall be construed to limit the authority of the Administrator to take action pursuant to section 309 of this Act.

(o) A copy of each permit application and each permit issued under this section shall be available to the public. Such permit application or portion thereof, shall further be available on request for the purpose of reproduction.

(p) Compliance with a permit issued pursuant to this section, including any activity carried out pursuant to a general permit issued under this section, shall be deemed compliance, for purposes of sections 309 and 505, with sections 301, 307, and 403.

(q) Not later than the one-hundred-eightieth day after the date of enactment of this subsection, the Secretary shall enter into agreements with the Administrator, the Secretaries of the Departments of Agriculture, Commerce, Interior, and Transportation, and the heads of other appropriate Federal agencies to minimize, to the maximum extent practicable, duplication, needless paperwork, and delays in the issuance of permits under this section. Such agreements shall be developed to assure that, to the maximum extent practicable, a decision with respect to an application for a permit under subsection (a) of this section will be made not later than the ninetieth day after the date the notice of such application is published under subsection (a) of this section.

(r) The discharge of dredged or fill material as part of the construction of a Federal project specifically authorized by Congress, whether prior to or on or after the date of enactment of this subsection, is not prohibited by or otherwise subject to regulation under this section, or a State program approved under this section, or section 301(a) or 402 of the Act (except for effluent standards or prohibitions under section 307), if information on the effects of such discharge, including consideration of the guidelines developed under subsection (b)(1) of this section, is included in an environmental impact statement for such project pursuant to the National Environmental Policy Act of 1969 and such environmental impact statement has been submitted to Congress before the actual discharge of dredged or fill material in connection with the construction of such project and prior to either authorization of such project or an appropriation of funds for each construction.

(s)(1) Whenever on the basis of any information available to him the Secretary finds that any person is in violation of any condition or limitation set forth in a permit issued by the Secretary under this section, the Secretary shall issue an order requiring such persons to comply with such condition or limitation, or the Secretary shall bring a civil action in accordance with paragraph (3) of this subsection.

(2) A copy of any order issued under this subsection shall be sent immediately by the Secretary to the State in which the violation occurs and other affected States. Any order issued under this subsection shall be by personal service and shall state with reasonable specificity the nature of the violation, specify a time for compliance, not to exceed thirty days, which the Secretary determines is reasonable, taking into account the seriousness of the violation and any good faith efforts to comply with applicable requirements. In any case in which an order under this subsection is issued to a corporation, a copy of such order shall be served on any appropriate corporate officers.

(3) The Secretary is authorized to commence a civil action for appropriate relief, including a permanent or temporary injunction for any violation for which he is authorized to issue a compliance order under paragraph (1) of this subsection. Any action under this paragraph may be brought in the district court of the United States for the district in which the defendant is located or resides or is doing business, and such court shall have jurisdiction to restrain such violation and to require compliance. Notice of the commencement of such acton [1] shall be given immediately to the appropriate State.

(4) Any person who violates any condition or limitation in a permit issued by the Secretary under this section, and any person who violates any order issued by the Secretary under paragraph (1) of this subsection, shall be subject to a civil penalty not to exceed $25,000 per day for each violation. In determining the amount of a civil penalty the court shall consider the seriousness of the violation or violations, the economic benefit (if any) resulting from the violaltion, any history of such violations, any good-faith efforts to comply with the applicable requirements, the economic impact of the penalty on the violator, and such other matters as justice may require.

(t) Nothing in the section shall preclude or deny the right of any State or interstate agency to control the discharge of dredged or fill material in any portion of the navigable waters within the jurisdiction of such State, including any activity of any Federal agency, and each such agency shall comply with such State or interstate requirements both substantive and procedural to control the discharge of dredged or fill material to the same extent that any person is subject to such requirements. This section shall not be construed as affecting or impairing the authority of the Secretary to maintain navigation.

(33 U.S.C. 1344)

---

[1] So in law. Probably should be "action".

SEC. 405. (a) Notwithstanding any other provision of this Act or of any other law, in the case where the disposal of sewage sludge resulting from the operation of a treatment works as defined in section 212 of this Act (including the removal of in-place sewage sludge from one location and its deposit at another location) would result in any pollutant from such sewage sludge entering the navigable waters, such disposal is prohibited except in accordance with a permit issued by the Administrator under section 402 of this Act.

(b) The Administrator shall issue regulations governing the issuance of permits for the disposal of sewage sludge subject to subsection (a) of this section and section 402 of this Act. Such regulations shall require the application to such disposal of each criterion, factor, procedure, and requirement applicable to a permit issued under section 402 of this title.

(c) Each State desiring to administer its own permit program for disposal of sewage sludge subject to subsection (a) of this section within its jurisdiction may do so in accordance with section 402 of this Act.

(d) REGULATIONS.—

(1) REGULATIONS.—The Administrator, after consultation with appropriate Federal and State agencies and other interested persons, shall develop and publish, within one year after the date of enactment of this subsection and from time to time thereafter, regulations poroviding guidelines for the disposal of sludge and the utilization of sludge for various purposes. Such regulations shall—

(A) identify uses for sludge, including disposal;

(B) specify factors to be taken into account in determining the measures and practices applicable to each such use or disposal (including publication of information on costs);

(C) identify concentrations of pollutants which interfere with each such use or disposal.

The Administrator is authorized to revise any regulation issued under this subsection.

(2) IDENTIFICATION AND REGULATION OF TOXIC POLLUTANTS.—

(A) ON BASIS OF AVAILABLE INFORMATION.—

(i) PROPOSED REGULATIONS.—Not later than November 30, 1986, the Administrator shall identify those toxic pollutants which, on the basis of available information on their toxicity, persistence, concentration, mobility, or potential for exposure, may be present in sewage sludge in concentrations which may adversely affect public health or the environment, and propose regulations specifying acceptable management practices for sewage sludge containing each such toxic pollutant and establishing numerical limitations for each such pollutant for each use identified under paragraph (1)(A).

(ii) FINAL REGULATIONS.—Not later than August 31, 1987, and after opportunity for public hearing, the

Administrator shall promulgate the regulations required by subparagraph (A)(i).

(B) OTHERS.—

(i) PROPOSED REGULATIONS.—Not later than July 31, 1987, the Administrator shall identify those toxic pollutants not identified under subparagraph (A)(i) which may be present in sewage sludge in concentrations which may adversely affect public health or the environment, and propose regulations specifying acceptable management practices for sewage sludge containing each such toxic pollutant and establishing numerical limitations for each pollutant for each such use identified under paragraph (1)(A).

(ii) FINAL REGULATIONS.—Not later than June 15, 1988, the Administrator shall promulate the regulations required by subparagraph (B)(i).

(C) REVIEW.—From time to time, but not less often than every 2 years, the Administrator shall review the regulations promulgated under this paragraph for the purpose of identifying additional toxic pollutants and promulgating regulations for such pollutants consistent with the requirements of this paragraph.

(D) MINIMUM STANDARDS; COMPLIANCE DATE.—The management practices and numerical criteria established under subparagraphs (A), (B), and (C) shall be adequate to protect public health and the environment from any reasonably anticipated adverse effects of each pollutant. Such regulations shall require compliance as expeditiously as practicable but in no case later than 12 months after their publication, unless such regulations require the construction of new pollution control facilities, in which case the regulations shall require compliance as expeditiously as practicable but in no case later than two years from the date of their publication.

(3) ALTERNATIVE STANDARDS.—For purposes of this subsection, if, in the judgment of the Administrator, it is not feasible to prescribe or enforce a numerical limitation for a pollutant identified under paragraph (2), the Administrator may instead promulgate a design, equipment, management practice, or operational standard, or combination thereof, which in the Administrator's judgment is adequate to protect public health and the environment from any reasonably anticipated adverse effects of such pollutant. In the event the Administrator promulgates a design or equipment standard under this subsection, the Administrator shall include as part of such standard such requirements as will assure the proper operation and maintenance of any such element of design or equipment.

(4) CONDITIONS ON PERMITS.—Prior to the promulgation of the regulations required by paragraph (2), the Administrator shall impose conditions in permits issued to publicly owned treatment works under section 402 of this Act or take such other measures as the Administrator deems appropriate to protect public health and the environment from any adverse effects which may occur from toxic pollutants in sewage sludge.

(5) LIMITATION ON STATUTORY CONSTRUCTION.—Nothing in this section is intended to waive more stringent requirements established by this Act or any other law.

(e) MANNER OF SLUDGE DISPOSAL.—The determination of the manner of disposal or use of sludge is a local determination, except that it shall be unlawful for any person to dispose of sludge from a publicly owned treatment works or any other treatment works treating domestic sewage for any use for which regulations have been established pursuant to subsection (d) of this section, except in accordance with such regulations.

(f) IMPLEMENTATION OF REGULATIONS.—

(1) THROUGH SECTION 402 PERMITS.—Any permit issued under section 402 of this Act to a publicly owned treatment works or any other treatment works treating domestic sewage shall include requirements for the use and disposal of sludge that implement the regulations established pursuant to subsection (d) of this section, unless such requirements have been included in a permit issued under the appropriate provisions of subtitle C of the Solid Waste Disposal Act, part C of the Safe Drinking Water Act, the Marine Protection, Research, and Sanctuaries Act of 1972, or the Clean Air Act, or under State permit programs approved by the Administrator, where the Administrator determines that such programs assure compliance with any applicable requirements of this section. Not later than December 15, 1986, the Administrator shall promulgate procedures for approval of State programs pursuant to this paragraph.

(2) THROUGH OTHER PERMITS.—In the case of a treatment works described in paragraph (1) that is not subject to section 402 of this Act and to which none of the other above listed permit programs nor approved State permit authority apply, the Administrator may issue a permit to such treatment works solely to impose requirements for the use and disposal of sludge that implement the regulations established pursuant to subsection (d) of this section. The Administrator shall include in the permit appropriate requirements to assure compliance with the regulations established pursuant to subsection (d) of this section. The Administrator shall establish procedures for issuing permits pursuant to this paragraph.

(g) STUDIES AND PROJECTS.—

(1) GRANT PROGRAM; INFORMATION GATHERING.—The Administrator is authorized to conduct or initiate scientific studies, demonstration projects, and public information and education projects which are designed to promote the safe and beneficial management or use of sewage sludge for such purposes as aiding the restoration of abandoned mine sites, conditioning soil for parks and recreation areas, agricultural and horticultural uses, and other beneficial purposes. For the purposes of carrying out this subsection, the Administrator may make grants to State water pollution control agencies, other public or nonprofit agencies, institutions, organizations, and individuals. In cooperation with other Federal departments and agencies, other public and private agencies, institutions, and organizations, the Administrator is authorized to collect and

disseminate information pertaining to the safe and beneficial use of sewage sludge.

(2) AUTHORIZATION OF APPROPRIATIONS.—For the purposes of carrying out the scientific studies, demonstration projects, and public information and education projects authorized in this section, there is authorized to be appropriated for fiscal years beginning after September 30, 1986, not to exceed $5,000,000.

(33 U.S.C. 1345)

SEC. 406. COASTAL RECREATION WATER QUALITY MONITORING AND NOTIFICATION.

(a) MONITORING AND NOTIFICATION.—

(1) IN GENERAL.—Not later than 18 months after the date of the enactment of this section, after consultation and in cooperation with appropriate Federal, State, tribal, and local officials (including local health officials), and after providing public notice and an opportunity for comment, the Administrator shall publish performance criteria for—

(A) monitoring and assessment (including specifying available methods for monitoring) of coastal recreation waters adjacent to beaches or similar points of access that are used by the public for attainment of applicable water quality standards for pathogens and pathogen indicators; and

(B) the prompt notification of the public, local governments, and the Administrator of any exceeding of or likelihood of exceeding applicable water quality standards for coastal recreation waters described in subparagraph (A).

(2) LEVEL OF PROTECTION.—The performance criteria referred to in paragraph (1) shall provide that the activities described in subparagraphs (A) and (B) of that paragraph shall be carried out as necessary for the protection of public health and safety.

(b) PROGRAM DEVELOPMENT AND IMPLEMENTATION GRANTS.—

(1) IN GENERAL.—The Administrator may make grants to States and local governments to develop and implement programs for monitoring and notification for coastal recreation waters adjacent to beaches or similar points of access that are used by the public.

(2) LIMITATIONS.—

(A) IN GENERAL.—The Administrator may award a grant to a State or a local government to implement a monitoring and notification program if—

(i) the program is consistent with the performance criteria published by the Administrator under subsection (a);

(ii) the State or local government prioritizes the use of grant funds for particular coastal recreation waters based on the use of the water and the risk to human health presented by pathogens or pathogen indicators;

(iii) the State or local government makes available to the Administrator the factors used to prioritize the use of funds under clause (ii);

(iv) the State or local government provides a list of discrete areas of coastal recreation waters that are subject to the program for monitoring and notification for which the grant is provided that specifies any coastal recreation waters for which fiscal constraints will prevent consistency with the performance criteria under subsection (a); and

(v) the public is provided an opportunity to review the program through a process that provides for public notice and an opportunity for comment.

(B) GRANTS TO LOCAL GOVERNMENTS.—The Administrator may make a grant to a local government under this subsection for implementation of a monitoring and notification program only if, after the 1-year period beginning on the date of publication of performance criteria under subsection (a)(1), the Administrator determines that the State is not implementing a program that meets the requirements of this subsection, regardless of whether the State has received a grant under this subsection.

(3) OTHER REQUIREMENTS.—

(A) REPORT.—A State recipient of a grant under this subsection shall submit to the Administrator, in such format and at such intervals as the Administrator determines to be appropriate, a report that describes—

(i) data collected as part of the program for monitoring and notification as described in subsection (c); and

(ii) actions taken to notify the public when water quality standards are exceeded.

(B) DELEGATION.—A State recipient of a grant under this subsection shall identify each local government to which the State has delegated or intends to delegate responsibility for implementing a monitoring and notification program consistent with the performance criteria published under subsection (a) (including any coastal recreation waters for which the authority to implement a monitoring and notification program would be subject to the delegation).

(4) FEDERAL SHARE.—

(A) IN GENERAL.—The Administrator, through grants awarded under this section, may pay up to 100 percent of the costs of developing and implementing a program for monitoring and notification under this subsection.

(B) NON-FEDERAL SHARE.—The non-Federal share of the costs of developing and implementing a monitoring and notification program may be—

(i) in an amount not to exceed 50 percent, as determined by the Administrator in consultation with State, tribal, and local government representatives; and

(ii) provided in cash or in kind.

(c) CONTENT OF STATE AND LOCAL GOVERNMENT PROGRAMS.—As a condition of receipt of a grant under subsection (b), a State or local government program for monitoring and notification under this section shall identify—

(1) lists of coastal recreation waters in the State, including coastal recreation waters adjacent to beaches or similar points of access that are used by the public;

(2) in the case of a State program for monitoring and notification, the process by which the State may delegate to local governments responsibility for implementing the monitoring and notification program;

(3) the frequency and location of monitoring and assessment of coastal recreation waters based on—

(A) the periods of recreational use of the waters;

(B) the nature and extent of use during certain periods;

(C) the proximity of the waters to known point sources and nonpoint sources of pollution; and

(D) any effect of storm events on the waters;

(4)(A) the methods to be used for detecting levels of pathogens and pathogen indicators that are harmful to human health; and

(B) the assessment procedures for identifying short-term increases in pathogens and pathogen indicators that are harmful to human health in coastal recreation waters (including increases in relation to storm events);

(5) measures for prompt communication of the occurrence, nature, location, pollutants involved, and extent of any exceeding of, or likelihood of exceeding, applicable water quality standards for pathogens and pathogen indicators to—

(A) the Administrator, in such form as the Administrator determines to be appropriate; and

(B) a designated official of a local government having jurisdiction over land adjoining the coastal recreation waters for which the failure to meet applicable standards is identified;

(6) measures for the posting of signs at beaches or similar points of access, or functionally equivalent communication measures that are sufficient to give notice to the public that the coastal recreation waters are not meeting or are not expected to meet applicable water quality standards for pathogens and pathogen indicators; and

(7) measures that inform the public of the potential risks associated with water contact activities in the coastal recreation waters that do not meet applicable water quality standards.

(d) FEDERAL AGENCY PROGRAMS.—Not later than 3 years after the date of the enactment of this section, each Federal agency that has jurisdiction over coastal recreation waters adjacent to beaches or similar points of access that are used by the public shall develop and implement, through a process that provides for public notice and an opportunity for comment, a monitoring and notification program for the coastal recreation waters that—

(1) protects the public health and safety;

(2) is consistent with the performance criteria published under subsection (a);

(3) includes a completed report on the information specified in subsection (b)(3)(A), to be submitted to the Administrator; and

(4) addresses the matters specified in subsection (c) .

(e) DATABASE.—The Administrator shall establish, maintain, and make available to the public by electronic and other means a national coastal recreation water pollution occurrence database that provides—

(1) the data reported to the Administrator under subsections (b)(3)(A)(i) and (d)(3); and

(2) other information concerning pathogens and pathogen indicators in coastal recreation waters that—

(A) is made available to the Administrator by a State or local government, from a coastal water quality monitoring program of the State or local government; and

(B) the Administrator determines should be included.

(f) TECHNICAL ASSISTANCE FOR MONITORING FLOATABLE MATERIAL.—The Administrator shall provide technical assistance to States and local governments for the development of assessment and monitoring procedures for floatable material to protect public health and safety in coastal recreation waters.

(g) LIST OF WATERS.—

(1) IN GENERAL.—Beginning not later than 18 months after the date of publication of performance criteria under subsection (a), based on information made available to the Administrator, the Administrator shall identify, and maintain a list of, discrete coastal recreation waters adjacent to beaches or similar points of access that are used by the public that—

(A) specifies any waters described in this paragraph that are subject to a monitoring and notification program consistent with the performance criteria established under subsection (a); and

(B) specifies any waters described in this paragraph for which there is no monitoring and notification program (including waters for which fiscal constraints will prevent the State or the Administrator from performing monitoring and notification consistent with the performance criteria established under subsection (a)).

(2) AVAILABILITY.—The Administrator shall make the list described in paragraph (1) available to the public through—

(A) publication in the Federal Register; and

(B) electronic media.

(3) UPDATES.—The Administrator shall update the list described in paragraph (1) periodically as new information becomes available.

(h) EPA IMPLEMENTATION.—In the case of a State that has no program for monitoring and notification that is consistent with the performance criteria published under subsection (a) after the last day of the 3-year period beginning on the date on which the Administrator lists waters in the State under subsection (g)(1)(B), the Administrator shall conduct a monitoring and notification program for the listed waters based on a priority ranking established by the Administrator using funds appropriated for grants under subsection (i)—

(1) to conduct monitoring and notification; and

(2) for related salaries, expenses, and travel.

(i) AUTHORIZATION OF APPROPRIATIONS.—There is authorized to be appropriated for making grants under subsection (b), including implementation of monitoring and notification programs by the Administrator under subsection (h), $30,000,000 for each of fiscal years 2001 through 2005.

(33 U.S.C. 1346)

TITLE V—GENERAL PROVISIONS

ADMINISTRATION

SEC. 501. (a) The Administrator is authorized to prescribe such regulations as are necessary to carry out his functions under this Act.

(b) The Administrator, with the consent of the head of any other agency of the United States, may utilize such officers and employees of such agency as may be found necessary to assist in carrying out the purposes of this Act.

(c) Each recipient of financial assistance under this Act shall keep such records as the Administrator shall prescribe, including records which fully disclose the amount and disposition by such recipient of the proceeds of such assistance, the total cost of the project or undertaking in connection with which such assistance is given or used, and the amount of that portion of the cost of the project or undertaking supplied by other sources, and such other records as will facilitate an effective audit.

(d) The Administrator and the Comptroller General of the United States, or any of their duly authorized representatives, shall have access, for the purpose of audit and examination, to any books, documents, papers, and records of the recipients that are pertinent to the grants received under this Act. For the purpose of carrying out audits and examinations with respect to recipients of Federal assistance under this Act, the Administrator is authorized to enter into noncompetitive procurement contracts with independent State audit organizations, consistent with chapter 75 of title 31, United States Code. Such contracts may only be entered into to the extent and in such amounts as may be provided in advance in appropriation Acts.

(e)(1) It is the purpose of this subsection to authorize a program which will provide official recognition by the United States Government to those industrial organizations and political subdivisions of States which during the preceding year demonstrated an outstanding technological achievement or an innovative process, method, or device in their waste treatment and pollution abatement programs. The Administrator shall, in consultation with the appropriate State water pollution control agencies, establish regulations under which such recognition may be applied for and granted, except that no applicant shall be eligible for an award under this subsection if such applicant is not in total compliance with all applicable water quality requirements under this Act, or otherwise does not have a satisfactory record with respect to environmental quality.

(2) The Administrator shall award a certificate or plaque of suitable design to each industrial organization or political subdivision which qualifies for such recognition under regulations established under this subsection.

(3) The President of the United States, the Governor of the appropriate State, the Speaker of the House of Representatives, and the President pro tempore of the Senate shall be notified of the award by the Administrator and the awarding of such recognition shall be published in the Federal Register.

(f) Upon the request of a State water pollution control agency, personnel of the Environmental Protection Agency may be detailed to such agency for the purpose of carrying out the provisions of this Act.

(33 U.S.C. 1361)

GENERAL DEFINITIONS

SEC. 502. Except as otherwise specifically provided, when used in this Act:

(1) The term "State water pollution control agency" means the State agency designated by the Governor having responsibility for enforcing State laws relating to the abatement of pollution.

(2) The term "interstate agency" means an agency of two or more States established by or pursuant to an agreement or compact approved by the Congress, or any other agency of two or more States, having substantial powers or duties pertaining to the control of pollution as determined and approved by the Adminstrator.

(3) The term "State" means a State, the District of Columbia, the Commonwealth of Puerto Rico, the Virgin Islands, Guam, American Samoa, the Commonwealth of the Northern Mariana Islands, and the Trust Territory of the Pacific Islands.

(4) The term "municipality" means a city, town, borough, county, parish, district, association, or other public body created by or pursuant to State law and having jurisdiction over disposal of sewage, industrial wastes, or other wastes, or an Indian tribe or an authorized Indian tribal organization, or a designated and approved management agency under section 208 of this Act.

(5) The term "person" means an individual, corporation, partnership, association, State, municipality, commission, or political subdivision of a State, or any interstate body.

(6) The term "pollutant" means dredged spoil, solid waste, incinerator residue, sewage, garbage, sewage sludge, munitions, chemical wastes, biological materials, radioactive materials, heat, wrecked or discarded equipment, rock, sand, cellar dirt and industrial, municipal, and agricultural waste discharged into water. This term does not mean (A) "sewage from vessels or a discharge incidental to the normal operation of a vessel of the Armed Forces" within the meaning of section 312 of this Act; or (B) water, gas, or other material which is injected into a well to facilitate production of oil or gas, or water derived in association with oil or gas production and disposed of in a well, if the well used either to facilitate production or for disposal purpose is approved by authority of the State in which the well is located, and if such State determines

that such injection or disposal will not result in the degradation of ground or surface water resources.

(7) The term "navigable waters" means the waters of the United States, including the territorial seas.

(8) The term "territorial seas" means the belt of the seas measured from the line of ordinary low water along that portion of the coast which is in direct contact with the open sea and the line marking the seaward limit of inland waters, and extending seaward a distance of three miles.

(9) The term "contiguous zone" means the entire zone established or to be established by the United States under article 24 of the Convention of the Territorial Sea and the Contiguous Zone.

(10) The term "ocean" means any portion of the high seas beyond the contiguous zone.

(11) The term "effluent limitation" means any restriction established by a State or the Administrator on quantities, rates, and concentrations of chemical, physical, biological, and other constituents which are discharged from point sources into navigable waters, the waters of the contiguous zone, or the ocean, including schedules of compliance.

(12) The term "discharge of a pollutant" and the term "discharge of pollutants" each means (A) any addition of any pollutant to navigable waters from any point source, (B) any addition of any pollutant to the waters of the contiguous zone or the ocean from any point source other than a vessel or other floating craft.

(13) The term "toxic pollutant" means those pollutants, or combinations of pollutants, including disease-causing agents, which after discharge and upon exposure, ingestion, inhalation or assimilation into any organism, either directly from the environment or indirectly by ingestion through food chains, will, on the basis of information available to the Administrator, cause death, disease, behavioral abnormalities, cancer, genetic mutations, physiological malfunctions (including malfunctions in reproduction) or physical deformations, in such organisms or their offspring.

(14) The term "point source" means any discernible, confined and discrete conveyance, including but not limited to any pipe, ditch, channel, tunnel, conduit, well, discrete fissure, container, rolling stock, concentrated animal feeding operation, or vessel or other floating craft, from which pollutants are or may be discharged. This term does not include agricultural stormwater discharges and return flows from irrigated agriculture.

(15) The term "biological monitoring" shall mean the determination of the effects on aquatic life, including accumulation of pollutants in tissue, in receiving waters due to the discharge of pollutants (A) by techniques and procedures, including sampling of organisms representative of appropriate levels of the food chain appropriate to the volume and the physical, chemical, and biological characteristics of the effluent, and (B) at appropriate frequencies and locations.

(16) The term "discharge" when used without qualification includes a discharge of a pollutant, and a discharge of pollutants.

(17) The term "schedule of compliance" means a schedule of remedial measures including an enforceable sequence of actions or

operations leading to compliance with an effluent limitation, other limitation, prohibition, or standard.

(18) The term "industrial user" means those industries identified in the Standard Industrial Classification Manual, Bureau of the Budget, 1967, as amended and supplemented, under the category "Division D—Manufacturing" and such other classes of significant waste producers as, by regulation, the Administrator deems appropriate.

(19) The term "pollution" means the man-made or man-induced alteration of the chemical, physical, biological, and radiological integrity of water.

(20) The term "medical waste" means isolation wastes; infectious agents; human blood and blood products; pathological wastes; sharps; body parts; contaminated bedding; surgical wastes and potentially contaminated laboratory wastes; dialysis wastes; and such additional medical items as the Administrator shall prescribe by regulation.

(21) COASTAL RECREATION WATERS.—

(A) IN GENERAL.—The term "coastal recreation waters" means—

(i) the Great Lakes; and

(ii) marine coastal waters (including coastal estuaries) that are designated under section 303(c) by a State for use for swimming, bathing, surfing, or similar water contact activities.

(B) EXCLUSIONS.—The term "coastal recreation waters" does not include—

(i) inland waters; or

(ii) waters upstream of the mouth of a river or stream having an unimpaired natural connection with the open sea.

(22) FLOATABLE MATERIAL.—

(A) IN GENERAL.—The term "floatable material" means any foreign matter that may float or remain suspended in the water column.

(B) INCLUSIONS.—The term "floatable material" includes—

(i) plastic;

(ii) aluminum cans;

(iii) wood products;

(iv) bottles; and

(v) paper products.

(23) PATHOGEN INDICATOR.—The term "pathogen indicator" means a substance that indicates the potential for human infectious disease.

(33 U.S.C. 1362)

WATER POLLUTION CONTROL ADVISORY BOARD

SEC. 503. (a)(1) There is hereby established in the Environmental Protection Agency a Water Pollution Control Advisory Board, composed of the Administrator or his designee, who shall be Chairman, and nine members appointed by the President, none of whom shall be Federal officers or employees. The appointed members, having due regard for the purposes of this Act, shall be selected from among representatives of various State, interstate, and local governmental agencies, of public or private interests contributing to, affected by, or concerned with pollution, and of other public and private agencies, organizations, or groups demonstrating an active interest in the field of pollution prevention and control, as well as other individuals who are expert in this field.

(2)(A) Each member appointed by the President shall hold office for a term of three years, except that (i) any member appointed to fill a vacancy occurring prior to the expiration of the term for which his predecessor was appointed shall be appointed for the remainder of such term, and (ii) the terms of office of the members first taking office after June 30, 1956, shall expire as follows: three at the end of one year after such date, three at the end of two years after such date, and three at the end of three years after such date, as designated by the President at the time of appointment, and (iii) the term of any member under the preceding provisions shall be extended until the date on which his successor's appointment is effective. None of the members appointed by the President shall be eligible for reappointment within one year after the end of his preceding term.

(B) The members of the Board who are not officers or employees of the United States, while attending conferences or meetings of the Board or while otherwise serving at the request of the Administrator, shall be entitled to receive compensation at a rate to be fixed by the Administrator, but not exceeding $100 per diem, including traveltime, and while away from their homes or regular places of business they may be allowed travel expenses, including per diem in lieu of subsistence, as authorized by law (5 U.S.C. 73b–2) for persons in the Government service employed intermittently.

(b) The Board shall advise, consult with, and make recommendations to the Administrator on matters of policy relating to the activities and functions of the Administrator under this Act.

(c) Such clerical and technical assistance as may be necessary to discharge the duties of the Board shall be provided from the personnel of the Environmental Protection Agency.

(33 U.S.C. 1363)

EMERGENCY POWERS

SEC. 504. (a) Notwithstanding any other provision of this Act, the Administrator upon receipt of evidence that a pollution source or combination of sources is presenting an imminent and substantial endangerment to the health of persons or to the welfare of persons where such endangerment is to the livelihood of such persons, such as inability to market shellfish, may bring suit on behalf of the United States in the appropriate district court to immediately restrain any person causing or contributing to the alleged pollution to stop the discharge of pollutants causing or contributing to such pollution or to take such other action as may be necessary.

[Subsection (b) repealed by §304(a) of P.L. 96–510, Dec. 11, 1980, 94 Stat. 2809]

(33 U.S.C. 1364)

SEC. 505. (a) Except as provided in subsection (b) of this section and section 309(g)(6), any citizen may commence a civil action on his own behalf—

(1) against any person (including (i) the United States, and (ii) any other governmental instrumentality or agency to the extent permitted by the eleventh amendment to the Constitution) who is alleged to be in violation of (A) an effluent standard or limitation under this Act or (B) an order issued by the Administrator or a State with respect to such a standard or limitation, or

(2) against the Administrator where there is alleged a failure of the Administrator to perform any act or duty under this Act which is not discretionary with the Administrator.

The district courts shall have jurisdiction, without regard to the amount in controversy or the citizenship of the parties, to enforce such an effluent standard or limitation, or such an order, or to order the Administrator to perform such act or duty, as the case may be, and to apply any appropriate civil penalties under section 309(d) of this Act.

(b) No action may be commenced—

(1) under subsection (a)(1) of this section—

(A) prior to sixty days after the plaintiff has given notice of the alleged violation (i) to the Administrator, (ii) to the State in which the alleged violation occurs, and (iii) to any alleged violator of the standard, limitation, or order, or

(B) if the Administrator or State has commenced and is diligently prosecuting a civil or criminal action in a court of the United States, or a State to require compliance with the standard, limitation, or order, but in any such action in a court of the United States any citizen may intervene as a matter of right.

(2) under subsection (a)(2) of this section prior to sixty days after the plaintiff has given notice of such action to the Administrator,

except that such action may be brought immediately after such notification in the case of an action under this section respecting a violation of sections 306 and 307(a) of this Act. Notice under this subsection shall be given in such manner as the Administrator shall prescribe by regulation.

(c)(1) Any action respecting a violation by a discharge source of an effluent standard or limitation or an order respecting such standard or limitation may be brought under this section only in the judicial district in which such source is located.

(2) In such action under this section, the Administrator, if not a party, may intervene as a matter of right.

(3) PROTECTION OF INTERESTS OF UNITED STATES.—Whenever any action is brought under this section in a court of the United States, the plaintiff shall serve a copy of the complaint on the Attorney General and the Administrator. No consent judgment shall be entered in an action in which the United States is not a party prior to 45 days following the receipt of

a copy of the proposed consent judgment by the Attorney General and the Administrator.

(d) The court, in issuing any final order in any action brought pursuant to this section, may award costs of litigation (including reasonable attorney and expert witness fees) to any prevailing or substantially prevailing party, whenever the court determines such award is appropriate. The court may, if a temporary restraining order or preliminary injunction is sought, require the filing of a bond or equivalent security in accordance with the Federal Rules of Civil Procedure.

(e) Nothing in this section shall restrict any right which any person (or class of persons) may have under any statute or common law to seek enforcement of any effluent standard or limitation or to seek any other relief (including relief against the Administrator or a State agency).

(f) For purposes of this section, the term "effluent standard or limitation under this Act" means (1) effective July 1, 1973, an unlawful act under subsection (a) of section 301 of this Act; (2) an effluent limitation or other limitation under section 301 or 302 of this Act; (3) standard or performance under section 306 of this Act; (4) prohibition, effluent standard or pretreatment standards under section 307 of this Act; (5) certification under section 401 of this Act; (6) a permit or condition thereof issued under section 402 of this Act, which is in effect under this Act (including a requirement applicable by reason of section 313 of this Act); or (7) a regulation under section 405(d) of this Act,.[1]

(g) For the purposes of this section the term "citizen" means a person or persons having an interest which is or may be adversely affected.

(h) A Governor of a State may commence a civil action under subsection (a), without regard to the limitations of subsection (b) of this section, against the Administrator where there is alleged a failure of the Administrator to enforce an effluent standard or limitation under this Act the violation of which is occurring in another State and is causing an adverse effect on the public health or welfare in his State, or is causing a violation of any water quality requirement in his State.

(33 U.S.C. 1365)

APPEARANCE

SEC. 506. The Administrator shall request the Attorney General to appear and represent the United States in any civil or criminal action instituted under this Act to which the Administrator is a party. Unless the Attorney General notifies the Administrator within a reasonable time, that he will appear in a civil action, attorneys who are officers or employees of the Environmental Protection Agency shall appear and represent the United States in such action.

(33 U.S.C. 1366)

[1] So in law. See P.L. 100–4, sec. 406(d)(2), 101 Stat. 73.

EMPLOYEE PROTECTION

SEC. 507. (a) No person shall fire, or in any other way discriminate against, or cause to be fired or discriminated against, any employee or any authorized representative or employees by reason of the fact that such employee or representative has filed, instituted, or caused to be filed or instituted any proceeding under this Act, or has testified or is about to testify in any proceeding resulting from the administration or enforcement of the provisions of this Act.

(b) Any employee or a representative of employees who believes that he has been fired or otherwise discriminated against by any person in violation of subsection (a) of this section may, within thirty days after such alleged violation occurs, apply to the Secretary of Labor for a review of such firing or alleged discrimination. A copy of the application shall be sent to such person who shall be the respondent. Upon receipt of such application, the Secretary of Labor shall cause such investigation to be made as he deems appropriate. Such investigation shall provide an opportunity for a public hearing at the request of any party to such review to enable the parties to present information relating to such alleged violation. The parties shall be given written notice of the time and place of the hearing at least five days prior to the hearing. Any such hearing shall be of record and shall be subject to section 554 of title 5 of the United States Code. Upon receiving the report of such investigation, the Secretary of Labor shall make findings of fact. If he finds that such violation did occur, he shall issue a decision, incorporating an order therein and his findings, requiring the party committing such violation to take such affirmative action to abate the violation as the Secretary of Labor deems appropriate, including, but not limited to, the rehiring or reinstatement of the employee or representative of employees to his former position with compensation. If he finds that there was no such violation, he shall issue an order denying the application. Such order issued by the Secretary of Labor under this subparagraph shall be subject to judicial review in the same manner as orders and decisions of the Administrator are subject to judicial review under this Act.

(c) Whenever an order is issued under this section to abate such violation, at the request of the applicant, a sum equal to the aggregate amount of all costs and expenses (including the attorney's fees), as determined by the Secretary of Labor, to have been reasonably incurred by the applicant for, or in connection with, the institution and prosecution of such proceedings, shall be assessed against the person committing such violation.

(d) This section shall have no application to any employee who, acting without direction from his employer (or his agent) deliberately violates any prohibition of effluent limitation or other limitation under section 301 or 302 of this Act, standards of performance under section 306 of this Act, effluent standard, prohibition or pretreatment standard under section 307 of this Act, or any other prohibition or limitation established under this Act.

(e) The Administrator shall conduct continuing evaluations of potential loss or shifts of employment which may result from the issuance of any effluent limitation or order under this Act, including, where appropriate, investigating threatened plant closures or reductions in employment allegedly resulting from such limitation or order. Any employee who is discharged or laid off, threatened with discharge or lay-off, or otherwise discriminated against by any person because of the alleged results of any effluent limitation or order issued under this Act, or any representative of such employee, may request the Administrator to conduct a full investigation of the matter. The Administrator shall thereupon investigate the matter and, at the request of any party, shall hold public hearings on not less than five days notice, and shall at such hearings require the parties, including the employer involved, to present information relating to the actual or potential effect of such limitation or order on employment and on any alleged discharge, lay-off, or other discrimination and the detailed reasons or justification therefor. Any such hearing shall be of record and shall be subject to section 554 of title 5 of the United States Code. Upon receiving the report of such investigation, the Administrator shall make findings of fact as to the effect of such effluent limitation or order on employment and on the alleged discharge, lay-off, or discrimination and shall make such recommendations as he deems appropriate. Such report, findings, and recommendations shall be available to the public. Nothing in this subsection shall be construed to require or authorize the Administrator to modify or withdraw any effluent limitation or order issued under this Act.

(33 U.S.C. 1367)

FEDERAL PROCUREMENT

SEC. 508. (a) No Federal agency may enter into any contract with any person, who has been convicted of any offense under section 309(c) of this Act, for the procurement of goods, materials, and services if such contract is to be performed at any facility at which the violation which gave rise to such conviction occurred, and if such facility is owned, leased, or supervised by such person. The prohibition in the preceding sentence shall continue until the Administrator certifies that the condition giving rise to such conviction has been corrected.

(b) The Administrator shall establish procedures to provide all Federal agencies with the notification necessary for the purposes of subsection (a) of this section.

(c) In order to implement the purposes and policy of this Act to protect and enhance the quality of the Nation's water, the President shall, not more than one hundred and eighty days after enactment of this Act, cause to be issued an order (1) requiring each Federal agency authorized to enter into contracts and each Federal agency which is empowered to extend Federal assistance by way of grant, loan, or contract to effectuate the purpose and policy of this Act in such contracting or assistance activities, and (2) setting forth procedures, sanctions, penalties, and such other provisions, as the President determines necessary to carry out such requirement.

(d) The President may exempt any contract, loan, or grant from all or part of the provisions of this section where he determines such exemption is necessary in the paramount interest of the United States and he shall notify the Congress of such exemption.

(e) The President shall annually report to the Congress on measures taken in compliance with the purpose and intent of this section, including, but not limited to, the progress and problems associated with such compliance.

(f)(1) No certification by a contractor, and no contract clause, may be required in the case of a contract for the acquisition of commercial items in order to implement a prohibition or requirement of this section or a prohibition or requirement issued in the implementation of this section.

(2) In paragraph (1), the term "commercial item" has the meaning given such term in section 4(12) of the Office of Federal Procurement Policy Act (41 U.S.C. 403(12)).

(33 U.S.C. 1368)

ADMINISTRATIVE PROCEDURE AND JUDICIAL REVIEW

SEC. 509. (a)(1) For purposes of obtaining information under section 305 of this Act, or carrying out section 507(e) of this Act, the Administrator may issue subpenas for the attendance and testimony of witnesses and the production of relevant papers, books, and documents, and he may administer oaths. Except for effluent data, upon a showing satisfactory to the Administrator that such papers, books, documents, or information or particular part thereof, if made public, would divulge trade secrets or secret processes, the Administrator shall consider such record, report, or information or particular portion thereof confidential in accordance with the purposes of section 1905 of title 18 of the United States Code, except that such paper, book, document, or information may be disclosed to other officers, employees, or authorized representatives of the United States concerned with carrying out this Act, or when relevant in any proceeding under this Act. Witnesses summoned shall be paid the same fees and mileage that are paid witnesses in the courts of the United States. In case of contumacy or refusal to obey a subpena served upon any person under this subsection, the district court of the United States for any district in which such person is found or resides or transacts business, upon application by the United States and after notice to such person, shall have jurisdiction to issue an order requiring such person to appear and give testimony before the Administrator, to appear and produce papers, books, and documents before the Administrator, or both, and any failure to obey such order of the court may be punished by such court as a contempt thereof.

(2) The district courts of the United States are authorized, upon application by the Administrator, to issue subpenas for attendance and testimony of witnesses and the production of relevant papers, books, and documents, for purposes of obtaining information under sections 304 (b) and (c) of this Act. Any papers, books, documents, or other information or part thereof, obtained by reason of such a subpena shall be subject to the same requirements as are provided in paragraph (1) of this subsection.

(b)(1) Review of the Administrator's action (A) in promulgating any standard of performance under section 306, (B) in making any determination pursuant to section 306(b)(1)(C), (C) in promulgating any effluent standard, prohibition, or pretreatment standard under section 307, (D) in making any determination as to a State permit program submitted under section 402(b), (E) in approving or promulgating any effluent limitation or other limitation under sections 301, 302, 306, or 405, (F) in issuing or denying any permit under section 402, and (G) in promulgating any individual control strategy under section 304(l), may be had by any interested person in the Circuit Court of Appeals of the United States for the Federal judicial district in which such person resides or transacts business which is directly affected by such action upon application by such person. Any such application shall be made within 120 days from the date of such determination, approval, promulgation, issuance or denial, or after such date only if such application is based solely on grounds which arose after such 120th day.

(2) Action of the Administrator with respect to which review could have been obtained under paragraph (1) of this subsection shall not be subject to judicial review in any civil or criminal proceeding for enforcement.

(3) AWARD OF FEES.—In any judicial proceeding under this subsection, the court may award costs of litigation (including reasonable attorney and expert witness fees) to any prevailing or substantially prevailing party whenever it determines that such award is appropriate.

(c) In any judicial proceeding brought under subsection (b) of this section in which review is sought of a determination under this Act required to be made on the record after notice and opportunity for hearing, if any party applies to the court for leave to adduce additional evidence, and shows to the satisfaction of the court that such additional evidence is material and that there were reasonable grounds for the failure to adduce such evidence in the proceeding before the Administrator, the court may order such additional evidence (and evidence in rebuttal thereof) to be taken before the Administrator, in such manner and upon such terms and conditions as the court may deem proper. The Administrator may modify his findings as to the facts, or make new findings, by reason of the additional evidence so taken and he shall file such modified or new findings, and his recommendation, if any, for the modification or setting aside of his original determination with the return of such additional evidence.

(33 U.S.C. 1369)

STATE AUTHORITY

SEC. 510. Except as expressly provided in this Act, nothing in this Act shall (1) preclude or deny the right of any State or political subdivision thereof or interstate agency to adopt or enforce (A) any standard or limitation respecting discharges of pollutants, or (B) any requirement respecting control or abatement of pollution; except that if an effluent limitation, or other limitation, effluent standard, prohibition, pretreatment standard, or standard of performance is in effect under this Act, such State or political subdivision or interstate agency may not adopt or enforce any effluent limitation, or other limitation, effluent standard, prohibition, pretreatment standard, or standard of performance which is less stringent than the effluent limitation, or other limitation, effluent standard prohibition, pretreatment standard, or standard of performance under this Act; or (2) be construed as impairing or in any manner affecting any right or jurisdiction of the States with respect to the waters (including boundary waters) of such States.

(33 U.S.C. 1370)

OTHER AFFECTED AUTHORITY

SEC. 511. (a) This Act shall not be construed as (1) limiting the authority or functions of any officer or agency of the United States under any other law or regulation not inconsistent with this Act; (2) affecting or impairing the authority of the Secretary of the Army (A) to maintain navigation or (B) under the Act of March 3, 1899 (30 Stat. 1112); except that any permit issued under section 404 of this Act shall be conclusive as to the effect on water quality of any discharge resulting from any activity subject to section 10 of the Act of March 3, 1899, or (3) affecting or impairing the provisions of any treaty of the United States.

(b) Discharges of pollutants into the navigable waters subject to the Rivers and Harbors Act of 1910 (36 Stat. 593; 33 U.S.C. 421) and the Supervisory Harbors, Act of 1888 (25 Stat. 209; 33 U.S.C. 441–451b) shall be regulated pursuant to this Act, and not subject to such Act of 1910 and the Act of 1888 except as to effect on navigation and anchorage.

(c)(1) Except for the provision of Federal financial assistance for the purpose of assisting the construction of publicly owned treatment works as authorized by section 201 of this Act, and the issuance of a permit under section 402 of this Act for the discharge of any pollutant by a new source as defined in section 306 of this Act, no action of the Administrator taken pursuant to this Act shall be deemed a major Federal action significantly affecting the quality of the human environment within the meaning of the National Environmental Policy Act of 1969 (83 Stat. 852); and

(2) Nothing in the National Environmental Policy Act of 1969 (83 Stat. 852) shall be deemed to—

(A) authorize any Federal agency authorized to license or permit the conduct of any activity which may result in the discharge of a pollutant into the navigable waters to review any effluent limitation or other requirement established pursuant to this Act or the adequacy of any certification under section 401 of ths Act; or

(B) authorize any such agency to impose, as a condition precedent to the issuance of any license or permit, any effluent limitation other than any such limitation established pursuant to this Act.

(d) Notwithstanding this Act or any other provisions of law, the Administrator (1) shall not require any State to consider in the development of the ranking in order of priority of needs for the construction of treatment works (as defined in title II of this Act), any water pollution control agreement which may have been entered into between the United States and any other nation, and (2) shall not consider any such agreement in the approval of any such priority ranking.

(33 U.S.C. 1371)

SEPARABILITY

SEC. 512. If any provision of this Act, or the application of any provision of this Act to any person or circumstance, is held invalid, the application of such provision to other persons or circumstances, and the remainder of this Act shall not be affected thereby.

(33 U.S.C. 1251 note)

LABOR STANDARDS

SEC. 513. The Administrator shall take such action as may be necessary to insure that all laborers and mechanics employed by contractors or subcontractors on treatment works for which grants are made under this Act shall be paid wages at rates not less than those prevailing for the same type of work on similar construction in the immediate locality, as determied by the Secretry of Labor, in accordance with the Act of March 3, 1931, as amended, known as the Davis-Bacon Act (46 Stat. 1494; 40 U.S.C., sec. 276a through 276a–5). The Secretary of Labor shall have, with respect to the labor standards specified in this subsection, the authority and functions set forth in Reorganization Plan Numbered 14 of 1950 (15 F.R. 3176) and section 2 of the Act of June 13, 1934, as amended (48 Stat. 948; 40 U.S.C. 276c).

(33 U.S.C. 1372)

PUBLIC HEALTH AGENCY COORDINATION

SEC. 514. The permitting agency under section 402 shall assist the applicant for a permit under such section in coordinating the requirements of this Act with those of the appropriate public health agencies.

(33 U.S.C. 1373)

EFFLUENT STANDARDS AND WATER QUALITY INFORMATION ADVISORY COMMITTEE

SEC. 515. (a)(1) There is established on Effluent Standards and Water Quality Information Advisory Committee, which shall be composed of a Chairman and eight members who shall be appointed by the Administrator within sixty days after the date of enactment of this Act.

(2) All members of the Committee shall be selected from the scientific community, qualified by education, training, and experience to provide assess, and evaluate scientific and technical information on effluent standards and limitations.

(3) Members of the Committee shall serve for a term of four years, and may be reappointed.

(b)(1) No later than one hundred and eighty days prior to the date on which the Administrator is required to publish any proposed regulations required by section 304(b) of this Act, any proposed standard of performance for new sources required by section 306 of this Act, or any proposed toxic effluent standard required by section 307 of this Act, he shall transmit to the Committee a notice of intent to propose such regulations. The Chairman of the Committee within ten days after receipt of such notice may publish a notice of a public hearing by the Committee, to be held within thirty days.

(2) No later than one hundred and twenty days after receipt of such notice, the Committee shall transmit to the Administrator such scientific and technical information as is in its possession, including that presented at any public hearing, related to the subject matter contained in such notice.

(3) Information so transmitted to the Administrator shall constitute a part of the administrative record and comments on any proposed regulations or standards as information to be considered with other comments and information in making any final determinations.

(4) In preparing information for transmittal, the Committee shall avail itself of the technical and scientific services of any Federal agency, including the United States Geological Survey and any national environmental laboratories which may be established.

(c)(1) The Committee shall appoint and prescribe the duties of a Secretary, and such legal counsel as it deems necessary. The Committee shall appoint such other employees as it deems necessary to exercise and fulfill its powers and responsibilities. The compensation of all employees appointed by the Committee shall be fixed in accordance with chapter 51 and subchapter III of chapter 53 of title V of the United States Code.

(2) Members of the Committee shall be entitled to receive compensation at a rate to be fixed by the President but not in excess of the maximum rate of pay grade for GS–18, as provided in the General Schedule under section 5332 of title V of the United States Code.

(d) Five members of the Committee shall constitute a quorum, and official actions of the Committee shall be taken only on the affirmative vote of at least five members. A special panel composed of one or more members upon order of the Committee shall conduct any hearing authorized by this section and submit the transcript of such hearing to the entire Committee for its action thereon.

(e) The Committee is authorized to make such rules as are necessary for the orderly transaction of its business.

(33 U.S.C. 1374)

REPORTS TO CONGRESS

SEC. 516. The Administrator, in cooperation with the States, including water pollution control agencies and other water pollution control planning agencies, shall make (1) a detailed estimate of the cost of carrying out the provisions of this Act; (2) a detailed estimate, biennially revised, of the cost of construction of all needed publicly owned treatment works in all of the States and of the cost of construction of all needed publicly owned treatment works in each of the States; (3) a comprehensive study of the economic impact on affected units of government of the cost of installation of treatment facilities; and (4) a comprehensive analysis of the national requirements for and the cost of treating municipal, industrial, and other effluent to attain the water quality objectives as established by this Act or applicable State law. The Administrator shall submit such detailed estimate and such comprehensive study of such cost to the Congress no later than February 10 of each odd-numbered year. Whenever the Administrator, pursuant to this subsection, requests and receives an estimate of cost from a State, he shall furnish copies of such estimate together with such detailed estimate to Congress.

(33 U.S.C. 1375)

GENERAL AUTHORIZATION

SEC. 517. There are authorized to be appropriated to carry out this Act, other than sections 104, 105, 106(a), 107, 108, 112, 113, 114, 115, 206, 207, 208 (f) and (h), 209, 304, 311 (c), (d), (i), (l), and (k), 314, 315, and 317, $250,000,000 for the fiscal year ending June 30, 1973, $300,000,000 for the fiscal year ending June 30, 1974, $350,000,000 for the fiscal year ending June 30, 1975, $100,000,000 for the fiscal year ending September 30, 1977, $150,000,000 for the fiscal year ending September 30, 1978, $150,000,000 for the fiscal year ending September 30, 1979, $150,000,000 for the fiscal year ending September 30, 1980, $150,000,000 for the fiscal year ending September 30, 1981, $161,000,000 for the fiscal year ending September 30, 1982, such sums as may be necessary for fiscal years 1983 through 1985, and $135,000,000 per fiscal year for each of the fiscal years 1986 through 1990.

(33 U.S.C. 1376)

SEC. 518. INDIAN TRIBES.

(a) POLICY.—Nothing in this section shall be construed to affect the application of section 101(g) of this Act, and all of the provisions of this section shall be carried out in accordance with the provisions of such section 101(g). Indian tribes shall be treated as States for purposes of such section 101(g).

(b) ASSESSMENT OF SEWAGE TREATMENT NEEDS; REPORT.—The Administrator, in cooperation with the Director of the Indian Health Service, shall assess the need for sewage treatment works to serve Indian tribes, the degree to which such needs will be met through funds allotted to States under section 205 of this Act and priority lists under section 216 of this Act, and any obstacles which prevent such needs from being met. Not later than one year after the date of the enactment of this section, the Administrator shall submit a report to Congress on the assessment under this subsection, along with recommendations specifying (1) how the Administrator intends to provide assistance to Indian tribes to develop waste treatment management plans and to construct treatment works under this Act, and (2) methods by which the participation in and administration of programs under this Act by Indian tribes can be maximized.

(c) RESERVATION OF FUNDS.—The Administrator shall reserve each fiscal year beginning after September 30, 1986, before allotments to the States under section 205(e), one-half of one percent of the sums appropriated under section 207. Sums reserved under this subsection shall be available only for grants for the develoment of waste treatment management plans and for the construction of sewage treatment works to serve Indian tribes, as defined in subsection (h) and former Indian reservations in Oklahoma (as determined by the Secretary of the Interior) and Alaska Native Villages as defined in Public Law 92–203.

(d) COOPERATIVE AGREEMENTS.—In order to ensure the consistent implementation of the requirements of this Act, an Indian tribe and the State or States in which the lands of such tribe are located may enter into a cooperative agreement, subject to the review and approval of the Administrator, to jointly plan and administer the requirements of this Act.

(e) TREATMENT AS STATES.—The Administrator is authorized to treat an Indian tribe as a State for purposes of title II and sections 104, 106, 303, 305, 308, 309, 314, 319, 401, 402, 404, and 406 of this Act to the degree necessary to carry out the objectives of this section, but only if—

(1) the Indian tribe has a governing body carrying out substantial governmental duties and powers;

(2) the functions to be exercised by the Indian tribe pertain to the management and protection of water resources which are held by an Indian tribe, held by the United States in trust for Indians, held by a member of an Indian tribe if such property interest is subject to a trust restriction on alienation, or otherwise within the borders of an Indian reservation; and

(3) the Indian tribe is reasonably expected to be capable, in the Administrator's judgment, of carrying out the functions to be exercised in a manner consistent with the terms and purposes of this Act and of all applicable regulations.

Such treatment as a State may include the direct provision of funds reserved under subsection (c) to the governing bodies of Indian tribes, and the determination of priorities by Indian tribes, where not determined by the Administrator in cooperation with the Director of the Indian Health Service. The Administrator, in cooperation with the Director of the Indian Health Service, is authorized to make grants under title II of this Act in an amount not to exceed 100 percent of the cost of a project. Not later than 18 months after the date of the enactment of this section, the Administrator shall, in consultation with Indian tribes, promulgate final regulations which specify how Indian tribes shall be treated as States for purposes of this Act. The Administrator shall, in promulgating such regulations, consult affected States sharing common water bodies and provide a mechanism for the resolution of any unreasonable consequences that may arise as a result of differing water quality standards that may be set by States and Indian tribes located on common bodies of water. Such mechanism shall provide for explicit consideration of relevant factors including, but not limited to, the effects of differing water quality permit requirements on upstream and downstream dischargers, economic impacts, and present and historical uses and quality of the waters subject to such standards. Such mechanism should provide for the avoidance of such unreasonable consequences in a manner consistent with the objective of this Act.

(f) GRANTS FOR NONPOINT SOURCE PROGRAMS.—The Administrator shall make grants to an Indian tribe under section 319 of this Act as though such tribe was a State. Not more than one-third of one percent of the amount appropriated for any fiscal year under section 319 may be used to make grants under this subsection. In addition to the requirements of section 319, an Indian tribe shall be required to meet the requirements of paragraphs (1), (2), and (3) of subsection (d)[1] of this section in order to receive such a grant.

(g) ALASKA NATIVE ORGANIZATIONS.—No provision of this Act shall be construed to—

(1) grant, enlarge, or diminish, or in any way affect the scope of the governmental authority, if any, of any Alaska Native organization, including any federally-recognized tribe, traditional Alaska Native council, or Native council organized pursuant to the Act of June 18, 1934 (48 Stat. 987), over lands or persons in Alaska;

(2) create or validate any assertion by such organization or any form of governmental authority over lands or persons in Alaska; or

(3) in any way affect any assertion that Indian country, as defined in section 1151 of title 18, United States Code, exists or does not exist in Alaska.

(h) DEFINITIONS.—For purposes of this section, the term—

(1) "Federal Indian reservation" means all land within the limits of any Indian reservation under the jurisdiction of the United States Government, notwithstanding the issuance of any patent, and including rights-of-way running through the reservation; and

(2) "Indian tribe" means any Indian tribe, band, group, or community recognized by the Secretary of the Interior and exercising governmental authority over a Federal Indian reservation.

(33 U.S.C. 1377)

SHORT TITLE

SEC. 519. This Act may be cited as the "Federal Water Pollution Control Act" (commonly referred to as the Clean Water Act).

(33 U.S.C. 1251 note)

TITLE VI—STATE WATER POLLUTION CONTROL REVOLVING FUNDS

SEC. 601. GRANTS TO STATES FOR ESTABLISHMENT OF REVOLVING FUNDS.

(a) GENERAL AUTHORITY.—Subject to the provisions of this title, the Administrator shall make capitalization grants to each State for the purpose of establishing a water pollution control revolving fund for providing assistance (1) for construction of treatment works (as defined in section 212 of this Act) which are publicly owned, (2) for implementing a management program under section 319, and (3) for developing and implementing a conservation and management plan under section 320.

(b) SCHEDULE OF GRANT PAYMENTS.—The Administrator and each State shall jointly establish a schedule of payments under which the Administrator will pay to the State the amount of each

[1] Probably should be subsection (e).

grant to be made to the State under this title. Such schedule shall be based on the State's intended use plan under section 606(c) of this Act, except that—

(1) such payments shall be made in quarterly installments, and

(2) such payments shall be made as expeditiously as possible, but in no event later than the earlier of—

(A) 8 quarters after the date such funds were obligated by the State, or

(B) 12 quarters after the date such funds were allotted to the State.

(33 U.S.C. 1381)

SEC. 602. CAPITALIZATION GRANT AGREEMENTS.

(a) GENERAL RULE.—To receive a capitalization grant with funds made available under this title and section 205(m) of this Act, a State shall enter into an agreement with the Administrator which shall include but not be limited to the specifications set forth in subsection (b) of this section.

(b) SPECIFIC REQUIREMENTS.—The Administrator shall enter into an agreement under this section with a State only after the State has established to the satisfaction of the Administrator that—

(1) the State will accept grant payments with funds to be made available under this title and section 205(m) of this Act in accordance with a payment schedule established jointly by the Administrator under section 601(b) of this Act and will deposit all such payments in the water pollution control revolving fund established by the State in accordance with this title;

(2) the State will deposit in the fund from State moneys an amount equal to at least 20 percent of the total amount of all capitalization grants which will be made to the State with funds to be made available under this title and section 205(m) of this Act on or before the date on which each quarterly grant payment will be made to the State under this title;

(3) the State will enter into binding commitments to provide assistance in accordance with the requirements of this title in an amount equal to 120 percent of the amount of each such grant payment within 1 year after the receipt of such grant payment;

(4) all funds in the fund will be expended in an expeditious and timely manner;

(5) all funds in the fund as a result of capitalization grants under this title and section 205(m) of this Act will first be used to assure maintenance of progress, as determined by the Governor of the State, toward compliance with enforceable deadlines, goals, and requirements of this Act, including the municipal compliance deadline;

(6) treatment works eligible under section 603(c)(1) of this Act which will be constructed in whole or in part before fiscal year 1995 with funds directly made available by capitalization grants under this title and section 205(m) of this Act will meet the requirements of, or otherwise be treated (as determined by the Governor of the State) under sections 201(b), 201(g)(1),

201(g)(2), 201(g)(3), 201(g)(5), 201(g)(6), 201(n)(1), 201(o), 204(a)(1), 204(a)(2), 204(b)(1), 204(d)(2), 211, 218, 511(c)(1), and 513 of this Act in the same manner as treatment works constructed with assistance under title II of this Act;

(7) in addition to complying with the requirements of this title, the State will commit or expend each quarterly grant payment which it will receive under this title in accordance with laws and procedures applicable to the commitment or expenditure of revenues of the State;

(8) in carrying out the requirements of section 606 of this Act, the State will use accounting, audit, and fiscal procedures conforming to generally accepted government accounting standards;

(9) the State will require as a condition of making a loan or providing other assistance, as described in section 603(d) of this Act, from the fund that the recipient of such assistance will maintain project accounts in accordance with generally accepted government accounting standards; and

(10) the State will make annual reports to the Administrator on the actual use of funds in accordance with section 606(d) of this Act.

(33 U.S.C. 1382)

SEC. 603. WATER POLLUTION CONTROL REVOLVING LOAN FUNDS. [1]

(a) REQUIREMENTS FOR OBLIGATION OF GRANT FUNDS.—Before a State may receive a capitalization grant with funds made available under this title and section 205(m) of this Act, the State shall first establish a water pollution control revolving fund which complies with the requirements of this section.

(b) ADMINISTRATOR.—Each State water pollution control revolving fund shall be administered by an instrumentality of the State with such powers and limitations as may be required to operate such fund in accordance with the requirements and objectives of this Act.

(c) PROJECTS ELIGIBLE FOR ASSISTANCE.—The amounts of funds available to each State water pollution control revolving fund shall be used only for providing financial assistance (1) to any municipality, intermunicipal, interstate, or State agency for construction of publicly owned treatment works (as defined in section 212 of this Act), (2) for the implementation of a management program established under section 319 of this Act, and (3) for development and implementation of a conservation and management plan under section 320 of this Act. The fund shall be established, maintained, and credited with repayments, and the fund balance shall be available in perpetuity for providing such financial assistance. [2]

---

[1] See section 104B of the Marine Protection, Research and Sanctuaries Act of 1972 (33 U.S.C. 1414G) for additional amounts that are to be deposited into a State's fund and treatment of such deposits.

[2] Section 1006 of the Ocean Dumping Ban Act of 1988 (P.L. 100–688) is as follows:

SEC. 1066. USE OF STATE WATER POLLUTION CONTROL REVOLVING FUND GRANTS FOR DEVELOPING ALTERNATIVE SYSTEMS.

(a) GENERAL REQUIREMENT.—Notwithstanding the provisions of title VI of the Federal Water Pollution Control Act, each of the States of New York and New Jersey shall use 10 percent of the amount of a grant payment made to such State under such title for each of the fiscal years 1990 and 1991 and 10 percent of the State's contribution associated with such grant payment in the 6-month period beginning on the date of receipt of such grant payment for making loans

(d) TYPES OF ASSISTANCE.—Except as otherwise limited by State law, a water pollution control revolving fund of a State under this section may be used only—

(1) to make loans, on the condition that—

(A) such loans are made at or below market interest rates, including interest free loans, at terms not to exceed 20 years;

(B) annual principal and interest payments will commence not later than 1 year after completion of any project and all loans will be fully amortized not later than 20 years after project completion;

(C) the recipient of a loan will establish a dedicated source of revenue for repayment of loans; and

(D) the fund will be credited with all payments of principal and interest on all loans;

(2) to buy or refinance the debt obligation of municipalities and intermunicipal and interstate agencies within the State at or below market rates, where such debt obligations were incurred after March 7, 1985;

(3) to guarantee, or purchase insurance for, local obligations where such action would improve credit market access or reduce interest rates;

(4) as a source of revenue or security for the payment of principal and interest on revenue or general obligation bonds issued by the State if the proceeds of the sale of such bonds will be deposited in the fund;

(5) to provide loan guarantees for similar revolving funds established by municipalities or intermunicipal agencies;

(6) to earn interest on fund accounts; and

(7) for the reasonable costs of administering the fund and conducting activities under this title, except that such amounts shall not exceed 4 percent of all grant awards to such fund under this title.

(e) LIMITATION TO PREVENT DOUBLE BENEFITS.—If a State makes, from its water pollution revolving fund, a loan which will finance the cost of facility planning and the preparation of plans, specifications, and estimates for construction of publicly owned treatment works, the State shall ensure that if the recipient of such loan receives a grant under section 201(g) of this Act for construction of such treatment works and an allowance under section 201(l)(1) of this Act for non-federal funds expended for such planning and preparation, such recipient will promptly repay such loan to the extent of such allowance.

---

and providing other assistance as described in section 603(d) of the Federal Water Pollution Control Act to any governmental entity in such State which has entered into a compliance agreement or enforcement agreement under section 104B of the Marine Protection, Research, and Sanctuaries Act of 1972 for identifying, developing, and implementing pursuant to such section alternative systems for management of sewage sludge.

(b) LIMITATION.—If, after the last day of the 6-month period beginning on the date of receipt of a grant payment by the State of New York or New Jersey under title VI of the Federal Water Pollution Control Act for each of fiscal years 1990 and 1991, 10 percent of the amount of such grant payment and the State's contribution associated with such grant payment has not been used for providing assistance described in subsection (a) as a result of insufficient applications for such assistance from persons eligible for such assistance, the 10 percent limitations set forth in subsection (a) shall not be applicable with respect to such grant payment and associated State contribution.

(f) CONSISTENCY WITH PLANNING REQUIREMENTS.—A State may provide financial assistance from its water pollution control revolving fund only with respect to a project which is consistent with plans, if any, developed under sections 205(j), 208, 303(e), 319, and 320 of this Act.

(g) PRIORITY LIST REQUIREMENT.—The State may provide financial assistance from its water pollution control revolving fund only with respect to a project for construction of a treatment works described in subsection (c)(1) if such project is on the State's priority list under section 216 of this Act. Such assistance may be provided regardless of the rank of such project on such list.

(h) ELIGIBILITY OF NON-FEDERAL SHARE OF CONSTRUCTION GRANT PROJECTS.—A State water pollution control revolving fund may provide assistance (other than under subsection (d)(1) of this section) to a municipality or intermunicipal or interstate agency with respect to the non-Federal share of the costs of a treatment works project for which such municipality or agency is receiving assistance from the Administrator under any other authority only if such assistance is necessary to allow such project to proceed.

(33 U.S.C. 1383)

SEC. 604. ALLOTMENT OF FUNDS.

(a) FORMULA.—Sums authorized to be appropriated to carry out this section for each of fiscal years 1989 and 1990 shall be allotted by the Administrator in accordance with section 205(c) of this Act.

(b) RESERVATION OF FUNDS FOR PLANNING.—Each State shall reserve each fiscal year 1 percent of the sums allotted to such State under this section for such fiscal year, or $100,000, whichever amount is greater, to carry out planning under sections 205(j) and 303(e) of this Act.

(c) ALLOTMENT PERIOD.—

(1) PERIOD OF AVAILABILITY FOR GRANT AWARD.—Sums allotted to a State under this section for a fiscal year shall be available for obligation by the State during the fiscal year for which sums are authorized and during the following fiscal year.

(2) REALLOTMENT OF UNOBLIGATED FUNDS.—The amount of any allotment not obligated by the State by the last day of the 2-year period of availability established by paragraph (1) shall be immediately reallotted by the Administrator on the basis of the same ratio as is applicable to sums allotted under title II of this Act for the second fiscal year of such 2-year period. None of the funds reallotted by the Administrator shall be reallotted to any State which has not obligated all sums allotted to such State in the first fiscal year of such 2-year period.

(33 U.S.C. 1384)

SEC. 605. CORRECTIVE ACTION.

(a) NOTIFICATION OF NONCOMPLIANCE.—If the Administrator determines that a State has not complied with its agreement with the Administrator under section 602 of this Act or any other requirement of this title, the Administrator shall notify the State of such noncompliance and the necessary corrective action.

(b) WITHHOLDING OF PAYMENTS.—If a State does not take corrective action within 60 days after the date a State receives notification of such action under subsection (a), the Administrator shall withhold additional payments to the State until the Administrator is satisfied that the State has taken the necessary corrective action.

(c) REALLOTMENT OF WITHHELD PAYMENTS.—If the Administrator is not satisfied that adequate corrective actions have been taken by the State within 12 months after the State is notified of such actions under subsection (a), the payments withheld from the State by the Administrator under subsection (b) shall be made available for reallotment in accordance with the most recent formula for allotment of funds under this title.

(33 U.S.C. 1385)

SEC. 606. AUDITS, REPORTS, AND FISCAL CONTROLS; INTENDED USE PLAN.

(a) FISCAL CONTROL AND AUDITING PROCEDURES.—Each State electing to establish a water pollution control revolving fund under this title shall establish fiscal controls and accounting procedures sufficient to assure proper accounting during appropriate accounting periods for—

(1) payments received by the fund;

(2) disbursements made by the fund; and

(3) fund balances at the beginning and end of the accounting period.

(b) ANNUAL FEDERAL AUDITS.—The Administrator shall, at least on an annual basis, conduct or require each State to have independently conducted reviews and audits as may be deemed necessary or appropriate by the Administrator to carry out the objectives of this section. Audits of the use of funds deposited in the water pollution revolving fund established by such State shall be conducted in accordance with the auditing procedures of the General Accounting Office, including chapter 75 of title 31, United States Code.

(c) INTENDED USE PLAN.—After providing for public comment and review, each State shall annually prepare a plan identifying the intended uses of the amounts available to its water pollution control revolving fund. Such intended use plan shall include, but not be limited to—

(1) a list of those projects for construction of publicly owned treatment works on the State's priority list developed pursuant to section 216 of this Act and a list of activities eligible for assistance under sections 319 and 320 of this Act;

(2) a description of the short- and long-term goals and objectives of its water pollution control revolving fund;

(3) information on the activities to be supported, including a description of project categories, discharge requirements under titles III and IV of this Act, terms of financial assistance, and communities served;

(4) assurances and specific proposals for meeting the requirements of paragraphs (3), (4), (5), and (6) of section 602(b) of this Act; and

(5) the criteria and method established for the distribution of funds.

(d) ANNUAL REPORT.—Beginning the first fiscal year after the receipt of payments under this title, the State shall provide an annual report to the Administrator describing how the State has met the goals and objectives for the previous fiscal year as identified in the plan prepared for the previous fiscal year pursuant to subsection (c), including identification of loan recipients, loan amounts, and loan terms and similar details on other forms of financial assistance provided from the water pollution control revolving fund.

(e) ANNUAL FEDERAL OVERSIGHT REVIEW.—The Administrator shall conduct an annual oversight review of each State plan prepared under subsection (c), each State report prepared under subsection (d), and other such materials as are considered necessary and appropriate in carrying out the purposes of this title. After reasonable notice by the Administrator to the State or the recipient of a loan from a water pollution control revolving fund, the State or loan recipient shall make available to the Administrator such records as the Administrator reasonably requires to review and determine compliance with this title.

(f) APPLICABILITY OF TITLE II PROVISIONS.—Except to the extent provided in this title, the provisions of title II shall not apply to grants under this title.

(33 U.S.C. 1386)

SEC. 607. AUTHORIZATION OF APPROPRIATIONS.

There is authorized to be appropriated to carry out the purposes of this title the following sums:

(1) $1,200,000,000 per fiscal year for each of fiscal year 1989 and 1990;

(2) $2,400,000,000 for fiscal year 1991;

(3) $1,800,000,000 for fiscal year 1992;

(4) $1,200,000,000 for fiscal year 1993; and

(5) $600,000,000 for fiscal year 1994.

(33 U.S.C. 1387)

Research Sources

The following EPA web pages provide substantial additional information about the various programs of the CWA:

- Combined Sewer Overflows: http://water.epa.gov/polwaste/npdes/cso/
- National Estuary Program: http://www.epa.gov/nep
- NPDES Program Home Page: http://cfpub.epa.gov/npdes/index.cfm
- Nonpoint-Source Program: http://www.epa.gov/owow/nps/index.html
- Oceans, Coasts, Estuaries & Beaches: http://water.epa.gov/type/oceb/
- Oil Spill Program: http://www.epa.gov/oilspill
- Office of Wastewater Management Home Page: http://www.epa.gov/owm
- Pretreatment Program: http://water.epa.gov/polwaste/npdes/pretreatment/index.cfm
- Sanitary Sewer Overflows: http://water.epa.gov/polwaste/npdes/sso/
- Storm-Water Program: http://water.epa.gov/polwaste/npdes/stormwater/index.cfm
- Total Maximum Daily Load Program: http://www.epa.gov/owow/tmdl/index.html
- Water Quality Standards for Surface Waters: http://water.epa.gov/scitech/swguidance/standards/index.cfm
- Watershed Management: http://www.epa.gov/owow/watershed/
- Wetlands Program: http://www.epa.gov/owow/wetlands/

Notes

CHAPTER 1

1. Pub. L. No. 92-500, 86 Stat. 816 (1972).

2. *NRDC v. Train*, 8 E.R.C. 2120 (D.D.C. 1976), *modified*, 12 E.R.C. 1833 (D.D.C. 1979), *aff'd, Environmental Defense Fund, Inc. v. Costle*, 636 F.2d 1229 (D.C. Cir. 1980).

3. This date was subsequently postponed to June 30, 1984. EPA completed issuance of the effluent guidelines required by the decree in 1987.

4. Pub. L. No. 95-217, 91 Stat. 1566 (1977).

5. Pub. L. No. 100-4, 101 Stat. 7 (1987).

6. Pub. L. 101-380, 104 Stat. 484 (1990).

7. 33 U.S.C. § 2701–2761.

8. Section 101(a), 33 U.S.C. § 1251(a).

9. Section 101(a)(2), 33 U.S.C. § 1251(a)(2).

10. Section 101(a)(1), 33 U.S.C. § 1251(a)(1).

11. Section 101(a)(3) & (7), 33 U.S.C. § 1251(a)(3) & (7).

12. Section 301(a), 33 U.S.C. § 1311(a).

13. Section 502(12), 33 U.S.C. § 1362(12).

14. *United States v. Wilson*, 133 F.3d 251(4th Cir. 1997); *see also Catskill Mountains Chapter of Trout Unlimited, Inc. v. New York City*, 273 F.3d 481 (2d Cir. 2001) (transfer of turbid water from reservoir into stream through tunnel requires NPDES permit); *Friends of Santa Fe County v. LAC Minerals, Inc.*, 892 F. Supp. 1333, 1354 (D.N.M. 1995) (migration of residual contamination resulting from previous release not an *addition*).

15. 40 C.F.R. § 403.15. This credit for pollutants in uptake water, referred to as a *net/gross credit*, is available only for technology-based limitations.

16. *L.A. Flood Control Dist. v. Natural Resources Def. Council, Inc.*, 133 S. Ct. 710 (2013).

17. *National Wildlife Fed'n v. Consumers Power Co.*, 862 F.2d 580 (6th Cir. 1988); *National Wildlife Federation v. Gorsuch*, 693 F.2d 156 (D.C. Cir. 1982). *But see Comm. to Save the Mokelumne River v. East Bay Mun. Util. Dist.*, 13 F.3d 305 (9th Cir. 1993), *cert. denied*, 115 S. Ct. 198 (1994) (discharges from dam used to collect acid mine drainage from abandoned mine subject to CWA permit requirement).

18. *In re General Motors Corp., CPC-Pontiac Fiero Plant*, 7 E.A.D. 465 (E.P.A. Dec. 24, 1997), *upheld*, *General Motors Corp. v. EPA*, 168 F.3d 1377 (D.C. Cir. 1999).

19. *Catskill Mountain Chapter of Trout Unlimited v. City of New York*, 451 F.3d 77 (2d Cir. 2006); *DuBois v. U.S. Dept. of Agriculture*, 102 F.3d 1273 (1st Cir. 1996), *cert. denied*, 117 S. Ct. 2510 (1997). Effective August 12, 2008, EPA adopted the National Pollutant Discharge Elimination System Water Transfers Rule, which is intended to clarify that "water transfers," defined as "activit[ies] that convey[] or connect[] waters of the United States without subjecting the transferred water to intervening industrial, municipal, or commercial uses," are excluded from regulation under the CWA because there is no "addition" of a pollutant to the waters being transferred. 73 Fed. Reg. 33697 (June 13, 2008) (codified at 40 C.F.R. pt. 122). Environmental groups, municipal water providers, states, and others filed a number of lawsuits challenging the new rule. Although some consolidated cases were dismissed for lack of subject matter jurisdiction, *see Friends of the Everglades v. EPA*, 699 F.3d 1280 (11th Cir. 2012), the Southern District of New York reached a decision on the merits in *Catskill Mountains Chapter of Trout Unlimited, Inc. v. EPA*, 8 F. Supp. 3d 500 (S.D.N.Y. 2014). The court found the Water Transfers Rule to be arbitrary and capricious and remanded the Rule to EPA. Multiple parties subsequently appealed the decision to the 2nd Circuit.

20. Section 502(6), 33 U.S.C. § 1362(6).

21. *E.g.*, *Natural Resources Defense Council v. EPA*, 859 F.2d 156 (D.C. Cir. 1988); *United States v. Hamel*, 551 F.2d 107, 110–12 (6th Cir. 1977); *see also United States v. Teixeira Foods Inc.*, C.R. 98-1015 (C.D. Cal. Jan. 13, 1999) (dumping of 100 ostrich carcasses into tributary constituted discharge of a "pollutant"); *United States v. West Indies Transp., Inc.*, 127 F.3d 299 (3d Cir. 1997), *cert. denied*, 118 S. Ct. 700 (1998) (dumping of 250-ton concrete and rebar block from barge was discharge of a pollutant).

22. Section 502(14), 33 U.S.C. § 1362(14).

23. In 2001, the Ninth Circuit held that the application of aquatic herbicides (e.g., to control weeds in irrigation ditches) required an NPDES permit. *Headwaters, Inc. v. Talent Irrigation District*, 243 F.3d 526 (9th Cir. 2001). In response to this decision, EPA issued a policy statement that the use of aquatic herbicides fell within the definition of return flows from irrigated agriculture, which are excluded from the "point source" definition. EPA thus takes

the position that such herbicide use results in only nonpoint-source discharges, which do not require an NPDES permit. U.S. EPA, "Interpretive Statement and Regional Guidance on the Clean Water Act's Exemption for Return Flows from Irrigated Agriculture" (March 2002).

24. *E.g., Concerned Area Residents for Env't v. Southview Farm*, 34 F.3d 114 (2d Cir. 1994), *cert. denied*, 115 S. Ct. 1793 (1995); *see also United States v. West Indies Transp., Inc.*, 127 F.3d 299 (3d Cir. 1997), *cert. denied*, 118 S. Ct. 700 (1998) (concrete barge is point source).

25. In other words, a person dumping pollutants into a water body, other than through a hose or pipe, for example, would not be in violation of the act's prohibition of discharges from point sources without a permit. The person may, however, be in violation of other laws and regulations. *United States v. Plaza Health Lab., Inc.*, 3 F.3d 643, 649 (2d Cir. 1993), *cert. denied*, 114 S. Ct. 2764 (1994) (dumping of vials of medical waste into Hudson River).

26. Although "point sources" are statutorily defined to include sources from which pollutants "may be discharged," EPA appears to have limited authority under the act to take regulatory action against "potential discharges" from a point source. In *Waterkeeper Alliance, Inc. v. EPA*, 399 F.3d 486 (2d Cir. 2005), environmental groups and farm groups brought multiple challenges to EPA-promulgated rules governing the emission of water pollutants from concentrated animal feeding operations (CAFOs). The court vacated various provisions of the CAFO rule, including a requirement that all CAFOs apply for an NPDES permit. *Id.* at 524. EPA argued that the provision requiring all CAFOs to apply for a permit or demonstrate that there is no potential to discharge was appropriate because all CAFOs presumptively present a "potential to discharge" pollutants. *Id.* at 505. The court found EPA's argument unavailing, concluding that the CWA gives EPA authority to regulate only "actual discharges" from point sources, not "potential discharges" or point sources themselves. *Id.* In response to the *Waterkeeper* decision, EPA adopted revised CAFO regulations, effective December 22, 2008. The revised regulations require only those CAFOs that "discharge" or "propose to discharge" wastewater to apply for NPDES permits. 73 Fed. Reg. 70418 (Nov. 20, 2008).

27. Section 502(7), 33 U.S.C. § 1362(7).

28. Clean Water Rule: Definition of "Waters of the United States," 80 Fed. Reg. 37054, 37104–05.

29. *Id.* at 37105.

30. *Id.*

31. *Solid Waste Agency of Northern Cook County v. U.S. Army Corps of Engineers*, 531 U.S. 159, 121 S. Ct. 675 (2001). Legislation has been introduced in the House of Representatives multiple times that would overturn this decision by broadly defining "waters of the United States," but none of the legislation has garnered sufficient support to be enacted into law.

32. 547 U.S. 715, 126 S. Ct. 2208 (2006).

33. *Id.* at 739.

34. *Id.* at 742 (emphasis in the original).

35. *Id.* at 759.

36. *Id.* at 779.

37. *United States v. Robison,* 505 F.3d 1208 (11th Cir. 2007) (holding that Justice Kennedy's test is controlling because it is the view of the member who concurred in the judgment on the narrowest grounds), *reh'g denied,* 521 F.3d 1319 (11th Cir. 2008); *United States v. Gerke Excavating, Inc.*, 464 F.3d 723 (7th Cir. 2006) (finding that Justice Kennedy's test is controlling in most cases because it is the least restrictive of federal authority to regulate); *Northern California River Watch v. City of Healdsburg*, 496 F.3d 993 (9th Cir. 2007) (stating that Justice Kennedy's test provides the controlling rule of law).

38. *See, e.g., United States v. Johnson*, 467 F.3d 56 (1st Cir. 2006); *United States v. Donovan*, 661 F.3d 174 (3rd Cir. 2011); *United States v. Cundiff,* 555 F.3d 200 (6th Cir. 2009); *United States v. Bailey*, 571 F.3d 791 (8th Cir. 2009).

39. *See* 80 Fed. Reg. 37054.

40. *Precon Dev. Corp. v. Army Corps of Engineers*, No. 13-2499, 2015 WL 1020693, at *5 (4th Cir. Mar. 10, 2015).

41. 80 Fed. Reg. 37054.

42. *In re: Environmental Protection Agency and Department of Defense, Final Rule: Clean Water Rule: Definition of "Waters of the United States,"* No. 15-3751 (6th Cir. 2016).

43. *Belle Co. v. U.S. Army Corps of Engineers*, 761 F.3d 383, 394 (5th Cir. 2014).

44. *Hawkes Co. v. U.S. Army Corps of Engineers*, 782 F.3d 994, 1001–02 (8th Cir. 2015).

45. No. 15-290 (May 31, 2016).

CHAPTER 2

1. Certain discharges of pollutants are exempted from the NPDES permit requirement. These include discharges from some agricultural and silvicultural activities, discharges into publicly owned treatment works, and discharges in compliance with instructions from an on-scene coordinator responding to a spill incident. 40 C.F.R. § 122.3. Discharges incidental to the normal operation of vessels also enjoyed a decades-long exemption, but this was ultimately invalidated by the courts. *Northwest Envtl. Advocates v. EPA*, No. 03-05760 (N.D. Cal. Sept. 18, 2006), *aff'd*, 537 F.3d 1006 (9th Cir. 2008). In response, EPA created a NPDES Vessel General Permit to cover these discharges. 73 Fed. Reg. 79473 (Dec. 29, 2008). Congress also passed the Clean Boating Act of 2008, Pub. L. No. 110-288, which amended the Clean Water Act and exempts recreational vehicles from the requirement to obtain a NPDES permit. *See* Section 402(r), 33 U.S.C. § 1342(r).

2. State-issued permits are sometimes referred to as State Pollutant Discharge Elimination System (SPDES) permits. *See* 40 C.F.R. Part 123 for EPA's regulations regarding the approval process for state programs.

3. Currently, only the following do *not* have permitting authority: District of Columbia, Idaho, Massachusetts, New Hampshire, New Mexico, and six U.S. territories. Indian tribes are also considered to be "states" for CWA purposes and have authority under the act to apply their own water quality standards to the activities of nonmembers operating on tribal lands. *State of Montana v. EPA*, 941 F. Supp. 945 (D. Mont. 1996), *aff'd*, 137 F.3d 1135 (9th Cir. 1998).

4. 40 C.F.R. § 123.24.

5. 40 C.F.R. § 123.25.

6. 40 C.F.R. § 123.1(i); *see also Atlantic States Legal Found., Inc. v. Eastman Kodak Co.*, 12 F.3d 353, 358–59 (2d Cir.), *cert. denied*, 115 S. Ct. 62 (1994) (state requirement more stringent than federal requirement may be enforced by the state or EPA but not through citizen suit provisions of act).

7. Section 402(d), 33 U.S.C. § 1342(d). EPA has 90 days in which to make its comments, objections, or recommendations to the state permitting agency. 40 C.F.R. § 123.44(a)(1).

8. Section 402(d)(4), 33 U.S.C. § 1342(d)(4).

9. Section 402(c)(3), 33 U.S.C. § 1342(c)(3). Citizens may also petition EPA to withdraw a state's authorization to issue NPDES permits. Such petitions have served as an effective aid in EPA's oversight of state programs.

10. Section 401(a), 33 U.S.C. § 1341(a); 40 C.F.R. Part 121. This state certification authority under § 401(a) also applies to licensing decisions made by other federal agencies, such as hydropower project licenses issued by the Federal Energy Regulatory Commission. *See PUD No. 1 of Jefferson County v. Washington Dept. of Ecology*, 511 U.S. 700, 114 S. Ct. 1900 (1994) (state certification may include requirement to maintain certain level of flow from hydropower facility); *American Rivers, Inc. v. FERC*, 129 F.3d 99 (2d Cir. 1997) (FERC may not exclude conditions in state certification from hydropower project license).

11. Section 401(a)(1), 33 U.S.C. § 1341(a)(1).

12. *See* 40 C.F.R. § 122.21 for a full discussion of the information required in NPDES permit application forms.

13. 40 C.F.R. § 123.25(a)(4).

14. 40 C.F.R. § 122.22(a).

15. 40 C.F.R. § 122.22(d).

16. 40 C.F.R. § 122.21(c) and (d).

17. This permit continuance policy is in effect for EPA-issued permits and in most authorized states. *See Natural Resources Defense Council v. EPA*, 859 F.2d 156, 213–14 (D.C. Cir. 1988) (upholding EPA policy to continue expired permits, reasoning that policy is necessary to prevent undue hardships on permittees).

18. 40 C.F.R. §§ 124.7, 124.8, 124.56. In some states, this document is referred to as a "Rationale."

19. 40 C.F.R. §§ 124.10(c), 124.11.

20. 40 C.F.R. § 124.12. There must be at least 30 days' notice of a hearing on a draft permit. Thus, the grant of a hearing extends the comment period at least until the end of the hearing.

21. 40 C.F.R. § 124.13.

22. *See, e.g., Mueller v. EPA*, 993 F.2d 1354, 1357 (8th Cir. 1993).

23. 40 C.F.R. § 124.15(b). If no comments requesting a change in the permit are received during the public comment period, the final permit will become effective immediately upon issuance. 40 C.F.R. § 124.15(b)(3).

24. 40 C.F.R. § 124.19(a).

25. *Id.*

26. 40 C.F.R. § 124.19(l)(2).

27. Section 509(b)(1)(F), 33 U.S.C. § 1369(b)(1)(F).

28. States must allow citizens as well as permit applicants to challenge final permit decisions in court. 40 C.F.R. § 123.30, 61 Fed. Reg. 20972 (May 8, 1996).

29. 5 U.S.C. § 706.

30. *See* Section 11.3.

31. 40 C.F.R. § 122.41.

32. The NPDES regulations were substantially revised in September 1984. The *Federal Register* preamble to that revision, found at 49 Fed. Reg. 37998 (Sept. 26, 1984), contains useful discussions of EPA policies concerning the NPDES program, including some boilerplate requirements. The D.C. Circuit Court of Appeals' discussion of these regulations also provides useful information about the permit program. *Natural Resources Defense Council, Inc. v. EPA*, 859 F.2d 156 (D.C. Cir. 1988).

33. 40 C.F.R. § 123.25(a).

34. *See* Section 304(e), 33 U.S.C. § 1314(e).

35. EPA's regulations require that permits contain monitoring requirements that are sufficient to yield representative data. 40 C.F.R. § 122.48. The regulations, however, establish

only minimal specific requirements for monitoring method and frequency, leaving most such decisions to the discretion of the permit writer. Regardless of monitoring frequency, monitoring results must be reported to the permitting agency at least once per year. 40 C.F.R. § 122.44(i)(2).

36. 65 Fed. Reg. 30886 (May 15, 2000).

37. 40 C.F.R. §§ 122.44(i)(1)(iii), 122.45(h). Note, however, that although EPA may require monitoring of an internal waste stream, it cannot impose effluent limitations on an internal stream. *American Iron and Steel Inst. v. EPA*, 115 F.3d 979, 995–96 (D.C. Cir. 1997).

38. EPA-approved analytical methods are contained in 40 C.F.R. Part 136.

39. Section 309(c)(4), 33 U.S.C. § 1319(c)(4). Such actions are frequently a target of enforcement action. For example, in 2007 a jury convicted a corporation and its president and principal owner of knowingly making false statements in required reports. *United States v. Hagerman*, 525 F. Supp. 2d 1058, 1060 (S.D. Ind. 2007), *aff'd*, 301 F. App'x 552 (7th Cir. 2008) published with modifications at 555 F.3d 553 (7th Cir. 2009).

40. Section 308(a), 33 U.S.C. § 1318(a).

CHAPTER 3

1. 40 C.F.R. § 122.44(d)(1)(iv). *See* U.S. EPA, *Technical Support Document for Water Quality-Based Toxics Control* at 1 (Mar. 1991).

2. Permittees measuring compliance with a pH limitation through the use of a continuous pH monitor are allowed to exceed the pH limitation on a limited basis. No individual pH excursion may exceed 60 minutes in length, and the total time of noncompliance during any calendar month may not exceed seven hours and 26 minutes. 40 C.F.R. § 401.17(a).

3. Despite the "guideline" nomenclature, the effluent guidelines are not merely guidance to permit writers in setting limitations in NPDES permits. The guidelines establish substantive requirements that must be met by facilities within the regulated industrial categories. *E.I. DuPont de Nemours & Co. v. Train*, 430 U.S. 112 (1977).

4. 40 C.F.R. Parts 405–471.

5. *Natural Resources Defense Council v. EPA*, No. 89-2980 (D.D.C. Jan. 31, 1992).

6. *See* U.S. EPA, "Effluent Guidelines Program Plan," 67 Fed. Reg. 55012 (Aug. 27, 2002). EPA continues to develop effluent guidelines for various industrial categories.

7. 33 U.S.C. § 1314(m).

8. *Natural Resource Defense Council v. EPA*, 437 F. Supp. 2d 1137 (C.D. Cal. 2006).

9. *See* 40 C.F.R. § 125.3(c).

10. A major difference between BPJ requirements and requirements based on effluent guidelines is that a permittee can challenge the propriety of a BPJ limitation when the permit is issued. Where a permit limitation is based on an effluent guideline, the validity of the guideline itself cannot be challenged in a permit appeal; rather, the permittee may challenge only the application of the guideline to the specific discharge and the permit writer's translation of the guideline into limitations.

11. Section 402(o), 33 U.S.C. § 1342(o); *see also Natural Resources Defense Council v. EPA*, 859 F.2d 156, 197–203 (D.C. Cir. 1988) (upholding EPA's pre-1987 nonstatutory antibacksliding policy).

12. Section 301(b)(1)(A), 33 U.S.C. § 1311(b)(1)(A).

13. The 1987 amendments to the act extended this deadline to March 31, 1989.

14. Section 301(b)(2)(A), 33 U.S.C. § 1311(b)(2)(A).

15. Section 301(b)(2)(E), 33 U.S.C. § 1311(b)(2)(E).

16. *See* discussion of *new source* in section 6.7.2.4 below.

17. Section 306, 33 U.S.C. § 1316.

18. *See, e.g.*, 52 Fed. Reg. 42522, 42525, 42533–38 (Nov. 5, 1987) (discussion of development of BPT limitations for OCPSF industrial category).

19. *BP Exploration & Oil, Inc. v. EPA*, 66 F.3d 784, 789 (6th Cir. 1995). EPA develops extensive supporting documentation for its categorical effluent guidelines. These "Development Documents" provide EPA's technical basis for the promulgated limitations and for its decision to regulate particular pollutants and waste streams under the guidelines. These Development Documents are often essential resources when challenging permit limitations.

20. *See, e.g.*, 40 C.F.R. Part 414; 79 Fed. Reg. 12661 (March 6, 2014).

21. *See Chemical Mfr. Ass'n v. EPA*, 870 F.2d 177, 207 (5th Cir. 1989), *cert. denied*, *PPG Indus. Inc. v. EPA*, 495 U.S. 910 (1990).

22. *BP Exploration & Oil v. EPA*, 66 F.3d at 796.

23. *See, e.g.*, 52 Fed. Reg. 42522, 42538–45 (Nov. 5, 1987) (discussion of development of BAT for OCPSF category).

24. 67 Fed. Reg. 38752 (June 5, 2002).

25. This *cost-reasonableness* test is discussed in detail at 51 Fed. Reg. 24974 (July 9, 1986). *See also BP Exploration & Oil, Inc. v. EPA*, 66 F.3d at 798; *American Paper Inst. v. EPA*, 660 F.2d 954 (4th Cir. 1981).

26. Section 306(a)(2), 33 U.S.C. § 1316(a)(2).

27. 40 C.F.R. § 122.2. It is almost impossible for EPA to issue a final NSPS within 120 days of its proposal. Thus, for practical purposes, new sources are sources for which construction begins after promulgation of the final NSPS.

28. *See Natural Resources Defense Council, Inc. v. EPA*, 822 F.2d 104, 114 (D.C. Cir. 1987) (120-day provision valid); *National Ass'n of Metal Finishers v. EPA*, 719 F.2d 624 (3d Cir. 1983), *rev'd on other grounds sub nom. Chemical Mfrs. Ass'n v. Natural Resources Defense Council*, 470 U.S. 116 (1985) (120-day provision not valid interpretation of congressional intent).

29. 40 C.F.R. § 122.29(b)(4).

30. 40 C.F.R. § 122.29(b)(1)(iii); 49 Fed. Reg. 37998, 38043–45 (Sept. 26, 1984); 45 Fed. Reg. 59343 (Sept. 9, 1980).

31. Section 306(a)(1), 33 U.S.C. § 1316(a)(1); *see also CPC Int'l, Inc. v. Train*, 540 F.2d 1329, 1341–42 (8th Cir. 1976), *cert. denied*, 430 U.S. 966 (1977) (setting NSPS does not require cost-benefit analysis; "[w]hat is required . . . is a thorough study of initial and annual costs and an affirmative conclusion that these costs can be reasonably borne by the industry").

32. *See* 40 C.F.R. § 423.15(g).

33. Section 306(d), 33 U.S.C. § 1316(d).

34. Section 301(n), 33 U.S.C. § 1311(n); 40 C.F.R. § 125.30–125.32. Only EPA may grant an FDF variance, even where the state is the permit-issuing authority.

35. *See* 40 C.F.R. § 125.31.

36. Section 301(n)(2), 33 U.S.C. § 1311(n)(2); 40 C.F.R. § 122.21(m)(1). For effluent guidelines promulgated before February 4, 1987, the request was required to be submitted by July 3, 1989. The statutory provisions regarding FDF variances apply only to variances from BAT limitations and pretreatment standards. FDF variances from BPT or BCT would presumably be handled under the regulatory provisions in 40 C.F.R. Part 125 Subpart D.

37. A federal court ruled in 1997 that state water quality standards cannot be effective until they are approved by EPA. *Alaska Clean Water Alliance v. Clarke*, 45 E.R.C. 1664, 1997 WL 446499 (W.D. Wash. 1997). EPA issued a final rule in April 2000 to bring the approval process in line with that decision. 65 Fed. Reg. 24641 (Apr. 27, 2000). The rule provides that new and revised state water quality standards are not effective until they are approved by EPA. Standards submitted to EPA before the effective date of the new rule, however, are considered effective whether or not they have been approved by EPA.

38. Section 303(c)(2), 33 U.S.C. § 1313(c)(2); 40 C.F.R. § 131.10(a); *see also* U.S. EPA, "Strategy for Water Quality Standards and Criteria" (Aug. 2003).

39. The 1987 amendments to the act require states to establish numerical criteria (rather than narrative criteria) for toxic pollutants. Pub. L. No. 100-4; *see also* Section 303(c)(2)(B),

33 U.S.C. § 1313(c)(2)(B). In December 1992, EPA issued what is referred to as the National Toxics Rule (NTR), which established numerical criteria for the 14 states that failed to meet the statutory deadline for developing their own numerical criteria. 57 Fed. Reg. 60848 (Dec. 22, 1992). The NTR was modified somewhat as the result of litigation challenging the rule. 60 Fed. Reg. 22229 (May 4, 1995) (rule modified to base criteria for metals on measurement of total dissolved metals rather than total recoverable metals).

40. 40 C.F.R. §§ 131.10(g), 131.12.

41. Section 304(a)(1), 33 U.S.C. § 1314(a)(1). EPA published a compilation of its criteria for 157 pollutants in December 1998, 63 Fed. Reg. 68354 (Dec. 10, 1998), and maintains a summary table of criteria on its website. *See* http://water.epa.gov/scitech/swguidance/standards/criteria/ current/index.cfm. Criteria are also updated periodically. *See, e.g.,* 79 Fed. Reg. 27303 (May 13, 2014) (draft revisions for criteria of 94 chemical pollutants).

42. U.S. EPA, *Water Quality Standards Handbook* at 3.7 (1983, updated 1994 and 2014).

43. *PUD No. 1 of Jefferson County v. Washington Dept. of Ecology*, 511 U.S. 700, 114 S. Ct. 1900 (1994); *Northwest Environmental Advocates v. City of Portland*, 56 F.3d 979 (9th Cir. 1995).

44. 40 C.F.R. § 122.44(d)(1)(i); see also *American Iron and Steel Inst. v. EPA*, 115 F.3d 979, 999–1001 (D.C. Cir. 1997) (upholding EPA procedures for determining reasonable potential).

45. The 7Q10 flow is the historical low flow for a period of seven days during a 10-year period, normally expressed as million gallons per day (MGD) or cubic feet per second (cfs). The 7Q10 flow may occur on nonconsecutive days during any one year.

46. *See* U.S. EPA, "Compilation of EPA Mixing Zone Documents" (July 2006).

47. This calculation is significantly more complicated for parameters such as pH, temperature, and biological oxygen demand. In these cases, permit limitations are often based on modeling of the effects of a discharge on the receiving waters.

48. *See American Iron and Steel Inst. v. EPA*, 115 F.3d at 998 (upholding EPA policy to limit mixing zones to the *discharge-induced mixing area*, defined as "the immediate vicinity of the discharge where mixing occurs due to the turbulence created by the momentum of the discharge" and limited to a quantity of water equal to 10 times the quantity of effluent).

49. *See* U.S. EPA, "Technical Support Document for Water Quality-Based Toxics Control" at 33–34 (Mar. 1991)

50. 33 U.S.C. § 1313(d).

51. The Ninth Circuit said in *dicta* that a state must implement TMDLs only to the extent that it seeks to avoid losing federal grant money, and that "there is no pertinent statutory provision otherwise requiring implementation of [Section] 303 plans or providing for their enforcement." *Pronsolino v. Nastri*, 291 F.3d 1123, 1140 (9th Cir. 2002).

52. *See* 33 U.S.C. § 1313(d); 40 C.F.R. § 130.7. The Eleventh Circuit held that EPA is not required to develop a formal TMDL implementation plan if a state fails to do so. *Sierra Club v. Meiburg*, 296 F.3d 1021 (11th Cir. 2002).

53. The Ninth Circuit held that waters contaminated exclusively by nonpoint sources of pollution must be included in state lists of impaired waters. *Pronsolino*, 291 F.3d at 1140–41.

54. 65 Fed. Reg. 43586 (July 13, 2000).

55. These cases were consolidated as *American Farm Bureau Fed. v. Whitman*, D.C. Cir. No. 00-1320.

56. 66 Fed. Reg. 53044 (Oct. 18, 2001).

57. U.S. EPA, "A Long-Term Vision for Assessment, Restoration, and Protection under the Clean Water Act Section 303(d) Program" (Dec. 2013).

58. *Friends of the Earth, Inc. v. EPA*, 446 F.3d 140 (D.C. Cir. Apr. 25, 2006).

59. *Id.* at *143.

60. *Id.* at *148.

61. Establishing TMDL "Daily" Loads in Light of the Decision by the U.S. Court of Appeals for the D.C. Circuit in Friends of the Earth, Inc. v. EPA, et al., No. 05-5015 (April 25, 2006) and Implications for NPDES Permits (Nov. 15, 2006) available at http://water.epa.gov/lawsregs/lawsguidance/cwa/tmdl/dailyloadsguidance.cfm.

62. 61 Fed. Reg. 29563 (June 11, 1996); *see also* EPA Proposed Water Quality Trading Policy, 67 Fed. Reg. 34709 (May 15, 2002).

63. *See* U.S. EPA, "Sharing the Load: Effluent Trading for Indirect Dischargers" (1998).

64. U.S. EPA, "Water Quality Trading Assessment Handbook" (Nov. 2004); U.S. EPA, "Water Quality Trading Toolkit for Permit Writers" (Aug. 2007, updated June 2009).

65. Watershed-based National Pollution Discharge Elimination System (NPDES) Permitting Implementation Guidance (December 2003) available at http://www.epa.gov/npdes/pubs/watershedpermitting_ finalguidance.pdf; Watershed-based National Pollution Discharge Elimination System (NPDES) Permitting Technical Guidance (August 2007) available at http://www.epa.gov/npdespub/pubs/watershed_ techguidance.pdf.

66. Toxicity-based permit limitations may also be imposed on a chemical-specific basis. This approach imposes concentration limitations on specific toxic pollutants based on established concentrations known to prevent toxic effects of the particular pollutants. EPA and many states have concentrated their toxicity-based permitting on WET rather than chemical-specific limitations.

67. EPA issued guidance for WET test methods in 2002. 67 Fed. Reg. 69952 (Nov. 19, 2002).

68. EPA has issued guidance documents providing useful information on WET testing: "National Whole Effluent Toxicity (WET) Implementation Guidance Under the NPDES Program" (Nov. 2004); "Method Guidance and Recommendations for Whole Effluent Toxicity (WET) Testing (40 C.F.R. Part 136)," 65 Fed. Reg. 46457 (July 28, 2000); and "Understanding and Accounting for Method Variability in Whole Effluent Toxicity (WET) Applications under the NPDES Program," 65 Fed. Reg. 44528 (June 30, 2000).

69. *See* 55 Fed. Reg. 30082, 30110 (July 24, 1990).

70. *See* U.S. EPA, "Estuarine and Coastal Marine Waters: Bioassessment and Biocriteria Technical Guidance" (Dec. 2000); "Policy on the Use of Biological Assessments and Criteria in the Water Quality Program" (June 19, 1991); U.S. EPA, "Lake and Reservoir Bioassessment and Biocriteria" (Aug. 1998); U.S. EPA, "Notice of Availability: Biological Criteria: Technical Guidance for Streams and Small Rivers," 61 Fed. Reg. 42610 (Aug. 16, 1996).

71. *TSD* at 41; *see also* 63 Fed. Reg. 36742 (July 7, 1998) (advance notice of proposed rulemaking regarding requirement to adopt biological criteria into state water quality standards).

72. 67 Fed. Reg. 9269 (Feb. 28, 2002). This was finalized in January 2003. 68 Fed. Reg. 557 (Jan. 6, 2003).

73. 66 Fed. Reg. 1671 (Jan. 9, 2001).

74. U.S. EPA, "State Adoption of Numeric Nutrient Standards (1998-2008)" (Dec. 2008).

75. *See* U.S. EPA, "Actions to Help States Address Barriers to Numeric Nutrient Criteria Implementation (2012-2014)" (Aug. 2013); U.S. EPA, "Working in Partnership with States to Address Phosphorus and Nitrogen Pollution through Use of a Framework for State Nutrient Reductions" (Mar. 16, 2011).

CHAPTER 4

1. In Natural Resources Defense Council, Inc. v. EPA, 859 F.2d 156 (D.C. Cir. 1988), the court held that EPA has the authority to express permit limitations in terms of toxicity as long as the limits reflect the appropriate requirements of the Act. The court concluded that, although toxicity appears to be an attribute of pollutants rather than a pollutant itself, the Act, by broadly defining "pollutant," authorizes the use of toxicity limitations.

2. EPA has proposed Part 136 test methods for conducting short-term acute and chronic WET testing for marine and freshwater species. 54 Fed. Reg. 50216 (Dec. 4, 1989).

3. For Ceriodaphnia, EPA's test protocol requires that the test be continued until at least 60 percent of the females produce three broods. This may require more or less than seven days to occur.

4. 49 Fed. Reg. 9016 (March 9, 1984).

5. See 54 Fed. Reg. 23868 (June 2, 1989).

6. See also U.S. EPA, Memorandum from Michael B. Cook, Director, Office of Wastewater Enforcement and Compliance, "Clarifications Regarding Certain Aspects of EPA's Surface Water Toxics Control Regulations" (Aug. 14, 1992); U.S. EPA, Technical Support Document for Water Quality-Based Toxics Control (March 1991) (hereinafter referred to as TSD); U.S. EPA, Memorandum from Rebecca W. Hanmer, Acting Assistant Administrator for Water, "Whole Effluent Toxicity Basic Permitting Principles and Enforcement Strategy" (Jan. 25, 1989). In October 1995, EPA issued new whole effluent toxicity test procedures guidelines. 60 Fed. Reg. 53529 (Oct. 16, 1995). These new guidelines are intended to reduce the costs and eliminate confusion caused by the multiple versions of test methods currently in use, and are expected to save up to 20 percent of the current costs of WET testing.

7. 40 C.F.R. § 122.44(d)(1)(iv).

8. 40 C.F.R. § 122.44(d)(1)(v).

9. U.S. EPA, "Whole Effluent Toxicity (WET) Control Policy" (July 1994).

10. See definition in Part 4.3.1 below.

11. A permit limit for a toxic pollutant stating "48-hour LC50 for Daphnia magna" means that the concentration of the pollutant in the discharge may not exceed that which would prove fatal to 50 percent of all of the test organisms exposed to the toxicant for a period of 48 hours. For WET, a permit limit may state, "The 96-hour LC50 of the effluent for bluegill fingerlings may not exceed 50%." This means that the effluent cannot require dilution by more than 50 percent to allow it to meet the 96-hour LC50 for bluegill fingerlings.

12. For example, if an effluent is lethal to 50 percent of test organisms when diluted to 25% concentration, the effluent would contain four acute toxic units (100/25).

13. TSD at 16.

14. *Id.* at 17.

15. *Id.* at 13.

16. *Id.* at 11.

17. *Id.* at 13.

18. *Id.* at 17–18.

CHAPTER 5

1. Pub. L. No. 100-4, 101 Stat. 7 (1987). The October 1, 1992, deadline was extended to October 1, 1993. *See* Section 402(p), 33 U.S.C. § 1342(p).

2. 40 C.F.R. § 122.26(b)(13).

3. *See* 40 C.F.R. § 122.26(b)(14). The scope of "associated with industrial activity" is discussed at length at 55 Fed. Reg. 47990, 48007–15 (Nov. 16, 1990).

4. 40 C.F.R. § 122.26(b)(14).

5. *Id.*

6. *See* 55 Fed. Reg. at 48007 (Nov. 16, 1990).

7. Section 402(p)(2)(E), 33 U.S.C. § 1342(p)(2)(E).

8. 40 C.F.R. § 122.26(g).

9. *Id.*

10. 40 C.F.R. § 122.26(b)(8).

11. Section 402(p)(3)(B)(iii), 33 U.S.C. § 1342(p)(3)(B)(iii).

12. *Defenders of Wildlife v. Browner*, 191 F.3d 1159 (9th Cir. 1999).

13. 64 Fed. Reg. 68722, 68840 (Dec. 8, 1999); 40 C.F.R. § 122.26.

14. *Envtl. Defense Center, Inc. v. EPA*, 344 F.3d 832 (9th Cir. 2003).

15. U.S. EPA, "Implementing the Partial Remand of the Stormwater Phase II Regulations Regarding Notices of Intent & NPDES General Permitting for Phase II MS4s" (Apr. 16, 2004); U.S. EPA, "MS4 Permit Improvement Guide" (Apr. 2010).

16. 55 Fed. Reg. 47990 (Nov. 16, 1990).

17. 40 C.F.R. § 122.26(c).

18. 60 Fed. Reg. 50804 (Sept. 29, 1995). The final modifications to this multi-sector permit were issued on September 30, 1998. 63 Fed. Reg. 52430. The revisions, among other things, require the notice of intent to be covered by the permit to include a certification that any endangered species in proximity to the storm-water discharge would not be adversely affected by the discharge. Facilities that cannot make this certification must obtain an individual permit. The multi-sector permit was reissued and expanded in October 2000. 65 Fed. Reg. 64746 (Oct. 30, 2000). The revised MSGP allows permitting authorities to apply the MSGP to facilities not covered by any of the 29 specific industrial sectors.

19. *See* EPA, *Guidance Manual for the Preparation of NPDES Permit Applications for Storm Water Discharges Associated with Industrial Activity* (April 1991).

20. 57 Fed. Reg. at 41243 (Sept. 9, 1992).

21. *Id.* at 41242.

CHAPTER 6

1. 59 Fed. Reg. 18688 (Apr. 19, 1994).

2. In January 2001, EPA proposed regulations that would clarify and expand permit requirements for municipal sanitary sewer systems in order to reduce SSOs, but the proposal was withdrawn. The proposed rules were designed to reduce the number of SSOs through increased recordkeeping; improvements in capacity, management, operation, and maintenance; and expanded permit coverage to satellite collection facilities. In June 2010, EPA announced "listening sessions" to seek input on potential revisions to the regulations. 75 Fed. Reg. 30395 (June 1, 2010).

3. U.S. EPA, "Compliance and Enforcement Strategy Addressing Combined Sewer Overflows and Sanitary Sewer Overflows" (Apr. 27, 2000). 175 33 U.S.C. § 1326.

4. *Entergy Corp. v. Riverkeeper, Inc.*, 556 U.S. 208 (2009).

5. Section 316(a) governs thermal discharges. Section 316(b) governs cooling water intake structures. In 2014, EPA finalized regulations under Section 316(b) that require new facilities with a design intake flow of more than 2 million gallons per day and which use at least 25% of water withdrawn exclusively for cooling to adopt certain design and location requirements to minimize adverse effects on aquatic life caused by cooling water intake. 79 Fed. Reg. 48300 (Aug. 15, 2014).

6. Section 403, 33 U.S.C. § 1343. *Territorial seas, contiguous zone*, and *ocean* are defined in CWA Section 502, 33 U.S.C. § 1362. On January 19, 2001, the EPA administrator signed a rulemaking that would have established minimum criteria for discharges; established a new designated use, Healthy Ocean Waters; and identified ocean areas as Special Ocean Sites. As part of the Bush administration transition, this proposed rule was withdrawn on January 20, 2001.

7. 40 C.F.R. § 125.121(e). The permitting authority may not issue a permit if it is uncertain whether the discharge will result in an unreasonable degradation, unless the permittee (1) agrees to provide additional data, such as a chemical analysis of the discharge, bioassays, or a dilution analysis, at a later date; and (2) can demonstrate that there are no reasonable alternatives to the disposal and that the discharge will not result in irreparable harm to the marine environment while the data are being obtained. 40 C.F.R. §125.123(c).

8. 33 U.S.C. § 1401–45.

9. 33 U.S.C. § 1411(a). The standard for permit issuance is whether the dumping will "unreasonably degrade or endanger" human health, welfare, or the marine environment.

10. 33 U.S.C. § 1413. No permit is required for dumping of fish wastes under the ocean dumping program, except for dumping in harbors or other areas where such wastes could endanger health, the environment, or ecological systems. 33 U.S.C. § 1412(d).

11. 33 U.S.C. § 1413; 33 C.F.R. Part 324.

12. 40 C.F.R. § 228.15.

13. 33 U.S.C. § 1414b.

14. 33 U.S.C. § 1901 *et seq.*

15. 33 U.S.C. § 1322; 40 C.F.R. Part 140.

CHAPTER 7

1. Section 307(b), 33 U.S.C. § 1317(b). Discharges *from* publicly owned treatment works must comply with NPDES permits for direct dischargers, including *secondary treatment standards* for biological oxygen demand (BOD), suspended solids, and pH. 40 C.F.R. Part 133.

2. EPA substantially revised its general pretreatment regulations in 1988. The preamble to that regulatory revision, found at 53 Fed. Reg. 40562 (Oct. 17, 1988), provides extensive discussions of several important aspects of the pretreatment program. EPA issued a proposed rule to amend its general pretreatment regulations and to relax the regulatory burden they impose in July 1999. 64 Fed. Reg. 39564 (July 22,1999). A final rule was issued in 2005. 70 Fed. Reg. 60134 (Oct. 14, 2005).

3. 40 C.F.R. § 403.5(a)(1).

4. 40 C.F.R. § 403.3(p).

5. 40 C.F.R. § 403.3(k).

6. 40 C.F.R. § 403.5(b).

7. 40 C.F.R. § 403.7.

8. 40 C.F.R. § 403.7(a)(i) (dilution); 52 Fed. Reg. 42522, 42547 (Nov. 5, 1987) (volatilization).

9. *Natural Resources Defense Council v. EPA*, 790 F.2d 289 (3d Cir. 1986), *cert. denied*, 479 U.S. 1084 (1987).

10. Pub. L. No. 100-4 § 402 (1987).

11. 56 Fed. Reg. 50978 (Oct. 9, 1991).

12. *Sierra Club v. EPA*, 992 F.2d 337 (D.C. Cir. 1993).

13. 40 C.F.R. Part 503; 58 Fed. Reg. 9248 (Feb. 19, 1993). Under the terms of a consent decree, *Gearhart v. Reilly*, Civil No. 89-6266-JO (D. Ore.), EPA was required to promulgate Phase II of the sludge regulations by December 15, 2001, although the deadline was later extended to October 17, 2003.

14. *See* 40 C.F.R. Part 403 App. G. In addition to the pollutants addressed in the Phase I sludge regulations, removal credits may be available for two additional metals and 14 organic pollutants when sludge is applied to the land, for seven additional metals and 14 organic pollutants when sludge is disposed in a surface disposal site, and for three other metals when sludge is incinerated, if the concentration of the pollutant in the sludge does not exceed levels established in Part 403.

15. *See* 70 Fed. Reg. 60199 (Oct. 14, 2005).

16. *See* Section 307(b)(4), 33 U.S.C. § 1317(b)(4); 40 C.F.R. § 403.4; *see also* U.S. EPA, "Local Limits Development Guidance" (July 2004).

17. Section 309, 33 U.S.C. § 1319; 40 C.F.R. § 403.10(f)(1).

18. Section 309(f), 33 U.S.C. § 1319(f); 40 C.F.R. § 403.5(e).

19. 40 C.F.R. § 403.10(f)(iv).

20. *See* U.S. EPA, "Intro to the National Pretreatment Program" (June 2011).

CHAPTER 8

1. Section 319(a)(1), 33 U.S.C. § 1329(a)(1).

2. Section 319(b), 33 U.S.C. § 1329(b).

3. Section 319(j), 33 U.S.C. § 1329(j).

4. In May 1996, EPA and the states issued an agreement titled "Nonpoint Source Program & Grants Guidance for Fiscal Year 1997 & Future Years." This agreement gives states more flexibility in administering their nonpoint-source control programs and streamlines the grants program. This guidance was updated most recently in 2013 and applies to 319-funded grant activities beginning in fiscal year 2014. U.S. EPA, "Nonpoint Source Program & Grants Guidelines for States and Territories" (April 12, 2013).

5. Coastal Zone Management Act, 16 U.S.C. § 1451–1466, as amended by the Coastal Zone Act Reauthorization Amendments of 1990 (Pub. L. 101-508); *see* 58 Fed. Reg. 5182 (Jan. 19, 1993).

6. U.S. EPA, Guidance Specifying Management Measures for Sources of Nonpoint Pollution in Coastal Waters (Jan. 1993).

7. Section 320, 33 U.S.C. § 1330; *see also* 54 Fed. Reg. 40798 (Oct. 3, 1989).

CHAPTER 9

1. 33 U.S.C. § 1344.

2. U.S. Army Corps of Engineers, Technical Rep. Y-87-1, *Corps of Engineers Wetlands Delineation Manual* (Jan. 1987), available at http://el.erdc.usace.army.mil/elpubs/pdf/wlman87.pdf.

3. *See* Clean Water Rule: Definition of "Waters of the United States," 80 Fed. Reg. 37054, 37088.

4. *See United States v. Riverside Bayview Homes*, 474 U.S. 121 (1985).

5. 80 Fed. Reg. 37104.

6. *See* 40 C.F.R. §§ 232.2(1), (2)(iii); 33 C.F.R. § 323.2(d)(1), (2)(iii); *see also Am. Mining Cong. v. U.S. Army Corps of Eng'rs*, 951 F. Supp. 267 (D.D.C. 1997), *aff'd sub nom. Nat'l Mining Ass'n v. U.S. Army Corps of Eng'rs*, 145 F.3d 1339 (D.C. 1998).

7. *See* 66 Fed. Reg. 4550 (Jan. 17, 2001); 33 C.F.R. § 323.2(d)(2)(i); 40 C.F.R. § 232.2(2)(i).

8. *United States v. Sinclair Oil Co.*, 767 F. Supp. 200, 203–05 (D. Mont. 1990).

9. *United States v. M.C.C. of Fla., Inc.*, 772 F.2d 1501, 1505–06 (11th Cir. 1985), *vacated and remanded on other grounds*, 107 S. Ct. 1968 (1987), *readopted in relevant part on remand*, 848 F.2d 1133 (11th Cir. 1988); *see also Avoyelles Sportsmen's League v. Marsh*, 715 F.2d 897, 923–24 (5th Cir. 1983).

10. *See* Section 404(f), 33 U.S.C. § 1344(f).

11. 486 F.3d 638 (9th Cir. 2007).

12. *Id.* at 641-43.

13. *Id.* at 640-41.

14. *Id.* at 644.

15. *Id.*

16. *Id.* at 655.

17. *Id.* at 655.

18. *Id.* at 647.

19. *Id.* at 649.

20. Petition for Writ of Certiorari at 11–12, *Couer Alaska Inc. v. S.E. Alaska Conservation Council*, No. 07-984, 2008 WL 244990 (Jan. 28, 2008).

21. *Coeur Alaska, Inc. v. S.E. Alaska Conservation Council*, 129 S. Ct. 2458 (2009).

22. 33 C.F.R. § 325.1.

23. 33 C.F.R. § 325.1(b).

24. 33 C.F.R. § 320.4(a).

25. *Id.*

26. *Mall Properties Inc. v. Marsh*, 672 F. Supp. 561, 566 (D. Mass. 1987).

27. 33 C.F.R. §§ 320.4(a)(1), 325.2(a)(6).

28. 40 C.F.R. § 230.10(a).

29. 40 C.F.R. § 230.3(q).

30. 40 C.F.R. § 230.10(a)(3).

31. *Id.* ("[W]here a discharge is proposed for a special aquatic site, all practicable alternatives to the proposed discharge which do not involve a discharge into a special aquatic site are presumed to have less adverse impact on the aquatic ecosystem, unless *clearly demonstrated otherwise*.") (emphasis added).

32. 40 C.F.R. § 230.10(c); *see also* 40 C.F.R. §§ 230.1(c).

33. Section 401, 33 U.S.C. § 1341(a)(1); 33 C.F.R. § 325.2(b)(1)(ii); *see also United States v. Marathon Dev. Co.*, 867 F.2d 96, 98–99 (1st Cir. 1989). The certifying agency may be deemed to have waived its certification if it does not act within 60 days. 33 C.F.R. § 325.2(b)(1)(ii).

34. Section 404(c), 33 U.S.C. § 1344(c). EPA may veto a Section 404 permit if it determines that the discharge would have "an unacceptable adverse effect on municipal water supplies, shellfish beds and fishery areas . . . , wildlife, or recreation areas." *Id.*; *see also Alameda Water & Sanitation Dist. v. Reilly*, 930 F. Supp. 486 (D. Colo. 1996).

35. Section 404(m), 33 U.S.C. § 1344(m).

36. *Mingo Logan Coal Co. v. EPA*, 714 F.3d 608 (D.C. Cir. 2013).

37. 42 U.S.C. §§ 4321–4370h.

38. 33 C.F.R. § 325.1(b).

39. *But see Wetlands Action Network v. U.S. Army Corps of Eng'rs*, 222 F.3d 1105, 1122 (9th Cir. 2000) (Corps may conduct abbreviated environmental assessment rather than an environmental impact statement in some cases).

40. Memorandum of Agreement between the Environmental Protection Agency and the Department of the Army concerning the Determination of Mitigation under Clean Water Act Section 404(b)(1) Guidelines (Feb. 6, 1990); *see also* 67 Fed. Reg. 2020 (Jan. 15, 2002) (reissuance of nationwide permits, reinforcing commitment to "no net loss" policy).

41. The February 1990 MOA contains a controversial provision limiting mitigation requirements where a high proportion of the land is wetlands. The MOA also states that the sequencing set forth in the MOA (avoidance, minimization, compensation) may not be required where wetland alterations constitute "insignificant environmental losses."

42. *Compensatory Mitigation for Losses of Aquatic Resources*, 73 Fed. Reg. 19594, 19671–72 (Apr. 10, 2008).

43. *Id.*

44. *Id.* at 19673.

45. *See* Section 404(e)(1), 33 U.S.C. § 1344(e)(1).

46. *See* 67 Fed. Reg. at 2089–94 (Jan. 15, 2002).

47. *See* 33 C.F.R. §§ 330.1(c), 330.2(c).

48. *See* C.F.R. § 330.1(b).

49. *See* 77 Fed. Reg. 10184 (Feb. 21, 2012).

50. *See* 33 C.F.R. §§ 330.1(c), 330.2(c).

51. *See* 33 C.F.R. § 330.6(a) ("Nationwide permittees may, and in some cases must, request from a [district engineer] confirmation that an activity complies with the terms and conditions of an NWP").

52. Section 309(d), 33 U.S.C. § 1319(d); Section 404(s), 33 U.S.C. § 1344(s); 40 C.F.R. § 19.4. EPA issued a Section 404 penalty policy in December 2001. "Clean Water Act Section 404 Settlement Penalty Policy" (Dec. 21, 2001).

53. *See, e.g., United States v. Cumberland Farms*, No. 91-10051-MLW (D. Mass. July 25, 1996), 61 Fed. Reg. 40248 (Aug. 1, 1996) (settlement under which company will establish 30-acre wildlife and wetlands corridor on site it allegedly damaged and will turn over additional 225 acres to state for permanent conservation); *United States v. Larkins*, 657 F. Supp. 76, 86 (W.D. Ky. 1987), *aff'd*, 852 F.2d 189 (6th Cir. 1988); *United States v. Bd. of Trs., Fla. Keys Cmty. Coll.*, 531 F. Supp. 267, 275–76 (S.D. Fla. 1981).

54. *See United States v. Ellen*, 961 F.2d 462, 467 (4th Cir. 1992), *cert. denied*, 506 U.S. 875 (1992); *United States v. Pozsgai*, 897 F.2d 524 (3d Cir.) (unpublished table decision), *cert. denied*, 498 U.S. 812 (1990).

CHAPTER 10

1. Section 311(b)(1); 33 U.S.C. § 1321(b)(1).

2. In determining whether a facility could reasonably discharge oil into navigable waters or adjoining shorelines, an operator may not take into consideration man-made features such as dikes, equipment, or other structures. 40 C.F.R. § 112.1(d)(1).

3. 40 C.F.R. § 112.1(b). Although the act requires EPA to issue regulations requiring SPCC plans for discharges of oil or hazardous substances, 33 U.S.C. § 1321(j)(5)(A), EPA has issued such regulations only for discharges of oil (40 C.F.R. Part 112). States, however, often require SPCC plans for hazardous substances under state programs.

4. 40 C.F.R. § 112.1(d)(2). In calculating these totals, only containers of oil with a capacity of 55 gallons or greater are counted.

5. 40 C.F.R. § 112.1(d)(4).

6. 40 C.F.R. § 112.3(e). SPCC plans must be maintained at a facility if the facility is normally attended at least four hours a day. If the facility is normally attended fewer than eight hours a day, the plan may be maintained at a nearby field office. 40 C.F.R. § 112.3(e).

7. 40 C.F.R. § 112.3(b).

8. 40 C.F.R. § 112.5.

9. 40 C.F.R. § 112.3(d). The engineer must certify that the plan has been prepared in accordance with good engineering practices.

10. 40 C.F.R. § 112.4(a).

11. 40 C.F.R. § 112.4(d). Section 112.4(e) and (f) set forth the procedures for appealing a decision by the regional administrator requiring amendment of an SPCC plan.

12. 40 C.F.R. § 112.7.

13. 40 C.F.R. § 112.7(a) and (b). This information is required if experience indicates that there is a reasonable potential for equipment failure (such as tank overflow, rupture, or leakage).

14. With the exception of secondary containment requirements, EPA's 2002 final rule allows deviations from most rule provisions when equivalent environmental protection is provided.

15. 40 C.F.R. § 112.7(f)(2).

16. *Id.* at (f)(1).

17. Section 311(j)(5)(C)(iv), 33 U.S.C. § 1321(j)(5)(C)(iv).

18. 40 C.F.R. § 112.20(e).

19. As in the SPCC program, EPA now exempts from the Facility Response Plan program completely buried underground tanks that are subject to the underground storage tank technical requirements and containers of oil with a capacity of less than 55 gallons.

20. 40 C.F.R. § 112.20(f).

21. 40 C.F.R. § 112.20(b), (f)(2).

22. 40 C.F.R. Part 112, App. D.

23. 40 C.F.R. Part 112, App. F.

24. Section 311(j)(5)(D), 33 U.S.C. § 1321(j)(5)(D).

25. 40 C.F.R. Part 300.

26. Section 311(j)(5)(F), 33 U.S.C. § 1321(j)(5)(F).

27. Section 311(j)(5)(G), 33 U.S.C. § 1321(j)(5)(G).

28. Section 311(b)(5), 33 U.S.C. § 1321(b)(5). For purposes of the Section 311 program, *discharge* is defined as including, but not limited to, "any spilling, leaking, pumping, pouring, emitting, emptying or dumping." Section 311(a)(2), 33 U.S.C. § 1321(a)(2).

29. Section 311(b)(5), 33 U.S.C. § 1321(b)(5).

30. *Oil* is defined as "oil of any kind or in any form, including, but not limited to, petroleum, fuel oil, sludge, oil refuse, and oil mixed with other than dredged spoil." Section 311(a)(1), 33 U.S.C. § 1321(a)(1). This definition includes animal fats and vegetable oils. 62 Fed. Reg. 54508 (Oct. 20, 1997).

31. 40 C.F.R. § 110.3; *see also Orgulf Transp. Co. v. United States*, 711 F. Supp. 344 (W.D. Ky. 1989) (EPA determination that any discharge of oil causing a sheen may be harmful is authorized by the act, even though all spills of oil create a sheen). The use of a dispersant or emulsifier to prevent a sheen to circumvent the notice requirement is prohibited. 40 C.F.R. § 110.4.

32. 40 C.F.R. Part 116.

33. 40 C.F.R. Part 117.

34. 40 C.F.R. § 117.21.

35. Section 311(a)(2), 33 U.S.C. § 1321(a)(2); 52 Fed. Reg. 10712 (Apr. 2, 1987).

36. Section 311(b)(3), 33 U.S.C. § 1321(b)(3).

37. For example, in 2011 Citgo Petroleum Corporation was assessed a penalty of $9 million under the CWA and a state environmental statute for discharging over 2 million gallons of slop oil. *United States. v. Citgo Petroleum Corp.*, No. 08-cv-0893 (W.D. La. Sept. 29, 2011), 2011 WL 10723934.

38. Section 311(f)(2), 33 U.S.C. § 1321(f)(2). These defenses include: (1) an act of God, (2) an act of war, (3) negligence on the part of the U.S. government, and (4) an act or omission of a third party. *Id.*

39. For discharges of hazardous substances, the maximum quantity-based penalty is $1,000 per reportable quantity unit. In other words, if the reportable quantity for a substance is ten pounds, and 10,000 pounds are discharged, the maximum penalty would be $1 million. Section 311(b)(7)(A), 33 U.S.C. § 1321(b)(7)(A); 40 C.F.R. § 19.4.

40. Section 311(b)(7)(D), 33 U.S.C. § 1321(b)(7)(D); 40 C.F.R. § 19.4.

CHAPTER 11

1. Section 402(b)(7), 33 U.S.C. § 1342(b)(7); 40 C.F.R. § 123.27(a).

2. 40 C.F.R. § 123.27(a)(3); *see also Natural Resources Defense Council v. EPA*, 859 F.2d 156, 179–180 (D.C. Cir. 1988).

3. 40 C.F.R. § 123.27(d).

4. Section 309(a), 33 U.S.C. § 1319(a).

5. A federal appeals court has held that EPA cannot *overfile* an RCRA enforcement case where a state environmental agency has resolved a violation through a consent decree. *Harmon Ind., Inc. v. Browner*, 191 F.3d 894 (8th Cir. 1999). In a CWA case raising the same issue, the Fourth Circuit held that EPA was *not* barred from bringing an action after the state had settled with a violator. *U.S. v. Smithfield Foods, Inc.*, 191 F.3d 516 (4th Cir. 1999), *cert. denied*, 531 U.S. 813 (2000).

6. *United States v. ITT Rayonier, Inc.*, 627 F.2d 996, 1001 (9th Cir. 1980).

7. Section 309(a)(2), 33 U.S.C. § 1319(a)(2).

8. 40 C.F.R. § 122.41(l)(4) and (7).

9. 40 C.F.R. § 122.41(l)(6).

10. 40 C.F.R. § 122.41(l)(2).

11. 40 C.F.R. § 122.41(l)(7).

12. 40 C.F.R. § 122.41(l)(8).

13. 33 U.S.C. § 1311(a).

14. E.g., *Am. Canoe Ass'n v. Murphy Farms*, 412 F.3d 536, 540 (4th Cir. 2005); United *States v. Texas Pipeline Co.*, 611 F.2d 345 (10th Cir. 1979); *United States v. Amoco Oil Co.*, 580 F. Supp. 1042 (W.D. Mo. 1984).

15. U.S. EPA, "Clean Water Act Action Plan" (Oct. 15, 2009).

16. *United States v. Amoco Oil Co.*, 580 F. Supp. 1042 (W.D. Mo. 1984) (good faith); *Student Public Interest Research Group of N.J. v. Monsanto*, 600 F. Supp. 1479 (D.N.J. 1985) (permit modification pendency); *Public Interest Research Group of N.J., Inc. v. Elf Atochem N.A., Inc.*, 817 F. Supp. 1164 (D. N.J. 1993) (where there was laboratory error, permittee will not be held liable for discharge violations but will be liable for each measurement proved to be erroneous). In 1998, a New Mexico hot sauce manufacturer pleaded guilty to a CWA violation for dumping hot pepper rinse water into a ditch that flowed into the Rio Grande. The manufacturer was charged with a criminal violation of the act although it had erroneously applied for and received a storm-water discharge permit rather than an NPDES permit. EPA officials had reviewed the issuance of the storm-water permit, and EPA had

informed the company that its NPDES permit was a low priority and could not expect to be issued for about 15 years. *United States v. Cervantes Enterprises, Inc.* (D.N.M., plea filed Aug. 6, 1998).

17. *Compare Hughey v. JMS Dev. Corp.*, 78 F.3d 1523, 1530 (11th Cir.), *cert. denied*, 117 S. Ct. 482 (1996) (where storm-water permit not available under state's program, developer not liable for storm-water runoff from construction site despite act's strict prohibition on discharges without a permit; based on finding that "Congress could not have intended a strict application of the zero discharge standard in section 1311[a] when compliance is factually impossible") *with Driscoll v. Adams*, 181 F.3d 1285 (11th Cir. 1999), *cert. denied*, 120 S. Ct. 1961 (2000) (facility operator could be subject to a citizen suit for violation of storm-water requirements even though state agency was not prepared to issue storm-water permits).

18. *See Marathon Oil Co. v. EPA*, 564 F.2d 1253 (9th Cir. 1977); *FMC Corp. v. Train*, 539 F.2d 973 (4th Cir. 1976). These decisions extend only to federally issued NPDES permits. As discussed below, states are not required to include the upset defense in state-issued permits.

19. 40 C.F.R. § 122.41(n)(1); *see also* 40 C.F.R. § 403.16(a); *Chesapeake Bay Found., Inc. v. Bethlehem Steel Corp.*, 652 F. Supp. 620 (D. Md. 1987) (violations of permit nearly every day during two-month period were not upsets).

20. 40 C.F.R. § 122.41(n)(1).

21. 40 C.F.R. § 122.41(n)(2). In 1988, the U.S. Court of Appeals for the D.C. Circuit ruled that EPA's refusal to extend the upset defense to water quality–based permit limits was arbitrary and capricious. *Natural Resources Defense Council v. EPA*, 859 F.2d 156, 209–10 (D.C. Cir. 1988). Upon remanding the regulation to the agency, the court specifically stated that it did not mean to imply that EPA must allow the defense for water quality–based limitations, only that if the agency decides not to extend the defense, it must provide a reasoned basis for its decision. EPA appears to have taken no action in response to the court's holding but appears to consider the issue on a case-by-case basis.

22. 40 C.F.R. § 122.41(n)(3)(iii); *Public Interest Research Group of N.J., Inc. v. U.S. Metals Refining Co.*, 681 F. Supp. 237 (D.N.J. 1987).

23. 40 C.F.R. § 122.41(n)(3).

24. 40 C.F.R. § 122.41(n)(4).

25. *Sierra Club v. Union Oil Co. of Cal.*, 813 F.2d 1480, 1487 (9th Cir. 1987), *vacated on other grounds*, 485 U.S. 931, *reinstated on remand*, 853 F.2d 667 (9th Cir. 1988).

26. 40 C.F.R. § 123.25(a)[Note].

27. 40 C.F.R. § 122.41(m)(1)(i); *see also* 40 C.F.R. § 403.17(a).

28. 40 C.F.R. § 122.41(m)(2); *see also Natural Resources Defense Council v. EPA*, 822 F.2d 104 (D.C. Cir. 1987) (upholding bypass prohibition even where limitations not exceeded).

29. *Id.*

30. 40 C.F.R. § 122.41(m)(4).

31. 40 C.F.R. § 122.41(m)(4)(i)(B); *see also United States v. Municipality of Penn Hills*, 6 F. Supp. 2d 432 (W.D. Pa. 1998) (bypass was illegal where facility could have installed equalization tanks or other alternatives to bypassing).

32. 40 C.F.R. § 122.41(m)(3)(i).

33. 40 C.F.R. § 122.41(m)(3)(ii).

34. *See, e.g., United States v. CPS Chem. Co.*, 779 F. Supp. 437, 454 (E.D. Ark. 1991); *Student Public Interest Research Group of N.J., Inc. v. AT&T Bell Lab.*, 617 F. Supp. 1190, 1204 (D.N.J. 1985).

35. 33 U.S.C. § 1342(k).

36. U.S. EPA, "Revised Policy Statement on Scope of Discharge Authorization and Shield Associated with NPDES Permits," Memorandum from Robert Perciasepe, Assistant Administrator for Water, et al., to Regional Administrators and Regional Counsels (April 11, 1995).

37. *E. I. du Pont de Nemours & Co. v. Train*, 430 U.S. 112, 138 n. 28 (1977).

38. *Atlantic States Legal Found.*, 12 F.3d at 357; *see also Piney Run Preservation Assn. v. County Commissioners of Carroll County, Maryland,* 268 F.3d 255 (4th Cir. 2001), *cert. denied*, 122 S. Ct. 1960 (2002); *In re Ketchikan Pulp Co.*, 7 E.A.D. 605 (1998) (EPA Appeals Board holding that permit shield covers all pollutants that have been adequately disclosed to the permitting agency during the application process, whether or not listed in the permit).

39. *S. Appalachian Mountain Stewards v. A&G Coal Corp.*, 758 F.3d 560 (4th Cir. 2014).

40. *Alaska Cmty. Action on Toxics v. Aurora Energy Servs., LLC*, 765 F.3d 1169 (9th Cir. 2014), *cert. denied*, 135 S. Ct. 2830 (2015).

41. U.S. EPA, "Guidance on Choosing Among Clean Water Act Administrative, Civil and Criminal Enforcement Remedies" (Aug. 28, 1987).

42. *Id.*

43. Section 309(g)(2)(A); 33 U.S.C. § 1319(g)(2)(A); 40 C.F.R. § 19.4.

44. For guidance on Class I penalty procedures, *see* 52 Fed. Reg. 30730 (Aug. 17, 1987).

45. Section 309(g)(2)(B), 33 U.S.C. § 1319(g)(2)(B); 40 C.F.R. § 19.4.

46. Section 309(g), 33 U.S.C. § 1319(g); 40 C.F.R. Part 22.

47. Section 309(a), 33 U.S.C. § 1319(a).

48. *Sackett v. EPA*, 132 S. Ct. 1367 (2012).

49. Section 309(a), 33 U.S.C. § 1319(a).

50. Section 309(d), 33 U.S.C. § 1319(d); 40 C.F.R. § 19.4.

51. *Tull v. United States*, 481 U.S. 412 (1987); *N. Carolina Envtl. Justice Network v. Taylor*, 2014 WL 7384970 (E.D.N.C. Dec. 29, 2014).

52. *United States v. Tull*, 481 U.S. 412 (1987).

53. U.S. EPA, "Interim Clean Water Act Settlement Penalty Policy" (March 1, 1995).

54. The Penalty Policy states that it is not to be used by a court in determining a penalty at trial.

55. EPA implemented changes to the BEN model in 2005. 70 Fed. Reg. 50326 (Aug. 26, 2005).

56. Section 309(c), 33 U.S.C. § 1319(c). A number of courts have held this standard only requires ordinary, rather than gross, negligence. *See, e.g.*, *United States v. Pruett*, 681 F.3d 232 (5th Cir. 2012).

57. Section 309(c), 33 U.S.C. § 1319(c). In 1991, the president of a drum reconditioning company in Florida was sentenced to 13 years in prison for intentionally and continuously dumping toxic waste into the Tampa sewer system. *United States v. Benkovitz*, Nos. 97-331, 98-349 (M.D. Fla. Aug. 16, 1999).

58. Section 309(c), 33 U.S.C. § 1319(c).

59. *United States v. Weitzenhoff*, 1 F.3d 1523 (9th Cir.), *amended and superseded on denial of reh'g en banc*, 35 F.3d 1275 (9th Cir. 1993), *cert. denied*, 513 U.S. 1128 (1995); *accord United States v. Sinskey*, 119 F.3d 712 (8th Cir. 1997) (government did not have to prove that defendants knew their conduct was illegal, only that they knew of their relevant conduct); *United States v. Hopkins*, 53 F.3d 533 (2d Cir. 1995) (defendant only had to know of the commission of the act, not that the act violated the company's discharge permit); *United States v. Wilson*, 133 F.3d 251 (4th Cir. 1997) (to prove felony violation of act, government must prove defendant's knowledge of facts meeting each essential element of offense but need not prove defendant knew conduct was illegal).

60. Section 309(c)(3), 33 U.S.C. § 1319(c)(3).

61. *See United States v. Borowski*, 977 F.2d 27 (1st Cir. 1992) (prosecution for knowing endangerment cannot be premised on danger that occurs before pollutant reaches water); *United States v. Villegas*, 784 F. Supp. 6 (E.D.N.Y. 1991), *rev'd in part and remanded*, *United States v. Plaza Health Lab., Inc.*, 3 F.3d 643 (2d Cir. 1993), *cert. denied*, 512 U.S. 1245 (1994) (defendant who placed vials of blood into river bulkhead could not be convicted of knowing

endangerment because there was insufficient evidence that he knew there was a high probability that by placing the vials in the river he was placing another person in imminent danger).

62. Section 309(c)(4), 33 U.S.C. § 1319(c)(4); *see also United States v. Carnival Corp.* (S.D. Fla. Apr. 19, 2002) (cruise ship operator ordered to pay $18 million after pleading guilty to falsifying records on oil discharges at sea); *United States v. Warner-Lambert Inc.*, Crim. No. 97-2394 (D.P.R. Sept. 19, 1997) (company fined $3.6 million for filing false discharge reports, despite lack of evidence of environmental harm from violations); *United States v. Rettig* (D. Va. Jan. 16, 1997) (30-month prison sentence imposed for filing false water quality reports and discarding or destroying laboratory reports).

63. Section 309(c)(6), 33 U.S.C. § 1319(c)(6). A corporate officer may be criminally liable if he has authority to exercise control over the activity that is causing the illegal discharges, even if such responsibility is not expressly vested in him by the company. *United States v. Iverson,* 162 F.3d 1015 (9th Cir. 1998). Moreover, a corporate officer can be convicted of "knowing endangerment" to others even if he has no actual knowledge of the violation. *United States v. Hansen,* 262 F.3d 1217 (11th Cir. 2001), *cert. denied,* 122 S. Ct. 2326 and 122 S. Ct. 2327 (2002).

64. The standing requirement of this section has been broadly construed. Normally, a citizen group can demonstrate standing by showing that one or more of its members makes use of the water body that is affected by the discharge at issue. *See Natural Resources Defense Council v. Texaco Ref. and Mktg., Inc.*, 2 F.3d 493 (3d Cir. 1993); *Public Interest Research Group of N.J., Inc. v. Powell-Duffryn Terminals, Inc.*, 913 F.2d 64 (3d Cir. 1990), *cert. denied*, 498 U.S. 1109 (1991). *But see Public Interest Research Group of N.J., Inc. v. Magnesium Elektron, Inc.*, 123 F.3d 111 (3d Cir. 1997) (group had no standing where violation of permit did not cause actual or threatened injury to river, although group argued knowledge of violations caused them to reduce their use and enjoyment of the river); *Friends of the Earth Inc. v. Crown Central Petroleum Corp.*, 95 F.3d 358 (5th Cir. 1996) (citizen group does not have standing where members used waters three tributaries and 18 miles downstream from facility and no evidence that pollutants had migrated from facility to those waters).

65. Section 505(g), 33 U.S.C. § 1365(g).

66. *American Canoe Assoc., Inc. v. City of Louisa Water & Sewer Commission*, 389 F.3d 536 (6th Cir. 2004).

67. *See Soundkeeper, Inc. v. A & B Auto Salvage Inc.*, 19 F. Supp. 3d 426 (D. Conn. 2014); *San Francisco Baykeeper v. Cargill Salt Div.*, 481 F.3d 700, 706 (9th Cir. 2007); *Sierra Club, Lone Star Chapter v. Cedar Point Oil Co. Inc.*, 73 F.3d 546 (5th Cir. 1996).

68. Section 505(b), 33 U.S.C. § 1365(b).

69. *Natural Resources Defense Council v. Fina Oil & Chem. Co.*, 806 F. Supp. 145 (E.D. Tex. 1992); *see also California Sportfishing Prot. Alliance v. Chico Scrap Metal, Inc.*, 728

F.3d 868 (9th Cir. 2013) (citizen suit barred only if federal or state action involves the same standard, limitation, or order as the citizen suit, not merely a comparable one); *Citizens for a Better Environment v. Union Oil Co. of Cal.*, 83 F.3d 1111 (9th Cir. 1996), *cert. denied*, 117 S. Ct. 789 (1997) (citizen suit not barred by company's settlement with state under which it paid $780,000 to push back a compliance date under state law not comparable to federal enforcement provisions); *Knee Deep Cattle Co. v. Bindana Inv. Co.*, 94 F.3d 514 (9th Cir. 1996), *cert. denied*, 117 S. Ct. 1027 (1997) (state prosecution for solely past violations did not constitute diligent prosecution for ongoing violations).

70. Section 505(b), 33 U.S.C. § 1365(b); *but see Black Warrior Riverkeeper, Inc. v. Black Warrior Minerals, Inc.*, 734 F.3d 1297 (11th Cir. 2013) (citizen suit cannot evade 60-day waiting period by asserting violations of new source performance standards and ignoring the NPDES permit which incorporates those standards).

71. *Natural Resources Defense Council v. Texaco Ref. and Mktg., Inc.*, 2 F.3d 493, 499 (3d Cir. 1993). The notice need not identify the specific outfall, however. *Atlantic States Legal Found., Inc. v. Stroh Die Casting Co.*, 116 F.3d 814 (7th Cir. 1997), *cert. denied*, 118 S. Ct. 442 (1997).

72. *Hudson Riverkeeper Fund Inc. v. Putnam Hosp. Ctr. Inc.*, 891 F. Supp. 152 (S.D.N.Y. 1995).

73. Any civil penalty recovered goes to the U.S. Treasury. *New Jersey Public Interest Research Group v. Powell-Duffryn Terminals, Inc.*, 913 F.2d 64 (3d Cir. 1990). If a suit is settled without a finding or admission of a violation of the act, the money recovered is not required to be paid to the Treasury, although it often will be. *Sierra Club v. Electronic Controls Design, Inc.*, 909 F.2d 1350 (9th Cir. 1990).

74. A 2001 Supreme Court case decided under the Americans with Disabilities Act suggests that attorney fee awards under CWA citizen suits may be limited to cases in which there is an official judicial ruling or court-sanctioned consent decree. *Buckhannon Board & Care Home v. West Virginia*, 532 U.S. 598, 121 S. Ct. 1835 (2001).

75. Section 505(c)(3), 33 U.S.C. § 1365(c)(3). Often, EPA or the U.S. Department of Justice will require that any environmental activity receiving settlement funds be related to the alleged violation (e.g., money paid in settlement of an alleged water permit violation would not be allowed to be designated for an air monitoring project).

76. *In re the Exxon Valdez*, 270 F.3d 1215 (9th Cir. 2001).

77. *Gwaltney of Smithfield Ltd. v. Chesapeake Bay Foundation, Inc.*, 484 U.S. 49 (1987).

78. *Id.* at 66, quoting *United States v. Phosphate Export Ass'n, Inc.*, 393 U.S. 199 (1968). The Fourth Circuit, on remand in the *Gwaltney* case, stated, "Intermittent or sporadic violations do not cease to be ongoing until the date when there is no real likelihood of repetition." *Gwaltney of Smithfield Ltd. v. Chesapeake Bay Foundation, Inc.*, 844 F.2d 170, 172 (4th Cir. 1988). *But see Allen County Citizens for the Environment v. BP Oil Co.*, 762 F. Supp. 733

(N.D. Ohio 1991), *aff'd*, 966 F.2d 1451 (6th Cir. 1992) (violation not "continuing" where no exceedances had occurred during the 42 months preceding the date the complaint was filed, even though two exceedances occurred after complaint was filed).

79. *Public Interest Research Group v. Hercules, Inc.*, 830 F. Supp. 1549, 1556 (D.N.J. 1993); *see also Paolino v. JF Realty, LLC,* 710 F.3d 31, 33 (1st Cir. 2013) (notice sufficient where "notice identifies the potential plaintiffs, provides basic contact information, and allows the putative defendants to identify and remedy the alleged violations").

Index

www.ingramcontent.com/pod-product-compliance
Ingram Content Group UK Ltd.
Pitfield, Milton Keynes, MK11 3LW, UK
UKHW051417150926
475561UK00005B/214